COMPLEX ANALYSIS

COMPLEX ANALYSIS

Guantie Deng
Beijing Normal University, China

World Scientific

NEW JERSEY · LONDON · SINGAPORE · BEIJING · SHANGHAI · HONG KONG · TAIPEI · CHENNAI

Published by

World Scientific Publishing Co. Pte. Ltd.
5 Toh Tuck Link, Singapore 596224
USA office: 27 Warren Street, Suite 401-402, Hackensack, NJ 07601
UK office: 57 Shelton Street, Covent Garden, London WC2H 9HE

Library of Congress Cataloging-in-Publication Data
Names: Deng, Guantie, author.
Title: Complex analysis / Guantie Deng, Beijing Normal University, China.
Description: New Jersey : World Scientific, [2025] | Series: Graduate textbooks for the
    21st century higher education | Reprint of Complex analysis : published by
    Beijing Normal University Press, 2010. | Includes bibliographical references and index.
Identifiers: LCCN 2024035244 | ISBN 9789811298813 (hardcover) |
    ISBN 9789819800339 (ebook for institutions) | ISBN 9789819800346 (ebook for individuals)
Subjects: LCSH: Analytic functions--Textbooks. | Mathematical analysis--Textbooks.
Classification: LCC QA331 .D47 2025 | DDC 515/.9--dc23/eng/20241102
LC record available at https://lccn.loc.gov/2024035244

British Library Cataloguing-in-Publication Data
A catalogue record for this book is available from the British Library.

Copyright © 2025 by World Scientific Publishing Co. Pte. Ltd.

All rights reserved. This book, or parts thereof, may not be reproduced in any form or by any means, electronic or mechanical, including photocopying, recording or any information storage and retrieval system now known or to be invented, without written permission from the publisher.

For photocopying of material in this volume, please pay a copying fee through the Copyright Clearance Center, Inc., 222 Rosewood Drive, Danvers, MA 01923, USA. In this case permission to photocopy is not required from the publisher.

For any available supplementary material, please visit
https://www.worldscientific.com/worldscibooks/10.1142/14001#t=suppl

Desk Editors: Kannan Krishnan/Rok Ting Tan

Typeset by Stallion Press
Email: enquiries@stallionpress.com

Author's Note

Complex Analysis is one of the degree foundation courses for master's students at the Mathematics Science School of Beijing Normal University, as listed in the 2000 curriculum. In June 2007, the Graduate School of Beijing Normal University revised the graduate program based on suggestions from some expert professors and supplemented the content of *Complex Analysis* with important advances in complex analysis since the 1970s, making it a degree foundation course for master's students at the Mathematics Science School of Beijing Normal University.

Since 2000, the authors have taught this course to six cohorts of master's students at the Mathematics Science School of Beijing Normal University. In our teaching, we have used the famous book *Real and Complex Analysis* by American mathematician Rudin, *Topics in Complex Analysis* by Andersson, and *Selected Topics in Complex Function Theory* by Zhang Nanyue and Chen Huaihui as the main reference books for this course. This book is based on the authors *Complex Analysis* lectures, supplemented and modified by referencing some important monographs and literature both domestically and internationally.

The book consists of six chapters. The knowledge introduced in Chapter 1 is the most basic and essential in complex analysis, and its application runs through the whole book. Chapter 2 mainly introduces normal families, Riemann mapping theorem, and conformal mapping. Chapter 3 introduces zero points of analytic functions, and Chapter 4 introduces the basic content and important properties of harmonic and subharmonic functions. Chapter 5 introduces H^p space and the entire Fourier transform and their connection. In the final chapter, we introduce some content on uniform approximation of rational functions.

In selecting the content, this book emphasizes the basic ideas, theories, and methods in modern complex analysis, while also introducing some cutting-edge research problems and recent advances. Readers who have already acquired knowledge of undergraduate courses such as *Real Analysis*, *Functions of One Complex Variable* and *Functional Analysis* should not have difficulty in studying this course.

This course is one of the bilingual courses for master's students established by the Graduate School of Beijing Normal University from 2005 to 2006. This book is one of the series of textbooks for degree foundation courses for master's students at the Mathematics Science School of Beijing Normal University from 2005 to 2008.

The authors sincerely thank the Graduate School and the Mathematics Science School of Beijing Normal University for their support during the teaching and writing of this course. We also thank the teachers and graduate students at the Mathematics Science School of Beijing Normal University for their valuable opinions during the trial use of the original manuscript, especially Gao Zhiqiang, Dr. Ke Siyu, and Dr. Li Zhen, who carefully proofread and corrected some errors in the original manuscript, for which we express our most sincere gratitude. In addition, we are very grateful to the Beijing Normal University Press for their hard work in editing and publishing this book. Due to the limitations of our own knowledge, there may still be errors and inadequacies in this book, and we sincerely ask experts and readers for their advice and corrections.

Guantie Deng
December, 2024

Notations

$\mathbb{R}^n$ denotes an n-dimensional Euclidean space. A connected open set of $\mathbb{R}^n$ is called a region. $\operatorname{supp}(f) = \overline{\{x \in \mathbb{R}^n : f(x) \neq 0\}}$ is defined as the closure of the set of points in $\{x \in \mathbb{R}^n : f(x) \neq 0\}$, which is called the support of a function of f. If $\operatorname{supp}(f)$ is a compact set in $\mathbb{R}^n$, f is said to have compact support. $\mathbb{C} = \mathbb{R}^2$ represents the complex plane, Ω represents an open set (or domain) in $\mathbb{R}^2$ (the complex plane $\mathbb{C}$), ω represents a bounded open set with (when necessary) piecewise smooth C^1 boundary $\partial\omega$, which is always supposed to be positively oriented, i.e., one has ω on the left-hand side when passing along $\partial\omega$. The notation $\omega \subset\subset \Omega$ means that the closure of ω is a compact subset of Ω, and $d(K, E)$ denotes the distance the sets K and E. Moreover, $D(a, r)$ is an open disk with center a and radius r, and U denotes the unit disk, and T is its boundary of the unit disk. The space of all analytic functions on Ω is denoted as $H(\Omega)$, $C(\Omega)$ represents the space of all continuous functions on Ω, $C^k(\Omega)$ is the space of k times (real) differentiable (complex-valued) functions in Ω whose derivatives up to order k (partial or ordinary) are continuous (however, we write $C(\Omega)$ rather than $C^0(\Omega)$) and $C^\infty(\Omega) = \bigcap_{k=1}^\infty C^k(\Omega)$ is the space of functions on Ω that are infinitely differentiable. $C_c^k(\Omega)$ is the subspace of functions in $C^k(\Omega)$ that have compact support in Ω, $C_c^\infty(\Omega) = \bigcap_{k=1}^\infty C_c^k(\Omega)$. Moreover, $C^k(\overline{\omega})$ is the subspace of functions in $C^k(\omega)$ whose derivatives up to the kth order (partial or ordinary) have continuous extensions to the closure $\overline{\omega}$, $C^\infty(\overline{\omega}) = \bigcap_{k=1}^\infty C^k(\overline{\omega})$ represents the space of functions that are infinitely differentiable and have continuous extensions of all derivatives to ω. $d\lambda$ represents the Lebesgue measure on $\mathbb{R}^2$. We use the standard abbreviation a.e. for "almost every(where)." We also use u.c. for "uniformly on compact sets." If f and g are functions, then "$f = o(g)$ when $z \to a$" means that $f/g \to 0$ when $z \to a$. For $0 < p \leqslant \infty$, $L^p(\mathbb{R}^n)$ denotes the Lebesgue space on $\mathbb{R}^n$ and is defined as the set of all functions:

$$L^p(\mathbb{R}^n) = \left\{ f : \|f\|_p = \left(\int_{\mathbb{R}^n} |f(x)|^p dx \right)^{1/p} < \infty \right\}, \quad 0 < p < \infty$$

and defined as the set of all functions f, for which

$$L^\infty(\mathbb{R}^n) = \left\{ f : \|f\|_\infty = \inf_{\substack{|E|=0 \\ E \subset \mathbb{R}^n}} \left(\sup_{x \in \mathbb{R}^n \setminus E} |f(x)| \right) < \infty \right\}$$

denotes the space of essentially bounded functions on $E \subset \mathbb{R}$ defined as the set of all functions that are integrable up to order. $L^p_{\text{loc}}(\mathbb{R}^n)$ $(0 < p < \infty)$ represents the space of functions that are integrable up to order p for $1 \leqslant p \leqslant \infty$, and p' denotes the conjugate exponent of p satisfying $\frac{1}{p} + \frac{1}{p'} = 1$. $\square$ indicates the end of a proof.

Other notations and symbols in the book have their usual meanings and do not require further explanation.

Contents

Author's Note   v
Notations   vii

1. **Basic Knowledge of Analytic Functions**   1
   - 1.1 Preliminaries 1
   - 1.2 Basic Properties of Analytic Functions 6
   - 1.3 Global Cauchy Theorem 26
   - Exercises I 34

2. **Normal Family and Conformal Mapping**   41
   - 2.1 Normal Family 41
   - 2.2 Conformal Mapping for Simply Connected Regions 51
   - 2.3 Boundary Correspondence Theorem 59
   - 2.4 Univalent Function 65
   - 2.5 Picard's Theorem 83
   - Exercises II 94

3. **Zeros of Holomorphic Functions**   99
   - 3.1 Infinite Products 99
   - 3.2 The Weierstrass Factorization Theorem 102
   - 3.3 Order and Type of Entire Functions 106

3.4 Exponent of Convergence, Genus and Canonical Product 114
3.5 The Gamma Function, the Beta Function and the Riemann Zeta Function 126
Exercises III 142

4. Harmonic and Subharmonic Functions — 147

4.1 Property of Harmonic Functions 147
4.2 Upper Semicontinuous Functions 161
4.3 Subharmonic Functions 163
4.4 The Dirichlet Problem and Green's Function 194
4.5 Harmonic Functions in the Unit Disk 205
4.6 Harmonic Functions in the Upper Half-Plane 217
Exercises IV 243

5. H^p Spaces — 247

5.1 H^p Spaces in the Unit Disk 247
5.2 H^p Space in the Upper Half-Plane 274
5.3 Holomorphic Fourier Transform 291
Exercises V 310

6. Uniform Approximation by Polynomials — 315

6.1 Uniform Approximation by Rational Function and Simply Connected Region 315
6.2 Uniform Approximation by Polynomials 324
Exercises VI 331

References — 333

Index — 335

Chapter 1

Basic Knowledge of Analytic Functions

1.1 Preliminaries

In addition to the complex function theory, some basic knowledge of calculus, Lebesgue integration and functional analysis is required. For convenience, we describe them and give proofs for some of them.

Definition 1.1.1 (Definition of Real Differentiability). Let f be a function from the open set Ω to $\mathbb{C}$ and $a \in \Omega$. If

$$f(a+z) = f(a) + Df|_a(z) + o(|z|) \ (z \to 0), \tag{1.1.1}$$

where $Df|_a(z) = Ax + By, z = x + iy$ with constants A and B dependent on a only, f is said to be real differentiable at a. Denote A by $\frac{\partial f}{\partial x}|_a$ and B by $\frac{\partial f}{\partial y}|_a$. If $f \in C^1(D(a,r))$ for a neighborhood $D(a,r)$ of a, then (1.1.1) holds, and thus, f is real differentiable at a.

Proposition 1.1.2 (Existence of Test Functions). *For any compact set $K \subset \Omega$, there is a $\phi \in C_c^\infty(\Omega)$ such that ϕ equals 1 in a neighborhood of K and $0 \leqslant \phi \leqslant 1$.*

Proof. For $\varepsilon > 0$, define the function $f_{0,\varepsilon}(t)$ on $\mathbb{R}$ as follows. If $t \in (-\varepsilon, \varepsilon)$, $f_{0,\varepsilon}(t) = \exp\{-(t+\epsilon)^{-2} - (t-\varepsilon)^{-2}\}$, and if $t \notin (-\varepsilon, \varepsilon)$, $f_{0,\varepsilon}(t) = 0$, then $f_{0,\varepsilon}(t)$ is a function of $C^\infty(\mathbb{R})$ and is positive on $(-\varepsilon, \varepsilon)$ and 0 otherwise. Set $f_{1,\varepsilon}(t) = \int_0^{2t} f_{0,\varepsilon}(s - \varepsilon)ds$. Then

$f_{2,\varepsilon}(t) = \frac{f_{1,\varepsilon}(t)}{f_{1,\varepsilon}(\varepsilon)}$ is a function of $C^\infty(\mathbb{R})$, assumes positive and monotonically increases on $(0, \varepsilon)$, becomes 0 when $t \leq 0$, and 1 when $t \geq \varepsilon$. If $a \in \mathbb{C}$, the function $g_{a,\varepsilon}(z) = f_{0,\varepsilon^2}(|z-a|^2)$ is a function of $C^\infty(\mathbb{C})$, assumes positive on $D(a, \varepsilon) = \{z : |z-a| < \varepsilon\}$ and 0 otherwise. If $\Omega \subset \mathbb{C}$ is open and $K \subset \Omega$ is compact, set $0 < \varepsilon < d(K, \Omega^c)$, where $d(K, \Omega^c)$ represents the distance between K and the complement of Ω. Then there exist a finite number of balls $D(a_1, \varepsilon), \ldots, D(a_N, \varepsilon)$ that cover K, where $a_1, \ldots, a_N \in K$. Consequently, the non-negative infinitely differentiable function $f_0(z) = \sum_{k=1}^{N} g_{a_k,\varepsilon}(z)$ satisfies the following properties: $f_0(z) > 0$ for $z \in K$ and $f_0 = 0$ outside $\bigcup_{k=1}^{N} \overline{D(a_k, \varepsilon)}$. Let $\delta = \frac{1}{2}\min\{f_0(z) : z \in K\}$. Then the function $\phi(z) = f_{2,\delta}(f_0(z)) \in C_c^\infty(\Omega)$. In addition, ϕ equals 1 in a neighborhood of K and $0 \leq \phi \leq 1$. □

Proposition 1.1.3 (Fubini's Theorem). *Suppose (X, μ) and (Y, λ) are σ-finite spaces and f is measurable on $X \times Y$.*

(a) *If f is a complex-valued function, and*

$$\int_X \varphi^*(x) d\mu(x) < \infty, \quad \text{where } \varphi^*(x) = \int_Y |f(x,y)| d\lambda(y),$$

then $f \in L^1(\mu \times \lambda)$.

(b) *If $f \in L^1(\mu \times \lambda)$, then for almost everywhere on X, $f_x(y) = f(x,y)$ as a function of y belongs to $L^1(\lambda)$ and for almost everywhere on Y, $f^y(x) = f(x,y)$ as a function of x belongs to $L^1(\mu)$. In addition,*

$$\int_X d\mu \int_Y f d\lambda = \int_{X \times Y} f d(\mu \times \lambda) = \int_Y d\lambda \int_X f d\mu.$$

Proposition 1.1.4 (Minkowski's Inequality). *Suppose (X, μ) and (Y, λ) are σ-finite spaces, f is measurable on $X \times Y$, and $1 \leq p < \infty$. then*

$$\left[\int_Y \left|\int_X f(x,y) d\mu(x)\right|^p d\lambda(y)\right]^{\frac{1}{p}} \leq \int_X \left[\int_Y |f(x,y)|^p d\lambda(y)\right]^{\frac{1}{p}} d\mu(x).$$

Proposition 1.1.5 (Differentiating under the Integral Sign). *Suppose (X, μ) is a σ-finite space and $I = (a, b)$ is an interval.*

Let $f(x,t)$ be measurable on $X \times I$ such that for almost all $x \in X$, $f_x(t) = f(x,t)$ as a function of t belongs to $C^1(I)$. If $f, \frac{\partial}{\partial t} f \in L^1(\mu \times \lambda)$ and there is a function $\psi(x) \in L^1(\mu)$ such that $|\frac{\partial}{\partial t} f(x,t)| \leq \psi(x)$ holds for almost all $x \in X$, then $g(t) = \int_X f(x,t) d\mu(x) \in C^1(I)$ and $g'(t) = \int_X \frac{\partial}{\partial t} f(x,t) d\mu(x)$.

Proposition 1.1.6 (Differentiating under the Summation). Suppose $I = (a,b)$ is an interval and $f_n(t) \in C^1(I)$. If there is a $t_0 \in I$ such that $\sum_{n=1}^{\infty} |f_n(t_0)| < \infty$ and $\sum_{n=1}^{\infty} \sup\{|f_n'(t)| : t \in I\} < \infty$, then $g(t) = \sum_{n=1}^{\infty} f_n(t)$ is uniformly convergent on any compact set of I, $g \in C^1(I)$ and $g'(t) = \sum_{n=1}^{\infty} f_n'(t)$.

Theorem 1.1.7 (Lebesgue Differentiation Theorem). Let $f \in L_{\text{loc}}(\mathbb{R})$.

(a) For almost all $x \in \mathbb{R}$,

$$\lim_{r \to 0} \frac{1}{r} \int_{|t| < r} [f(x-t) - f(x)] dt = 0. \tag{1.1.2}$$

(b) For almost all $x \in \mathbb{R}$,

$$\lim_{r \to 0} \frac{1}{r} \int_{|t| < r} |f(x-t) - f(x)| dt = 0. \tag{1.1.3}$$

(c) If $f \in L^p(\mathbb{R})$ $(1 \leq p < \infty)$, then for almost all $x \in \mathbb{R}$,

$$\lim_{\varepsilon \to 0} \frac{1}{\varepsilon} \int_{|t| \leq \varepsilon} |f(x-t) - f(x)|^p dt = 0. \tag{1.1.4}$$

The set of all points x such that (1.1.2) holds is called the set of differentiable points of the f integral. A point x at which (1.1.3) holds is called a Lebesgue point of f. The set of all Lebesgue points is called the Lebesgue point set. A point x at which (1.1.4) holds is called a p-Lebesgue point of f. Clearly, a 1-Lebesgue point is simply a Lebesgue point. The notion of Lebesgue points describes a local property, which can be defined only for functions that are locally integrable (or locally p-integrable). If $f \in L^p_{\text{loc}}(\mathbb{R})$ $(1 < p < \infty)$, then a p-Lebesgue point is necessarily a Lebesgue point. Lebesgue differential theorem is a very basic and important theorem in analysis.

Definition 1.1.8 (Definition of a Curve). A curve γ in a topological space X refers to a continuous mapping from a compact interval $[\alpha, \beta] \subset \mathbb{R}$ to X, where $\alpha < \beta$. The interval $[\alpha, \beta]$ is called the parameter interval of γ. The image $\gamma([\alpha, \beta])$ is denoted by γ^*. The curve γ is called a simple curve or Jordan curve if the following is true: $\gamma(t_1) = \gamma(t_2)$ if and only if $t_1 = t_2$ or $t_1, t_2 \in \{\alpha, \beta\}$. The curve γ is called a closed curve if $\gamma(\alpha)$ is equal to $\gamma(\beta)$. A simple closed curve is also called a Jordan closed curve. A path in a plane refers to a piecewise C^1 curve. A polygonal line in a topological vector space X is a curve consisting of a finite number of straight line segments joined end to end.

Proposition 1.1.9 (Jordan Closed Curve Theorem). *A plane Jordan closed curve γ divides the plane into two regions. One is bounded, called the interior of γ, and the other is unbounded, called the exterior of γ. γ^* is the common boundary of the two regions.*

The following special case of the Jordan Closed Curve Theorem is easy to prove (see [16] and [24] for a proof of the general scenario).

Proposition 1.1.10 (Jordan Closed Polygonal Line Theorem). *A plane Jordan closed polygonal line γ divides the plane into two regions. One is bounded, called the interior of γ, and the other is unbounded, called the exterior of γ. γ^* is the common boundary of the two regions.*

Proof. Jordan Closed Polygonal Line Theorem can be proved by induction. The theorem clearly holds for the Jordan closed polygonal line represented by the boundary of a triangle. For a Jordan closed polygonal line with n ($n \geqslant 4$) line segments, we can add a new line segment to obtain two Jordan closed polygonal lines, such that they share a common line segment and each consists of less than or equal to $n - 1$ line segments. Thus by induction we can prove the theorem. $\square$

Definition 1.1.11 (Definition of A Simply Connected Region). A region Ω in the extended complex plane is simply connected if for any Jordan closed polygonal line γ ($\gamma^* \subset \Omega$) its interior or exterior is contained in Ω.

Let E be a set in the extended complex plane. A maximal connected subset of E is called a connected component of E. Let Ω be a region in the extended complex plane. Ω is said to be n-connected if $\mathbb{C}_\infty \setminus \Omega$ consists of n connected components.

Theorem 1.1.12. *Let Ω be a region in the plane. If $D \subset \Omega$ is a non-empty open set and the closure $\overline{D}$ of D is such that $\overline{D} \cap \Omega = D$, then $D = \Omega$.*

Proof. Let $a \in D$ and D_a be the connected component in D that contains a. If $D_a \neq \Omega$, then there is a $b \in \Omega$ such that $b \notin D_a$. Connecting a with b by a curve in Ω, we see that there is a a point $c \in \Omega \cap \partial D_a \subset \overline{D}$. However, $\overline{D} \cap \Omega = D$. Therefore there is a $r > 0$ such that $D(c, r) \subset D$, contradicting the definition of a connected component. $\square$

Definition 1.1.13 (Operators $\frac{\partial}{\partial z}$ and $\frac{\partial}{\partial \bar{z}}$). A complex number takes the form $z = x + iy$, where $(x, y) \in \mathbb{R}^2$. The mapping $\mathbb{C} \to \mathbb{R}^2$, $z = x + iy \to (x, y)$, is a bijection between $\mathbb{C}$ and $\mathbb{R}^2$. From $z = x + iy$, we have $dz = dx + idy$, $d\bar{z} = dx - idy$. Then $Pdx + Qdy$ in $\mathbb{R}^2$ may be written as $fdz + gd\bar{z}$ ($f = (P - iQ)/2$, $g = (P + iQ)/2$). We now introduce the following differential operators:

$$\frac{\partial}{\partial z} = \frac{1}{2}\left(\frac{\partial}{\partial x} - i\frac{\partial}{\partial y}\right), \quad \frac{\partial}{\partial \bar{z}} = \frac{1}{2}\left(\frac{\partial}{\partial x} + i\frac{\partial}{\partial y}\right).$$

Then for $f, g \in C^1(\Omega)$, $\alpha, \beta \in \mathbb{C}$,

$$\frac{\partial}{\partial z}(\alpha f + \beta g) = \alpha \frac{\partial f}{\partial z} + \beta \frac{\partial g}{\partial z}, \quad \frac{\partial}{\partial \bar{z}}(\alpha f + \beta g) = \alpha \frac{\partial f}{\partial \bar{z}} + \beta \frac{\partial g}{\partial \bar{z}},$$

$$\frac{\partial}{\partial z}(fg) = \frac{\partial f}{\partial z}g + f\frac{\partial g}{\partial z}, \quad \frac{\partial}{\partial \bar{z}}(fg) = \frac{\partial f}{\partial \bar{z}}g + f\frac{\partial g}{\partial \bar{z}},$$

$$df = \frac{\partial f}{\partial x}dx + \frac{\partial f}{\partial y}dy = \frac{\partial f}{\partial z}dz + \frac{\partial f}{\partial \bar{z}}d\bar{z}. \tag{1.1.5}$$

If $g \in C^1(\Omega)$ and $f \in C^1(\overline{g(\Omega)})$, then the composition $f \circ g \in C^1(\Omega)$ and

$$\frac{\partial (f \circ g)}{\partial z} = \frac{\partial f}{\partial \zeta}\frac{\partial g}{\partial z} + \frac{\partial f}{\partial \bar{\zeta}}\frac{\partial \bar{g}}{\partial z}, \quad \frac{\partial (f \circ g)}{\partial \bar{z}} = \frac{\partial f}{\partial \zeta}\frac{\partial g}{\partial \bar{z}} + \frac{\partial f}{\partial \bar{\zeta}}\frac{\partial \bar{g}}{\partial \bar{z}}.$$

1.2 Basic Properties of Analytic Functions

Definition 1.2.1 (Complex Differentiability and Analytic Functions). Let f be a complex-valued function defined on an open set Ω and $a \in \Omega$. If the limit

$$\lim_{z \to 0} \frac{f(a+z) - f(a)}{z} \qquad (1.2.1)$$

exists, f is said to be complex differentiable at $a \in \Omega$. When existing, the limit is denoted by $f'(a)$. The function f is said to be analytic or holomorphic on Ω if it is complex differentiable at every $a \in \Omega$. Denote by $H(\Omega)$ the set of all analytic functions on Ω. For any set E of the plane, denote by $H(\overline{E})$ the collection of all functions $f \in H(\Omega_f)$ for some open set Ω_f containing the closure of E.

From Definition 1.2.1, it is easy to see that if $f'(a)$ exists, then $f(z)$ is continuous at a. Therefore, an analytic function is always continuous. In fact, it follows from the definition that for any $\varepsilon > 0$, there is a $\delta_1 > 0$, such that if $|z| < \delta_1$, then

$$\left| \frac{f(a+z) - f(a)}{z} - f'(a) \right| < \varepsilon.$$

Thus,

$$|f(a+z) - f(a)| < |f'(a)||z| + \varepsilon|z| = (|f'(a)| + \varepsilon)|z|.$$

Choose $\delta = \min\{\delta_1, \varepsilon/(|f'(a)|+\varepsilon)\}$. When $|z| < \delta$, we have $|f(a+z) - f(a)| < \varepsilon$, showing $f(z)$ is continuous at a.

Proposition 1.2.2. *If a complex function f is real differentiable at $a \in \Omega$, then it is complex differentiable at a if and only if $\frac{\partial f}{\partial \bar{z}}|_a = 0$. If f is complex differentiable at $a \in \Omega$, then $f'(a)$ exists and $f'(a) = \frac{\partial f}{\partial z}|_a$.*

Proof. Assuming f is real differentiable at $a \in \Omega$, then for $z = x + iy, \bar{z} = x - iy$, we have

$$x \frac{\partial f}{\partial x}(a) + y \frac{\partial f}{\partial y}(a) = z \frac{\partial f}{\partial z}(a) + \bar{z} \frac{\partial f}{\partial \bar{z}}(a).$$

From the definition of real differentiability, (1.1.1) holds, and thus, f is complex differentiable at a if and only if the limit

$\lim_{z\to 0} \frac{\bar{z}}{z}\frac{\partial f}{\partial \bar{z}}(a)$ exists. Note that for $z = x + iy = re^{i\theta}$, where $\theta = \arg z \in (-\pi, \pi]$, $r = \sqrt{x^2 + y^2}$,

$$\frac{\bar{z}}{z}\frac{\partial f}{\partial \bar{z}}(a) = \frac{\mathrm{Re}^{-i\theta}}{\mathrm{Re}^{i\theta}}\frac{\partial f}{\partial \bar{z}}(a) = \frac{\partial f}{\partial \bar{z}}(a)e^{-2i\theta}.$$

It is seen that if $\frac{\partial f}{\partial \bar{z}}(a) \neq 0$, then as z approaches 0, the limit $\lim_{z\to 0}\frac{\bar{z}}{z}\frac{\partial f}{\partial \bar{z}}|_a$ does not exist. Therefore, $\frac{\partial f}{\partial \bar{z}}(a) = 0$ is a necessary and sufficient condition for $\lim_{z\to 0}\frac{\bar{z}}{z}\frac{\partial f}{\partial \bar{z}}(a)$ to exist. Hence, if f is complex differentiable at a, we have $\frac{\partial f}{\partial \bar{z}}(a) = 0$, $f'(a) = \frac{\partial f}{\partial z}(a)$. □

Proposition 1.2.2 shows that if $f \in C^1(\Omega)$ and $\frac{\partial f}{\partial \bar{z}} = 0$ on Ω, then $f \in H(\Omega)$. Conversely, if $f \in H(\Omega)$, then f is real differentiable and $\frac{\partial f}{\partial \bar{z}} = 0 (z \in \Omega)$. We call $\frac{\partial f}{\partial \bar{z}} = 0$ the Cauchy–Riemann equation.

For $f \in C^1(\Omega)$, we write $f(z) = u(x,y) + iv(x,y)$, where u and v are real-valued functions. Then we have

$$\frac{\partial f}{\partial \bar{z}} = \frac{1}{2}\left[\left(\frac{\partial u}{\partial x} + i\frac{\partial v}{\partial x}\right) + i\left(\frac{\partial u}{\partial y} + i\frac{\partial v}{\partial y}\right)\right]$$
$$= \frac{1}{2}\left[\left(\frac{\partial u}{\partial x} - \frac{\partial v}{\partial y}\right) + i\left(\frac{\partial v}{\partial x} + \frac{\partial u}{\partial y}\right)\right],$$

so

$$\frac{\partial f}{\partial \bar{z}} = 0 \text{ is equivalent to } \begin{cases} \frac{\partial u}{\partial x} = \frac{\partial v}{\partial y}, \\ \frac{\partial u}{\partial y} = -\frac{\partial v}{\partial x}. \end{cases}$$

Next, we study the integration of analytic functions. It is known that in $\mathbb{R}^2$, we have the Green formula:

$$\int_{\partial\omega} Pdx + Qdy = \int_{\omega}(Q_x - P_y)d\lambda,$$

where ω is a bounded domain whose boundary consists of one or several smooth curves and is positively oriented (hereafter), $P, Q \in C^1(\overline{\omega})$. With notations from complex analysis, the Green formula can be written as

$$\int_{\partial\omega} fdz + gd\bar{z} = 2i\int_{\omega}\left(\frac{\partial f}{\partial \bar{z}} - \frac{\partial g}{\partial z}\right)d\lambda(z), \quad f, g \in C^1(\overline{\omega}). \quad (1.2.2)$$

In fact, for $f, g \in C^1(\overline{\omega})$, we may assume without loss of generality that both f and g are real. Otherwise, we can consider their real and imaginary parts separately. Setting $P = f + g$ and $Q = i(f - g)$ in the Green formula, we have

$$\int_{\partial \omega} f dz + g d\bar{z} = \int_{\partial \omega} P dx + Q dy = \int_{\omega} (Q_x - P_y) d\lambda$$

$$= \int_{\omega} \left[i \left(\frac{\partial f}{\partial x} - \frac{\partial g}{\partial x} \right) - \left(\frac{\partial f}{\partial y} + \frac{\partial g}{\partial y} \right) \right] d\lambda$$

$$= 2i \int_{\omega} \left[\frac{1}{2} \left(\frac{\partial f}{\partial x} + i \frac{\partial f}{\partial y} \right) - \frac{1}{2} \left(\frac{\partial g}{\partial x} - i \frac{\partial g}{\partial y} \right) \right] d\lambda$$

$$= 2i \int_{\omega} \left(\frac{\partial f}{\partial \bar{z}} - \frac{\partial g}{\partial z} \right) d\lambda(z).$$

In particular, if $g = 0$ in (1.2.2), then

$$\int_{\partial \omega} f dz = 2i \int_{\omega} \frac{\partial f}{\partial \bar{z}} d\lambda(z). \tag{1.2.3}$$

If we choose $f = \bar{z}$ in the above, then the area of ω is $\lambda(\omega) = \frac{1}{2i} \int_{\partial \omega} \bar{z} dz$. It follows from (1.2.3) that we have the following.

Theorem 1.2.3 (Cauchy's Theorem). *If $f \in H(\omega) \cap C^1(\overline{\omega})$, then*

$$\int_{\partial \omega} f(z) dz = 0. \tag{1.2.4}$$

Cauchy's theorem can be used to show that the values of an analytic function inside a domain are determined by the values of that function on the boundary of the region. This is what Cauchy integral formula states. Here, however, we will relax the condition appropriately and study functions that have continuous first-order partial derivatives on a bounded closed region (not necessarily analytic on the region). This allows for a more general Green–Pompeiu formula.

Theorem 1.2.4 (Green–Pompeiu Formula). *If $f \in C^1(\overline{\omega})$, $z \in \omega$, then*

$$f(z) = \frac{1}{2\pi i} \int_{\partial \omega} \frac{f(\zeta)}{\zeta - z} d\zeta - \frac{1}{\pi} \int_{\omega} \frac{\partial f}{\partial \bar{\zeta}} \frac{d\lambda(\zeta)}{\zeta - z}. \tag{1.2.5}$$

In particular, if $f \in H(\omega) \cap C^1(\overline{\omega})$, then we have the Cauchy formula

$$f(z) = \frac{1}{2\pi i} \int_{\partial \omega} \frac{f(\zeta)}{\zeta - z} d\zeta. \tag{1.2.6}$$

Proof. For any $z \in \omega$, choose a small $\varepsilon > 0$, such that $D(z,\varepsilon) = \{\zeta : |\zeta - z| < \varepsilon\} \subset \omega$. Applying (1.2.3) to $f(\zeta)/(\zeta - z)$ as a function of ζ, we have

$$\int_{\partial(\omega\setminus D(z,\varepsilon))} \frac{f(\zeta)}{\zeta - z} d\zeta = 2i \int_{\omega\setminus D(z,\varepsilon)} \frac{\frac{\partial f(\zeta)}{\partial \bar\zeta}}{\zeta - z} d\lambda(\zeta)$$

$$= \left(\int_{\partial\omega} - \int_{\partial D(z,\varepsilon)}\right) \frac{f(\zeta) d\zeta}{\zeta - z}, \quad (1.2.7)$$

and $\partial D(a,\varepsilon)$ has a parametrization $z = z(\theta) = a + \varepsilon e^{i\theta}, 0 \leqslant \theta \leqslant 2\pi$, and consequently, $z'(\theta) = i\varepsilon e^{i\theta}$. Then

$$\int_{\partial D(a,\varepsilon)} \frac{f(z)}{z - a} dz = i \int_0^{2\pi} f(a + \varepsilon e^{i\theta}) d\theta.$$

Letting $\varepsilon \to 0$ in (1.2.7) leads to (1.2.5). If in addition f is analytic on ω, then $\partial f/\partial \bar\zeta = 0$ on ω, and thus, (1.2.6) holds. $\square$

From 1.2.4, we have the following.

Corollary 1.2.5 (Solution to $\bar\partial$ Problem). *Assume $\phi \in C_c^1(\mathbb{C})$ and let*

$$\psi(z) = -\frac{1}{\pi} \int_{\mathbb{C}} \phi(\zeta) \frac{d\lambda(\zeta)}{\zeta - z}, \quad (1.2.8)$$

then $\psi \in C^1(\mathbb{C})$ and $\frac{\partial \psi(z)}{\partial \bar z} = \phi(z)$.

Proof. Assume $\phi \in C_c^1(\mathbb{C})$. Then ϕ has compact support supp (ϕ), which is closed and bounded. For all $z \in \mathbb{C}$, choose $R > |z| + \max\{|\zeta| : \zeta \in \text{supp}(\phi)\}$. Then $\phi = 0$ and $\frac{\partial \phi}{\partial \bar\zeta} = 0$ on $\mathbb{C} \setminus \text{supp}(\phi)$. Therefore,

$$\psi(z) = -\frac{1}{\pi} \int_{\mathbb{C}} \phi(z + \zeta) \frac{d\lambda(\zeta)}{\zeta} = -\frac{1}{\pi} \int_{D(0,R)} \phi(z + \zeta) \frac{d\lambda(\zeta)}{\zeta}.$$

Since $\frac{1}{\zeta}$ is integrable over $D(0,R)$, we have $\psi \in C(\mathbb{C})$. By differentiating under the integral sign and the Green–Pompeiu formula, we have

$$\frac{\partial \psi(z)}{\partial \bar z} = -\frac{1}{\pi} \int_{D(0,R)} \frac{\partial \phi(z + \zeta)}{\partial \bar z} \frac{d\lambda(\zeta)}{\zeta} = -\frac{1}{\pi} \int_{\mathbb{C}} \frac{\partial \phi}{\partial \bar\zeta} \frac{d\lambda(\zeta)}{\zeta - z} = \phi(z).$$

Thus, $\psi \in C^1(\mathbb{C})$ and $\frac{\partial \psi(z)}{\partial \bar z} = \phi(z)$. $\square$

Theorem 1.2.6. *If $f \in H(\omega) \cap C^1(\overline{\omega})$, then for any positive integer m, the mth derivative $f^{(m)} \in H(\omega)$ and*

$$f^{(m)}(z) = \frac{m!}{2\pi i} \int_{\partial \omega} \frac{f(\zeta)d\zeta}{(\zeta - z)^{m+1}}, \quad z \in \omega. \qquad (1.2.9)$$

Proof. Take the derivative with respect to z from both sides of (1.2.6) and apply differentiating under the integral sign. □

The above theorem shows that a continuously differentiable analytic function f on a domain Ω must have derivatives of any order, i.e., $H(\Omega) \cap C^1(\Omega) \subset C^\infty(\Omega)$, and $f^{(m)} \in H(\Omega)$.

As mentioned before, an analytic function must be continuous. The following theorem tells us that when a continuous function satisfies certain conditions, it must be analytic.

Theorem 1.2.7 (Morera's Theorem). *Suppose f is continuous on Ω. If for any triangle $\triangle \subset \Omega$, $\int_{\partial \triangle} f(z)dz = 0$, then $f \in H(\Omega) \cap C^1(\Omega)$.*

Proof. For any $z_0 \in \Omega$, choose a convex domain Ω_{z_0} such that $z_0 \in \Omega_{z_0} \subset \Omega$. In the following, we prove f is analytic on Ω_{z_0}. Fix a point a in Ω_{z_0} and set

$$F(z) = \int_{[a,z]} f(\zeta)d\zeta,$$

where $[a, z]$ is the directed linear segment from a to z. Choose a sufficiently small $|\triangle z|$ such that $z + \triangle z \in \Omega_{z_0}$. Then the triangle with $a, z, z + \triangle z$ as vertices is contained in Ω_{z_0}. Thus, the integral of f along the boundary of the triangle is 0, i.e.,

$$\int_{[a,z]} f(\zeta)d\zeta + \int_{[z,z+\triangle z]} f(\zeta)d\zeta + \int_{[z+\triangle z,a]} f(\zeta)d\zeta = 0,$$

hence,

$$F(z + \triangle z) - F(z) = \int_{[z,z+\triangle z]} f(\zeta)d\zeta$$

$$= \int_{[z,z+\triangle z]} (f(\zeta) - f(z))d\zeta + \triangle z f(z).$$

For any $\varepsilon > 0$, there is a $\delta \in (0, \varepsilon)$, such that $\overline{D(z, \delta)} \subset \Omega_{z_0}$, $z + \triangle z \in \overline{D(z, \delta)}$ for $|\triangle z| < \delta$, and consequently, $[z, z + \triangle z] \subset \overline{D(z, \delta)}$, and

$|f(\zeta) - f(z)| < \varepsilon$ for any $\zeta \in \overline{D(z,\delta)}$. Therefore,

$$\left| \int_{[z,z+\triangle z]} (f(\zeta) - f(z))d\zeta \right| \leqslant \int_{[z,z+\triangle z]} |f(\zeta) - f(z)||d\zeta| \leqslant \varepsilon |\triangle z|.$$

Then we have

$$\lim_{|\triangle z| \to 0} \frac{1}{\triangle z} \int_{[z,z+\triangle z]} (f(\zeta) - f(z))\, d\zeta = 0,$$

and thus,

$$F(z + \triangle z) - F(z) = \triangle z f(z) + o(|\triangle z|).$$

This shows F is differentiable at z and $\frac{\partial F}{\partial z} = f(z)$, $\frac{\partial F}{\partial \bar{z}} = 0$. Thus, F is analytic on Ω_{z_0} and $F' = f$ on Ω_{z_0}, and so, $F \in H(\Omega_{z_0}) \cap C^1(\Omega_{z_0})$. From Theorem 1.2.6, we see that $f \in H(\Omega_{z_0}) \cap C^1(\Omega_{z_0})$. Since z_0 is arbitrary, we see that $f \in H(\Omega) \cap C^1(\Omega)$. □

Theorem 1.2.8 (Goursat's Theorem). $H(\Omega) = H(\Omega) \cap C^1(\Omega)$.

Proof. We will show $f \in H(\Omega) \cap C^1(\Omega)$ for $f \in H(\Omega)$. From Morera's Theorem, it suffices to show that for any $\triangle$ contained in Ω, $I = \int_{\partial \triangle} f(z)dz = 0$. By bisecting each side of the given triangle $\triangle$ and connecting these points two by two, the given triangle $\triangle$ is divided into four congruent triangles: $\triangle^1, \triangle^2, \triangle^3$, and $\triangle^4$. Clearly, $I = \sum_{k=1}^{4} \int_{\partial \triangle^k} f(z)dz$, since along each line segment connecting the points, the integrals are taken exactly twice in opposite directions and thus cancel each other out. It is obvious that among the four integrals along the boundary $\partial \triangle^k$, at least one is not less than $|I|/4$. Say, $\triangle_1$ is a triangle such that $|\int_{\partial \triangle_1} f(z)dz| \geqslant |I|/4$. As before, divide this triangle $\triangle_1$ into four congruent triangles, one of which, say, $\triangle_2$, satisfies $|\int_{\partial \triangle_2} f(z)dz| \geqslant \frac{|I|}{4^2}$. We continue this practice indefinitely. Then we obtain a sequence of triangles $\triangle_1, \triangle_2, \triangle_3, \ldots, \triangle_n, \ldots$, where $\triangle_n$ contains $\triangle_{(n+1)}$ and $|\int_{\partial \triangle_n} f(z)dz| \geqslant \frac{|I|}{4^n}$. Denote the perimeter of $\triangle$ by $|\partial \triangle|$. Since each triangle in the sequence contains all the subsequent triangles and the perimeter $\frac{|\partial \triangle|}{2^n}$ of the triangle $\triangle_n$ approaches zero as $n \to \infty$, there is an $a \in \bigcap_{n=1}^{\infty} \triangle_n$. In addition, since f is complex differentiable at a, we see that for $\varepsilon > 0$, there is a $\delta > 0$ such that $B(a, 2\delta) \subset \Omega$ and for $|z - a| < \delta$,

$$|f(z) - f(a) - f'(a)(z - a)| \leqslant \varepsilon |z - a|.$$

Obviously, $\triangle_n \subset B(a, \delta)$ when n is sufficiently large, so

$$\leqslant \varepsilon \int_{\partial \triangle_n} |z - a||dz| \leqslant \varepsilon |\partial \triangle_n|^2 = \varepsilon 4^{-n} |\partial \triangle|^2.$$

Then $|I| \leqslant \varepsilon |\partial \triangle|^2$, showing $I = 0$. □

In complex analysis, we sometimes need the notion of uniform convergence on compact subsets of a region. Let $f_n(z)$ be analytic on the open set Ω_n ($n = 1, 2, 3, \ldots$). $\{f_n(z)\}$ is said to be *uniformly convergent to f on compact subsets of Ω*, where

$$\Omega = \bigcup_{N=1}^{\infty} \operatorname{int} \left(\bigcap_{n=N}^{\infty} \Omega_n \right),$$

if $\{f_n\}$ converges to f uniformly on compact subsets of Ω (which will be abbreviated to $f_n \to f$ u.c. on Ω), i.e., for any compact set $K \subset \Omega$ and any $\varepsilon > 0$, there exist N and $M = M(K, N)$ (dependent on K, N), such that $K \subset \operatorname{int}(\bigcap_{n=N}^{\infty} \Omega_n)$ and $|f_n(z) - f(z)| < \varepsilon$ for any $n \geqslant N$ and z of K (int E denoting the set of interior points of E).

Theorem 1.2.9 (Weierstrass's Theorem). *If Ω_n is open and $f_n \in H(\Omega_n)$ and $f_n \to f$ u.c. on Ω, where*

$$\Omega = \bigcup_{N=1}^{\infty} \operatorname{int} \left(\bigcap_{n=N}^{\infty} \Omega_n \right),$$

then $f \in H(\Omega)$ and $f_n^{(m)} \to f^{(m)}$ u.c. on Ω.

Proof. We first prove that f is continuous on Ω. For any $a \in \Omega$, there is a $r > 0$, such that the closed disk $\overline{D(a,r)} \subset \Omega$. Then there is an N such that $\overline{D(a,r)} \subset \operatorname{int}(\bigcap_{n=N}^{\infty} \Omega_n)$ and $\{f_n : n \geqslant N\}$ is convergent uniformly to f on $\overline{D(a,r)}$. As $f_n \in H(\Omega_n)$, f_n is necessarily continuous on $\overline{D(a,r)}$, and so, f is continuous on $\overline{D(a,r)}$. Since a is arbitrary, f is continuous on Ω. Now, we choose any triangle $\triangle \subset \overline{D(a,r)} \subset \Omega$. Since f_n converges uniformly to f on $\partial \triangle$ and $f_n \in H(\Omega_n)$, from Cauchy's Theorem, we have

$$\int_{\partial \triangle} f(z)dz = \lim_{n \to \infty} \int_{\partial \triangle} f_n(z)dz = 0.$$

Then it follows from Morera's theorem that $f \in H(\Omega)$. For any compact set $K \subset \Omega$, there is a $\omega \subset\subset \Omega$, such that $K \subset \omega$. By Theorem 1.2.6, we have that for $z \in K$,

$$f_n^{(m)}(z) = \frac{m!}{2\pi i} \int_{\partial \omega} \frac{f_n(\zeta)}{(\zeta-z)^{m+1}} d\zeta, \quad f^{(m)}(z) = \frac{m!}{2\pi i} \int_{\partial \omega} \frac{f(\zeta)}{(\zeta-z)^{m+1}} d\zeta.$$

Since f_n converges uniformly to f on $\overline{\omega}$, $\frac{f_n(\zeta)}{(\zeta-z)^{m+1}}$ converges uniformly to $\frac{f(\zeta)}{(\zeta-z)^{m+1}}$ on $\partial \omega$. Then

$$\lim_{n\to\infty} f_n^{(m)}(z) = \lim_{n\to\infty} \frac{m!}{2\pi i} \int_{\partial \omega} \frac{f_n(\zeta)}{(\zeta-z)^{m+1}} d\zeta$$

$$= \frac{m!}{2\pi i} \int_{\partial \omega} \frac{f(\zeta)}{(\zeta-z)^{m+1}} d\zeta = f^{(m)}(z).$$

Therefore, $f_n^{(m)}(z)$ converges uniformly to $f^{(m)}(z)$ on K. This proves that $f_n^{(m)} \to f^{(m)}$ u.c. on Ω. □

Suppose $f(z)$ is analytic on Ω. If $z_0 \in \Omega$ and $\overline{D(z_0, r)} \subset \Omega$ for $r > 0$, then the Cauchy integral formula gives

$$f(z_0) = \frac{1}{2\pi i} \int_{\partial D(z_0, r)} \frac{f(\zeta) d\zeta}{\zeta - z_0}.$$

Note that the point ζ on $\partial D(z_0, r)$ can be represented by $\zeta = z_0 + re^{it}, 0 \leqslant t \leqslant 2\pi$. Therefore, the Cauchy integral formula becomes

$$f(z_0) = \frac{1}{2\pi i} \int_0^{2\pi} \frac{f(z_0 + re^{it}) i re^{it}}{re^{it}} dt = \frac{1}{2\pi} \int_0^{2\pi} f(z_0 + re^{it}) dt.$$

This is the mean value property of analytic functions. It states that the value of $f(z)$ at $z = z_0$ is equal to the average value of $f(z_0)$ over $\partial D(z_0, r)$.

Suppose $f(z)$ is analytic on ω and continuous to the boundary $\partial \omega$. Then clearly, $|f(z)|$ is continuous on $\overline{\omega}$, so that the maximum value of $|f(z)|$ is attained on $\overline{\omega} = \omega \cup \partial \omega$. The maximum modulus theorem states that unless f is a constant, the maximum value of $|f(z)|$ is obtained only on the boundary $\partial \Omega$.

Theorem 1.2.10. *Let $f(z)$ be analytic on a bounded region Ω and continuous to the boundary $\partial \Omega$. Set $M = \max\{|f(z)| : z \in \overline{\Omega}\}$. Then $|f(z)| < M$ on Ω unless $f(z) = Me^{i\alpha}$ for constants M, α.*

This is the maximum modulus theorem. We now prove it by using the mean value property of analytic functions.

Proof. If there is a $z_0 \in \Omega$, such that $|f(z_0)| = M$, then $D = \{z \in \Omega : |f(z)| = M\} \neq \varnothing$. Since f is continuous on $\overline{\Omega}$, we have $\overline{D} \cap \Omega = D$. In the following, we show D is open. For $a \in D$, choose $r > 0$, such that $D(a, r) \subset \Omega$. Take r' such that $0 < r' < r$. By the mean value property of analytic functions, we have

$$M = |f(a)| = \left| \frac{1}{2\pi} \int_0^{2\pi} f(a + r'e^{it}) dt \right|$$

$$\leqslant \frac{1}{2\pi} \int_0^{2\pi} |f(a + r'e^{it})| dt \leqslant M.$$

Since the left and right ends of the above are equal, all equal signs hold, i.e., $|f(a + r'e^{it})| = M$ for all t and $0 < r' < r$. Therefore, $D(a, r) \subseteq D$, showing that D is open. Hence, D is a non-empty open and closed set in Ω. Since Ω is connected, we see that from Theorem 1.1.12, $D = \Omega$. So, $|f(z)|$ is constant on Ω. It then follows from the Cauchy–Riemann equation that $f(z)$ is constant on Ω.

Given a power series

$$\sum_{n=0}^{\infty} c_n (z - a)^n,$$

its "radius of convergence" R is determined by Hadamard's formula

$$\frac{1}{R} = \limsup_{n \to \infty} |c_n|^{1/n}.$$

If $R > 0$, the series converges in $D(a, R)$. If $R < \infty$, the series diverges in $\{z : |z - a| > R\}$.

We say a function f defined on an open set Ω has a power series representation on Ω if for every disk $D(a, r) \subset \Omega$, there is a power series $\sum_{n=0}^{\infty} c_n (z-a)^n$ that is convergent to $f(z)$ for every $z \in D(a, r)$. □

Theorem 1.2.11. *If f has a power series representation on Ω, then $f \in H(\Omega)$ and f' has a power series representation on Ω. In fact, if*

$$f(z) = \sum_{n=0}^{\infty} c_n (z - a)^n \qquad (1.2.10)$$

holds for $z \in D(a, R) \subset \Omega$, then for these z, we have

$$f'(z) = \sum_{n=1}^{\infty} n c_n (z-a)^{n-1}. \tag{1.2.11}$$

Proof. Let $f_n(z) = \sum_{k=0}^{n} c_k (z-a)^k$. Then $f_n(z) \in H(D(a,R))$. Since $f_n \to f$ u.c. on $D(a,R)$, $\lim_{n \to \infty} f_n(z) = f(z) \in H(D(a,R))$, so (1.2.10) holds. By Weierstrass's theorem, $f'_n(z) \to f'(z)$ u.c. on $D(a,R)$, and thus, (1.2.11) holds. $\square$

Corollary 1.2.12. *If f has a power series representation on Ω, then f is infinitely complex differentiable on Ω and the kth derivative $f^{(k)}(z)$ of f has a power series representation on Ω, i.e.,*

$$f^{(k)}(z) = \sum_{n=k}^{\infty} n(n-1) \cdots (n-k+1) c_n (z-a)^{n-k}.$$

Therefore,

$$k! c_k = f^{(k)}(a) \quad (k = 0, 1, 2, \ldots), \tag{1.2.12}$$

and so, for every $a \in \Omega$, there is a sequence $\{c_n\}$ such that (1.2.10) holds.

Theorem 1.2.13 (Cauchy–Type Integral). *Let μ be a finite complex measure on a measurable space X and ϕ a measurable function on X. If Ω is an open set in the plane that is disjoint to $\phi(X)$ and*

$$f(z) = \int_X \frac{d\mu(\zeta)}{\phi(\zeta) - z} \quad (z \in \Omega),$$

then f has a power series representation on Ω. In fact, if $D(a, R) \subset \Omega$, then for $z \in D(a, R)$,

$$f(z) = \sum_{n=0}^{\infty} c_n (z-a)^n, \tag{1.2.13}$$

where

$$c_n = \int_X \frac{d\mu(\zeta)}{(\phi(\zeta) - z)^{n+1}} \quad (n = 0, 1, 2, \ldots).$$

Proof. If $D(a, R) \subset \Omega$, then for $|z - a| < r < R$,

$$\frac{1}{\phi(\zeta) - z} = \sum_{n=0}^{\infty} \frac{(z-a)^n}{(\phi(\zeta) - a)^{n+1}}$$

converges uniformly with respect to $\zeta \in X$, so
$$f(z) = \sum_{n=0}^{\infty} \int_X \frac{d\mu(\zeta)}{(\phi(\zeta)-a)^{n+1}}(z-a)^n.$$
Thus, (1.2.13) holds. □

Theorem 1.2.14 (Taylor's Theorem). *If $f \in H(\Omega)$, then f has a power series representation on Ω, i.e.,*
$$f(z) = \sum_{n=0}^{\infty} c_n (z-a)^n \qquad (1.2.14)$$
for $z \in D(a,R) \subset \Omega$, where
$$c_n = \frac{1}{2\pi i} \int_{\partial D(a,r)} \frac{f(\zeta)d\zeta}{(\zeta-a)^{n+1}} \quad (0 < r < R,\ n = 0, 1, 2, \ldots). \qquad (1.2.15)$$

Proof. If $D(a,R) \subset \Omega$, then for $|z-a| < r < R$, the Cauchy formula gives
$$f(z) = \frac{1}{2\pi i} \int_{\partial D(a,r)} \frac{f(\zeta)d\zeta}{\zeta - z}.$$
Now, it follows from Theorem 1.2.13 that Theorem 1.2.14 holds. □

Theorem 1.2.15 (Cauchy Inequality). *If $f \in H(D(a,R))$ and $|f(z)| \leq M$ for any $z \in D(a,R)$, then*
$$|f^{(n)}(a)| \leq \frac{n!M}{R^n} \quad (n = 1, 2, 3, \ldots). \qquad (1.2.16)$$

Proof. From (1.2.12) and (1.2.15), we have (1.2.16). □

Theorem 1.2.16 (Liouville's Theorem). *Every bounded entire function $f \in H(\mathbb{C})$ is constant.*

Proof. From (1.2.16), we see that for any $a \in \mathbb{C}$, $f'(a) = 0$, so for any $z \in \mathbb{C}$, $f(z) = f(0)$. □

Theorem 1.2.17 (Fundamental Theorem of Algebra). *Let n be a positive integer. If*
$$P(z) = z^n + a_{n-1}z^{n-1} + \cdots + a_1 z + a_0,$$

where $a_0, \ldots, a_{n-1}$ are complex numbers, then P has at least one root in the plane.

Proof. If P has no roots in the complex plane, then $\frac{1}{P}$ is a bounded and entire function. From Liouville's theorem, we see that $\frac{1}{P}$ is constant, and thus, P is constant, a contradiction. □

Theorem 1.2.18 (Isolation of Zeros). *If Ω is a domain, $f \in H(\Omega)$, $f \not\equiv 0$, and $Z(f) = \{a \in \Omega : f(a) = 0\}$, then $Z(f)$ has no limit points in Ω. For every $a \in Z(f)$, there is a unique positive integer $m = m(a)$ such that*

$$f(z) = (z-a)^m g(z) \quad (z \in \Omega),$$

where $g \in H(\Omega)$ and $g(a) \neq 0$. In addition, $Z(f)$ is at most countable.

Proof. Let $a \in Z(f)$. By Theorem 1.2.14, there is a $R > 0$ such that

$$f(z) = \sum_{n=1}^{\infty} c_n (z-a)^n, \quad n! c_n = f^{(n)}(a) \quad (n = 0, 1, 2, \ldots)$$

for $z \in D(a, R) \subset \Omega$.

If $f^{(n)}(a) = 0$ $(n = 0, 1, 2, \ldots)$, then $D = \{b \in \Omega : f^{(n)}(b) = 0, n = 0, 1, 2, \ldots\}$ is non-empty open set. Since $f^{(n)}(z)$ is continuous, we have $f^{(n)}(b) = 0$ $(n = 0, 1, 2, \ldots)$ for any $b \in \overline{D} \cap \Omega$. Therefore, $\overline{D} \cap \Omega \subset D$. By Theorem 1.1.12, $D = \Omega$, so $f(z) \equiv 0$, a contradiction.

If there is an $m = m(a)$ such that $c_k = 0$ $(k = 0, 1, \ldots, m-1)$, $c_m \neq 0$, then on $D(a, R)$ we have

$$f(z) = \sum_{n=m}^{\infty} c_n (z-a)^n = (z-a)^m g_1(z),$$

where $g_1 \in H(D(a, R))$ and $g_1(a) \neq 0$. Set $g(z) = (z-a)^{-m} f(z)$, $z \in \Omega \setminus \{a\}$ and $g(a) = g_1(0)$. Then $g(z)$ meets the requirement. □

Corollary 1.2.19 (Uniqueness Theorem). *If f and g are analytic on Ω and $f(z) = g(z)$ on a set of Ω that has a limit point, then $f(z) = g(z)$ for all $z \in \Omega$.*

Proof. By Theorem 1.2.18. □

Theorem 1.2.20 (Laurent's Theorem). *Let $f \in H(\Omega)$, where $\Omega = \{z : r < |z - a| < R\}$ with $0 \leqslant r < R$. Then on Ω the function f has a Laurent series representation, i.e.,*

$$f(z) = \sum_{n=-\infty}^{\infty} c_n (z-a)^n, \tag{1.2.17}$$

where

$$c_n = \frac{1}{2\pi i} \int_{\partial D(a,\rho)} \frac{f(\zeta) d\zeta}{(\zeta - a)^{n+1}} \quad (r < \rho < R, n \in \mathbb{Z}). \tag{1.2.18}$$

Proof. If $0 < 2\varepsilon < R - r$, then for $r + \varepsilon < |z - a| < R - \varepsilon$, the Cauchy formula gives

$$f(z) = \frac{1}{2\pi i} \int_{\partial D(a,R-\varepsilon)} \frac{f(\zeta) d\zeta}{\zeta - z} - \frac{1}{2\pi i} \int_{\partial D(a,r+\varepsilon)} \frac{f(\zeta) d\zeta}{\zeta - z}$$

$$= \frac{1}{2\pi i} \int_{\partial D(a,r+\varepsilon)} \sum_{n=-\infty}^{-1} \frac{(z-a)^n}{(\zeta - a)^{n+1}} f(\zeta) d\zeta$$

$$+ \frac{1}{2\pi i} \int_{\partial D(a,R-\varepsilon)} \sum_{n=0}^{\infty} \frac{(z-a)^n}{(\zeta - a)^{n+1}} f(\zeta) d\zeta.$$

From Fubini's theorem, we see that (1.2.17) holds, where

$$c_n = \frac{1}{2\pi i} \int_{\partial D(a,R-\varepsilon)} \frac{f(\zeta) d\zeta}{(\zeta - a)^{n+1}} \quad (n = 0, 1, 2, \ldots)$$

and

$$c_n = \frac{1}{2\pi i} \int_{\partial D(a,r+\varepsilon)} \frac{f(\zeta) d\zeta}{(\zeta - a)^{n+1}} \quad (n = -1, -2, -3, \ldots).$$

Applying the Cauchy formula again gives (1.2.18). □

Definition 1.2.21. *If $a \in \Omega$ and $f \in H(\Omega \setminus \{a\})$, then a is called an isolated singular point of f. Then there is a $R > 0$, such that $D(a, R) \subset \Omega$ and $f(z)$ has a Laurent series representation (1.2.17) on $\{z : 0 < |z - a| < R\}$. The coefficient*

$$c_{-1} = \frac{1}{2\pi i} \int_{\partial D(a,\rho)} f(\zeta) d\zeta \quad (0 < \rho < R)$$

independent of ρ is called the residue of f at a, denoted by $\text{Res}(f, a)$. The series

$$\sum_{n=-\infty}^{-1} c_n (z-a)^n$$

is called the principal part of f at a. If f can be redefined at a such that the extended function is analytic on Ω, the singularity a is called a removable singular point.

Theorem 1.2.22. *Suppose $f \in H(\Omega\setminus\{a\})$ and there is a $r > 0$, such that f is bounded in the punctured disk $D'(a,r) = \{z : 0 < |z-a| < r\} \subset \Omega$ with center a, then a is a removable singular point of f.*

Proof. Set $h(a) = 0$ and $h(z) = (z-a)^2 f(z)$ for $z \in \Omega \setminus \{a\}$. Then from the assumption of the theorem, we conclude that $h'(a) = 0$, $h \in H(\Omega)$. Thus, on $D(a,r)$, the function h can be represented by a power series

$$h(z) = \sum_{n=2}^{\infty} c_n (z-a)^n.$$

Letting $f(a) = c_2$, we see that on $D(a,r)$, the function f is represented by the power series

$$f(z) = \sum_{n=0}^{\infty} c_{n+2} (z-a)^n.$$

This completes the proof of the theorem. □

Theorem 1.2.23. *If $a \in \Omega$ and $f \in H(\Omega \setminus \{a\})$, then one of the following three cases must hold:*

(a) *a is a removable singular point of f;*
(b) *there are complex numbers $c_1, \ldots, c_m$, where m is a positive integer and $c_m \neq 0$, such that a is a removable singular point of*

$$f(z) - \sum_{k=1}^{m} \frac{c_k}{(z-a)^k}$$

(in this case, a is called a pole of order m of f);
(c) *If $r > 0$ and $D(a,r) \subset \Omega$, then $f(D'(a,r))$ is dense in the plane (in this case, a is called an essential singular point of f).*

Proof. If (c) does not hold, then there exist $r > 0, \delta > 0$ and w_0 such that $|f(z) - w_0| \geq \delta$ on the punctured neighborhood $D'(a,r) = \{z : 0 < |z - a| < r\} \subset \Omega$. Clearly, the function $g(z) = (f(z) - w_0)^{-1}$ is analytic on $D'(a,r)$. From Theorem 1.2.22, $g(z)$ can be extended to become analytic on $D(a,r)$. If $g(a) \neq 0$, then f is bounded on some punctured neighborhood $D'(a,\rho) = \{z : 0 < |z - a| < \rho\} \subset \Omega$. Then (a) is true because of Theorem 1.2.22. If $g(z)$ has a zero of order m at a, then it follows from Theorem 1.2.18 that there is a $g_1(z) \in H(D(a,r))$ such that $g_1(a) \neq 0$, and $g(z) = (z-a)^m g_1(z)$ for $z \in D(a,r)$. Then $g_1(z) \neq 0$ for $z \in D(a,r)$. Set $h(z) = \frac{1}{g_1(z)}$. Then $h(z) \in H(D(a,r))$, so $f(z) = w_0 + (z-a)^{-m} h(z)$. Since $h(z)$ can be represented by a power series $h(z) = \sum_{n=0}^{\infty} b_n (z-a)^n$ on $D(a,r)$, where $b_0 \neq 0$, (b) holds for $c_k = b_{m-k}, k = 1, 2, \ldots, m$. □

Theorem 1.2.24. *Suppose*
$$f(z) = \sum_{n=0}^{\infty} c_n (z-a)^n, \quad z \in D(a, R),$$
and $0 < r < R$. Then
$$\sum_{n=0}^{\infty} |c_n|^2 r^{2n} = \frac{1}{2\pi} \int_{-\pi}^{\pi} |f(a + re^{i\theta})|^2 d\theta. \tag{1.2.19}$$

Proof. For $0 < r < R$, we have
$$f(a + re^{i\theta}) = \sum_{n=0}^{\infty} c_n r^n e^{in\theta}.$$
Since this series converges uniformly on $[-\pi, \pi]$, we see that (1.2.19) is true. □

Lemma 1.2.25. *Let Ω be a region. The following conclusions hold:*

(i) *If $f \in H(\Omega)$ and for any closed polygonal line γ in Ω,*
$$\int_{\gamma} f(z) dz = 0, \tag{1.2.20}$$
then there is an $F \in H(\Omega)$, such that for any $z \in \Omega$, $F'(z) = f(z)$.

(ii) If $g \in H(\Omega)$ has no zeros in Ω, and for any closed polygonal line γ in Ω, $f(z) = \frac{g'(z)}{g(z)}$ satisfies (1.2.20), then there is an $h \in H(\Omega)$, such that $e^h = g$. Correspondingly, we write $h = \log g$.

(iii) Let m be a positive integer. If $g \in H(\Omega)$ has no zeros in Ω, and for any closed polygonal line γ in Ω, $f(z) = \frac{g'(z)}{g(z)}$ satisfies (1.2.20), then there is a $g_m \in H(\Omega)$, such that $g_m^m = g$. Correspondingly, we write $g_m = \sqrt[m]{g}$.

(iv) Let Ω be a simply connected domain. Suppose $f \in H(\Omega)$. Then for any closed polygonal line γ in Ω, $f(z) = \frac{g'(z)}{g(z)}$ satisfies (1.2.20). Therefore, there is an $F \in H(\Omega)$, such that for any $z \in \Omega$, $F'(z) = f(z)$. In addition, if m is a positive integer and $f \in H(\Omega)$ has no zeros in Ω, then there exist $h, g_m \in H(\Omega)$, such that $e^h = f$ and $g_m^m = f$.

Proof. (i) By Morera's theorem, we have $f \in H(\Omega) \cap C^1(\Omega)$. Fix a point $z_0 \in \Omega$ and define $F(z) = \int_{\Gamma_z} f(\zeta) d\zeta$, where Γ_z is a polygonal line in Ω that starts with z_0 and ends with z. From the assumption, we see that $F(z)$ is independent of the choice of Γ_z. According to our earlier proof of Morera's theorem, it is seen that $F \in H(\Omega)$ and $F'(z) = f(z)$.

(ii) By the assumption and (i), there is an $F_0 \in H(\Omega)$ such that $F'(z) = g'(z)/g(z)$. Then $(ge^{-F_0})' = g'e^{-F_0} - ge^{-F_0}F_0' = 0$, so $ge^{-F_0} \equiv c$, where c is constant. Set $h = F_0 + a$, were $e^a = c$. Obviously, $h' = F_0' = g'/g$ and $e^h = e^{F_0+a} = ce^{F_0} = g$.

(iii) Set $g_m = e^{h/m}$ for h from (ii).

(iv) If Ω is simply connected and $f \in H(\Omega)$, then Goursat's theorem and Cauchy's theorem show that (1.2.20) holds for any closed polygonal line γ in Ω. $\square$

Definition 1.2.26. A function f is said to be meromorphic in an open set Ω if these is a set $B \subset \Omega$, such that

(a) B has no limit points in Ω;
(b) $f \in H(\Omega \setminus B)$;
(c) every point in B is a pole of f.

Theorem 1.2.27 (Residue Theorem). *If $f \in H(\overline{\omega} \setminus A)$, where $A = \{a_1, a_2, \ldots, a_n\} \subset \omega$, then*

$$\frac{1}{2\pi i} \int_{\partial \omega} f(z) dz = \sum_{k=1}^{n} \text{Res}(f, a_k). \tag{1.2.21}$$

Proof. There is an $\varepsilon > 0$ such that $D(a_k, \varepsilon) \subset \omega$ and $\{D(a_k, \varepsilon) : k = a_1, a_2, \ldots, a_n\}$ is pairwise disjoint. By the Cauchy formula, we have

$$\frac{1}{2\pi i} \int_{\partial \omega} f(z) dz = \sum_{k=1}^{n} \frac{1}{2\pi i} \int_{\partial D(a_k, \varepsilon)} f(z) dz = \sum_{k=1}^{n} \text{Res}(f, a_k). \quad \square$$

Theorem 1.2.28 (Argument Principle). *Let $\omega \subset\subset \Omega$. If f is meromorphic on the open set ω and has neither a zero nor a pole on $\partial \omega$, then*

$$\frac{1}{2\pi i} \int_{\partial \omega} \frac{f'(z)}{f(z)} dz = N_f - P_f,$$

where N_f represents the number of zeros of f in ω (a zero of multiplicity k is counted k times) and P_f is the number of poles of f in ω (a pole of order k is counted k times).

Proof. Let B be the collection of zeros and poles of f in ω. Since ω is bounded, B is finite. Assume $B = \{a_1, a_2, \ldots, a_n\}$. Then there is an $\varepsilon > 0$ such that $D(a_k, \varepsilon) \subset \omega$ and $\{D(a_k, \varepsilon) : k = a_1, a_2, \ldots, a_n\}$ is pairwise disjoint. Using the Cauchy formula, we have

$$\frac{1}{2\pi i} \int_{\partial \omega} \frac{f'(z)}{f(z)} dz = \sum_{k=1}^{n} \frac{1}{2\pi i} \int_{\partial D(a_k, \varepsilon)} \frac{f'(z)}{f(z)} dz$$

$$= \sum_{k=1}^{n} \text{Res}\left(\frac{f'(z)}{f(z)}, a_k\right) = N_f - P_f.$$

This completes the proof of the theorem. $\square$

Theorem 1.2.29 (Rouché's Theorem). *(a) Let $\omega \subset\subset \Omega$, $f_t(z) = f(t, z)$, $\frac{\partial}{\partial z} f(t, z) \in C([0,1] \times \Omega)$, $f_t \in H(\Omega)$ for $t \in [0,1]$, and $f(t, z) \neq 0$ for $(t, z) \in [0,1] \times \partial \omega$. Let N_{f_t} be the number of zeros of f_t in ω, counting multiplicities. Then $N_{f_0} = N_{f_1}$.*

(b) Let $\omega \subset \Omega$. Suppose that ω is bounded and its boundary $C = \partial \omega \subset \Omega$ is a closed piecewise smooth Jordan curve. If $f(z)$ and $g(z)$ are analytic on the open set Ω and

$$|f(z) - g(z)| < |f(z)| + |g(z)|$$

on C. Then $f(z)$ and $g(z)$ have the same number of zeros, counting multiplicities, in ω.

Proof. From Theorem 1.2.28, the function

$$\varphi(t) = \frac{1}{2\pi i} \int_{\partial \omega} \frac{\frac{\partial}{\partial z} f(t, z)}{f(t, z)} dz$$

is continuous on $[0, 1]$ and takes values of integers. Thus, $\varphi(0) = \varphi(1)$. This proves (a).
(b) Since $|f(z) - g(z)| < |f(z)| + |g(z)|$ on C, $f(z)$ and $g(z)$ have no zeros on C. Set

$$h(z) = \frac{g(z)}{f(z)}, \quad f_t(z) = f(t, z) = (1-t)f(z) + tg(z), \quad (t, z) \in [0,1] \times C.$$

Then for any $z \in C$, $|h(z) - 1| < 1 + |h(z)|$. If there exists $(t, z) \in [0, 1] \times C$ such that $f(t, z) = 0$, then $0 < t < 1, (1 - t) + th(z) = 0$, so $h(z) = 1 - t^{-1} < 0$. Then there exists a $z \in C$ such that $|h(z) - 1| = t^{-1} - 1 + 1 = 1 + |h(z)|$, which contradicts the fact that $|h(z) - 1| < 1 + |h(z)|$ for any $z \in C$. Therefore, by (a), we have $N_{f_0} = N_{f_1}$. Since $f_0(z) = f(z), f_1(z) = g(z)$, we see that $f(z)$ and $g(z)$ have the same number of zeros in ω. $\square$

Theorem 1.2.30. Suppose that f is analytic on a region Ω and $b = f(a), a \in \Omega$. If $f(z) - b$ has a zero of order m at a, then there is an $\varepsilon_0 > 0$, such that for any $\varepsilon \in (0, \varepsilon_0]$, there is a $\delta > 0$, such that for every A satisfying $0 < |A - b| < \delta$, the function $f(z) - A$ has exactly m distinct zeros of order 1 in $|z - a| < \varepsilon$.

Proof. Assume a is a zero of order m of $f(z) - b$. By Theorem 3.3.5, there is an analytic function $h(z)$ on Ω such that $f(z) - b = (z - a)^m h(z)$ and $h(a) \neq 0$. Clearly, the derivative $f'(z) = m(z - a)^{m-1}h(z) + (z - a)^m h'(z)$. Since $h(z)$ and $mh(z) + (z - a)h'(z)$ are continuous at a, there are an $\varepsilon_0 > 0$ and an open disk $D(a, \varepsilon_0) \subset \overline{D(a, \varepsilon_0)} \subset \Omega$, such that $h(z)$ and

$mh(z) + (z-a)h'(z)$ have no zeros in $\overline{D(a,\varepsilon_0)}$, and thus, $f(z) - b$ and $f'(z)$ have no zeros in $\overline{D(a,\varepsilon_0)} \setminus \{a\}$. Since for any $\varepsilon \in (0, \varepsilon_0], f(z) - b$ is analytic on $D = \overline{D(a,\varepsilon)}$ and has no zeros on ∂D, $|f(z) - b|$ attains its minimum $\min\{|f(z) - b| : z \in \partial D\} = \delta > 0$ on ∂D. For any A satisfying $0 < |A - b| < \delta$, we have $f(z) - A = f(z) - b + b - A$ and $|A - b| < \delta \leqslant |f(z) - b|$ on ∂D. Then from Rouché's Theorem, $f(z) - A$ and $f(z) - b$ have the same number of zeros in $D(a,\varepsilon)$. Then $f(z) - A$ has m zeros in $D(a,\varepsilon)$. Since $f'(z)$ has no zeros in $\overline{D(a,\varepsilon)} \setminus \{a\}$, these m zeros are simple zeros, i.e., $f(z) - A$ has exactly m distinct zeros of order 1 in $|z - a| < \varepsilon$. $\square$

Theorem 1.2.31. *Let Ω be a region. If $f \in H(\Omega)$ and f is not constant, then $f(\Omega)$ is a region.*

Proof. For any $a \in \Omega$, a is an isolated zero of $f(z) - f(a)$ because of isolation of zeros. By Theorem 1.2.30, there are an $\varepsilon_0 > 0$, an open disk $D(a,\varepsilon_0) \subset \overline{D(a,\varepsilon_0)} \subset \Omega$, and a $\delta_0 > 0$, such that for any w satisfying $0 < |w - f(a)| < \delta_0$, $f(z) - w$ has at least one zero in $|z - a| < \varepsilon_0$. So, $D(f(a), \delta_0) \subset f(D(a,\varepsilon_0)) \subset f(\Omega)$, showing $f(\Omega)$ is open. For any $w_1, w_2 \in f(\Omega)$, there are $z_1, z_2 \in \Omega$ such that $w_1 = f(z_1), w_2 = f(z_2)$. Since Ω is connected, there is a continuous curve $L : z = \gamma(t)$ $(0 \leqslant t \leqslant 1)$ in Ω that joins z_1 and z_2. Then, the continuous curve $f(\gamma(t))$ lies in $f(\Omega)$ and joins w_1 and w_2. Therefore, $f(\Omega)$ is connected. $\square$

Theorem 1.2.32. *If $f \in H(\Omega)$ and is one to one (in this case, f is called univalent on Ω), then for any $z \in \Omega, f'(z) \neq 0$. Conversely, if $f \in H(\Omega), a \in \Omega, f'(a) \neq 0$, then there is a neighborhood V of a, such that f is univalent on V.*

Proof. First, we prove that if $f \in H(\Omega)$ and is one to one, then for any $z \in \Omega, f'(z) \neq 0$. If otherwise, there is an $a \in \Omega$, such that $f'(a) = 0$, then a is a zero of order $m (m \geqslant 2)$ of $f(z) - f(a)$. By Theorem 1.2.30, there are $\varepsilon > 0$ and $\delta > 0$, such that for every w satisfying $0 < |w - f(a)| < \delta$, the function $f(z) - w$ has at least $m (\geq 2)$ distinct zeros in $|z - a| < \varepsilon$. This contradicts the fact that $f(z)$ is univalent on Ω, and thus, $f'(z) \neq 0$ on Ω.

Conversely, if $f'(a) \neq 0$, then a is a zero of order 1 of $f(z) - f(a)$. From Theorem 1.2.30, there are an $\varepsilon > 0$ and a $\delta > 0$, such that for every w with $0 < |w - f(a)| < \delta$, the function $f(z) - w$ has exactly one zero in $|z-a|<\varepsilon$. By the continuity of $f(z)$, there is an $\varepsilon_1>0$ ($\varepsilon_1 \leqslant \varepsilon$), such that $f(D(a,\varepsilon_1)) \subset D(f(a),\delta)$. Then $f(z)$ is univalent on $D(a,\varepsilon_1)$. □

Theorem 1.2.33. *[Inverse Function Theorem] Let Ω be a region. If $f \in H(\Omega)$ is univalent, then $f(\Omega)$ is a region. Let $z = g(w)$ be the inverse function of $w = f(z)$, then g is univalent on $W = f(\Omega)$ and $g \in H(W)$, $g'(w)|_{w=f(z)} = 1/f'(z)$.*

Proof. From Theorem 1.2.31, $f(\Omega)$ is a region. Next, we show that $g(w)$ is continuous on $W = f(\Omega)$. Given $w_0 \in W$, there exists a unique $z_0 \in \Omega$, such that $g(w_0) = z_0$. For any $\varepsilon > 0$, there is an $\varepsilon_1 \leqslant \varepsilon (\varepsilon_1 > 0)$, such that $D(z_0,\varepsilon_1) \subset \Omega$. Theorem 1.2.31 shows that $f(D(z_0,\varepsilon_1))$ is open, and thus, there is a $\delta > 0$, such that $D(w_0,\delta) \subset f(D(z_0,\varepsilon_1))$, i.e., $g(D(w_0,\delta)) \subset D(z_0,\varepsilon_1)$. Then $|g(w) - g(w_0)| < \varepsilon_1 \leqslant \varepsilon$ for $|w-w_0| < \delta$. So, $g(w)$ is continuous at w_0. Since w_0 is arbitrary, $g(w)$ is continuous on $W = f(\Omega)$. Now, for any $w_0 \in W$, $z = g(w) \to z_0 = g(w_0)$ as $w \to w_0$, so

$$g'(w_0) = \lim_{w \to w_0} \frac{g(w) - g(w_0)}{w - w_0} = \lim_{z \to z_0} \frac{z - z_0}{f(z) - f(z_0)} = \frac{1}{f'(z_0)}.$$

Since w_0 is arbitrary, we see that $g \in H(W)$ and $g'(w) = 1/f'(z)$. □

A general non-constant analytic function f may be written in some form, as shown in the following theorem.

Theorem 1.2.34. *Let Ω be a domain. If $f \in H(\Omega)$, f is not constant, and $f(z) - f(a)$ has a zero of order m at $z = a$, then there are a neighborhood (of a) $V \subset \Omega$, a function $\phi \in H(V)$, and a $r > 0$, such that*

(a) $f(z) = f(a) + (\phi(z))^m$;
(b) $\phi'(z) \neq 0$ on V and $\phi: V \to D(0,r)$ is bijective.

Proof. Clearly, a is a zero of $f(z) - f(a)$. Since f is not constant, a is an isolated zero of $f(z) - f(a)$. Therefore, there is an open disk $D \subset \Omega$ that contains a, such that $f(z) - f(a)$ has no zeros in $D \setminus \{a\}$. On D, we have $f(z) - f(a) = (z-a)^m g(z)$, where $g(z)$ is analytic on D and has no zeros in D. Since D is convex, from Lemma 1.2.25 (b), we see that there is an $h \in H(D)$, such that $h^m = g$. Let $\phi(z) = (z-a)h(z)$. Then $f(z) = f(a) + (\phi(z))^m$. Note that $\phi(a) = 0$,

$\phi'(a) \neq 0$. By Theorem 1.2.32, there is a neighborhood $\tilde{V} \subset D$ of a, such that $\phi(\tilde{V})$ is a neighborhood of $\phi(a) = 0$ and ϕ is a bijection from $\tilde{V}$ to $\phi(\tilde{V})$. Choose $r > 0$ such that $D(0,r) \subset \phi(\tilde{V})$ and let $V = \phi^{-1}(D(0,r)) \subset \tilde{V}$. Note that $(\phi^{-1})'(0) = 1/\phi'(a) \neq 0$. Again, by Theorem 1.2.32, there is a neighborhood $V_1 \subset D(0,r)$ of 0, such that $\phi^{-1}(V_1)$ is a neighborhood of $\phi^{-1}(0) = a \subset V$. Hence, V is a neighborhood of a. The proofs of (a) and (b) are complete. □

1.3 Global Cauchy Theorem

Definition 1.3.1 (Closed Chain and Index). Suppose Γ_1, Γ_2, ..., Γ_N are directed piecewise smooth curves (referred to as paths). Each Γ_k may be treated as a bounded linear functional on $C(\Gamma_k^*)$, defined by $\Gamma_k(f) = \int_{\Gamma_k} f(z)dz$, $f \in C(\Gamma_k^*)$. Then we may define a bounded linear functional $\Gamma = \sum_1^N \Gamma_k$ on $C(\bigcup_{k=1}^N \Gamma_k^*)$ as follows: $\Gamma(f) = \sum_{k=1}^N \int_{\Gamma_k} f(z)dz$, $f \in C(\bigcup_{k=1}^N \Gamma_k^*)$. We call a Γ so defined a chain and denote $\Gamma^* = \bigcup_{k=1}^N \Gamma_k^*$. If $\Gamma_1, \Gamma_2, \ldots, \Gamma_N$ all are closed paths, we call Γ a closed chain.

We now define the index of a closed chain with respect to a point. Let Γ be a closed chain defined above and $z \notin \Gamma^* = \bigcup_{k=1}^N \Gamma_k^*$. The index of Γ with respect to z is defined to be

$$Ind_\Gamma(z) = \frac{1}{2\pi i} \int_\Gamma \frac{d\zeta}{\zeta - z} = \sum_{k=1}^N \frac{1}{2\pi i} \int_{\Gamma_k} \frac{d\zeta}{\zeta - z}.$$

In the following, we discuss the case where Γ is a closed path. Let $\Gamma : z = \Gamma(t) : [0,1] \to \mathbb{C}$ be a closed path and $0 \notin \Gamma^*$. Consider the index of Γ with respect to 0: $Ind_\Gamma(0) = \frac{1}{2\pi i}\int_\Gamma \frac{d\zeta}{\zeta}$. Since $0 \notin \Gamma^*$ and Γ is a closed path, there are $\rho(t) = |\Gamma(t)|$ and a piecewise smooth $\vartheta(t)$ on $[0,1]$, such that $\Gamma(t) = \rho(t)e^{i\vartheta(t)}$, and thus,

$$Ind_\Gamma(0) = \frac{1}{2\pi i}\int_\Gamma \frac{d\zeta}{\zeta} = \frac{1}{2\pi i}\int_0^1 \left(\frac{\rho'(t)}{\rho(t)} + i\vartheta'(t)\right) dt = \frac{1}{2\pi}(\vartheta(1) - \vartheta(0)).$$

Since $\Gamma(0) = \Gamma(1)$, we have $e^{i\vartheta(0)} = e^{i\vartheta(1)}$ and then $\vartheta(1) - \vartheta(0)$ is an integer multiple of 2π. It represents the amount of change in the argument when z moves on Γ along the given direction for one cycle.

Hence, $Ind_\Gamma(0)$ is an integer, showing the number of times that the closed path Γ winds around the origin. Similarly, for $z \notin \Gamma^*$, $Ind_\Gamma(z)$ is an integer, representing the number of times that the closed path Γ winds around the point z. An index has the following properties:

Theorem 1.3.2. *If Γ is a closed chain, then*

(a) *$Ind_\Gamma(z)$ takes integer values;*
(b) *$Ind_\Gamma(z)$ is constant on each connected component of $\mathbb{C} \setminus \Gamma^*$, and $Ind_\Gamma(z) = 0$ on each unbounded connected component.*

Proof. (a) Assume $\Gamma = \sum_1^N \Gamma_k$, where Γ_k is a closed path. From the above analysis, $Ind_{\Gamma_k}(z)$ is an integer. Therefore, $Ind_\Gamma(z) = \sum_1^N Ind_{\Gamma_k}(z)$ is an integer.

(b) By Theorem 1.2.13, $Ind_\Gamma(z) \in H(\mathbb{C} \setminus \Gamma^*)$. Therefore, $Ind_\Gamma(z)$ is constant on each connected component of $\mathbb{C} \setminus \Gamma^*$. In the following, we show that $Ind_\Gamma(z) = 0$ on each unbounded connected component. Since Γ^* is compact, there is $r > 0$, such that $\Gamma^* \subset D(0, r)$. For any $\varepsilon > 0$, choose $N > 0$, so that $\frac{|\Gamma|}{2\pi(N-r)} < \varepsilon$. Then if $|z| > N$,

$$\left| \frac{1}{2\pi i} \int_\Gamma \frac{d\zeta}{\zeta - z} \right| \leqslant \frac{1}{2\pi |z|} \int_\Gamma \left| \frac{1}{1 - \zeta/z} \right| |d\zeta| \leqslant \frac{|\Gamma|}{2\pi(N - r)} < \varepsilon,$$

where $|\Gamma|$ denotes the length of Γ. Note that for any $\zeta \in \Gamma$, we have $|\zeta| < r$, and so, $|1 - \zeta/z| \geqslant 1 - |\zeta|/|z| > 1 - r/N$. Then, $Ind_\Gamma(z) \to 0$ $(z \to \infty)$. Since $Ind_\Gamma(z)$ is constant on each connected component, it must be 0 on each unbounded connected component. This completes the proof. □

Definition 1.3.3 (Homology). Suppose that Γ_1 and Γ_2 are two closed chains in Ω. If $Ind_{\Gamma_1}(a) = Ind_{\Gamma_2}(a)$ for any $a \notin \Omega$, Γ_1 and Γ_2 are said to be homologous in Ω. In particular, a closed chain Γ in Ω is said to be homologous to zero in Ω if $Ind_\Gamma(a) = 0$ for any $a \notin \Omega$.

Theorem 1.3.4 (Homology Version of Cauchy's Theorem). *Suppose $f \in H(\Omega)$*

(a) *If a closed chain Γ is homologous to zero in Ω, then*

$$\int_\Gamma f(\zeta)d\zeta = 0 \qquad (1.3.1)$$

and

$$f(z)Ind_\Gamma(z) = \frac{1}{2\pi i}\int_\Gamma \frac{f(\zeta)}{\zeta - z}d\zeta \quad (z \in \Omega \setminus \Gamma^*). \qquad (1.3.2)$$

(b) *If closed chains Γ_1 and Γ_2 are homologous in Ω, then*

$$\int_{\Gamma_1} f(\zeta)d\zeta = \int_{\Gamma_2} f(\zeta)d\zeta.$$

Proof. (a) By Theorem 1.3.2, supp (Ind_Γ) is compact, and so, $H = \text{supp}\,(Ind_\Gamma) \cup \Gamma^*$ is a compact set in Ω. Then there is an ω such that $H \subset \omega \subset\subset \Omega$. Choose a $\phi \in C_c^\infty(\omega)$, such that $\phi \equiv 1$ in a neighborhood of H. Noting that $f\phi \in C_c^\infty(\omega)$ and using Theorem 1.2.14, we have, for $z \in H$,

$$f(z) = \phi(z)f(z) = -\frac{1}{\pi}\int_\Omega \frac{\partial \phi}{\partial \bar\zeta} \frac{f(\zeta)d\lambda(\zeta)}{\zeta - z}.$$

From Theorem 1.3.1 (Fubini's theorem),

$$\frac{1}{2\pi i}\int_\Gamma f(z)dz = \frac{1}{2\pi i}\int_\Gamma \left(-\frac{1}{\pi}\int_\Omega \frac{\partial\phi(\zeta)}{\partial\bar\zeta}\frac{f(\zeta)d\lambda(\zeta)}{\zeta - z}\right)dz$$

$$= \frac{1}{\pi}\int_\Omega \frac{\partial\phi(\zeta)}{\partial\bar\zeta}f(\zeta)\left(\frac{1}{2\pi i}\int_\Gamma \frac{dz}{z - \zeta}\right)d\lambda(\zeta)$$

$$= \frac{1}{\pi}\int_\Omega \frac{\partial\phi(\zeta)}{\partial\bar\zeta}f(\zeta)Ind_\Gamma(\zeta)d\lambda(\zeta)$$

$$= \frac{1}{\pi}\int_H \frac{\partial\phi(\zeta)}{\partial\bar\zeta}f(\zeta)Ind_\Gamma(\zeta)d\lambda(\zeta) = 0.$$

The last equality is due to the fact that the support of $Ind_\Gamma(\zeta)$ is contained in H and $\phi \equiv 1$ on H, and thus, $\frac{\partial\phi}{\partial\bar\zeta} \equiv 0$ on H. Therefore, (1.3.1) holds. In the following, we prove (1.3.2). Set $F(\zeta) = \frac{f(\zeta)-f(z)}{\zeta-z}$ for $z \in \Omega \setminus \Gamma^*$. Since $\lim_{\zeta \to z} F(\zeta) = f'(z)$, $\zeta = z$ is a removable singular point of $F(\zeta)$. Define $F(z) = f'(z)$. Then $F(\zeta) \in H(\Omega)$. Thus, (1.3.1) holds for $F(\zeta)$:

$$\int_\Gamma \frac{f(\zeta) - f(z)}{\zeta - z} d\zeta = \int_\Gamma F(\zeta) d\zeta = 0,$$

so

$$f(z) Ind_\Gamma(z) = \frac{1}{2\pi i} \int_\Gamma \frac{f(\zeta)}{\zeta - z} d\zeta.$$

(b) If closed chains Γ_1 and Γ_2 are homologous in Ω, then $\Gamma_1 - \Gamma_2$ is homologous to zero in Ω. From (1.3.1), we have

$$\int_{\Gamma_1 - \Gamma_2} f(\zeta) d\zeta = 0, \quad \text{that is,} \quad \int_{\Gamma_1} f(\zeta) d\zeta = \int_{\Gamma_2} f(\zeta) d\zeta.$$

$\square$

Theorem 1.3.5 (Residue Theorem). *Suppose that f is meromorphic on the open set Ω, $A \subset \Omega$, $f \in H(\Omega \backslash A)$, and every point in A is a pole of f. If Γ is a closed chain in $\Omega \backslash A$ and for any $\alpha \notin \Omega$, we have*

$$Ind_\Gamma(\alpha) = 0, \tag{1.3.3}$$

then

$$\frac{1}{2\pi i} \int_\Gamma f(z) dz = \sum_{a \in A} \text{Res}(f, a) Ind_\Gamma(a). \tag{1.3.4}$$

Proof. Let $B = \{a \in A : Ind_\Gamma(a) \neq 0\}$. Without loss of generality, we may assume that B is not empty. For any $c \in \partial \Omega$, let V_c be a connected component of $\mathbb{C} \backslash \Gamma^*$ that contains c. Since $Ind_\Gamma(z)$ is constant on V_c, (1.3.3) implies that $Ind_\Gamma(z)$ is zero on V_c. Let V_∞ be the unbounded connected component of $\mathbb{C} \backslash \Gamma^*$. Then $Ind_\Gamma(z)$ is zero on V_∞. Hence,

$$B \subset \left(\bigcup_{c \in \partial \Omega} V_c \right)^c \cap V_\infty^c, \quad \text{and thus} \quad \overline{B} \subset \left(\bigcup_{c \in \partial \Omega} V_c \right)^c \cap V_\infty^c.$$

Therefore, $\overline{B} \subset \Omega$ is compact. Since A has no limit points on Ω, $B \subset A$ is a finite set. Then there is a positive integer n, such that

$B = \{a_1, a_2, \ldots, a_n\}$ contains n points. Let $Q_k(z)$ be the principal part of f at a_k. Set $g(z) = f(z) - (Q_1(z) + Q_2(z) + \cdots + Q_n(z))$, $\Omega_0 = \Omega \backslash (A \backslash B)$. Since $a_1, a_2, \ldots, a_n$ are removable singular points of $g(z)$, applying Theorem 1.3.4 to g and Ω_0 leads to

$$\frac{1}{2\pi i} \int_\Gamma f(z)dz = \sum_{k=1}^n \frac{1}{2\pi i} \int_\Gamma Q_k(z)dz = \sum_{k=1}^n \mathrm{Res}(Q_k, a_k) \mathrm{Ind}_\Gamma(a_k).$$

Since $\mathrm{Res}(f, a_k) = \mathrm{Res}(Q_k, a_k)$, we see that (1.3.4) holds. □

Lemma 1.3.6. *Suppose that* $\gamma_0 : z = \gamma_0(t), \gamma_1 : z = \gamma_1(t)$ $(0 \leqslant t \leqslant 1)$ *are closed chains in* $\mathbb{C}$. *If for* $\alpha \in \mathbb{C}$ *and any* $t \in [0, 1]$, *we have*

$$|\gamma_1(t) - \gamma_0(t)| < |\alpha - \gamma_0(t)| + |\alpha - \gamma_1(t)|, \quad (1.3.5)$$

then

$$\mathrm{Ind}_{\gamma_0}(\alpha) = \mathrm{Ind}_{\gamma_1}(\alpha).$$

Proof. First, (1.3.5) implies $\alpha \notin \gamma_i^* (i = 0, 1)$. This is because if $\alpha \in \gamma_i^* (i = 0, 1)$, then there is a $t_0 \in [0, 1]$, such that $\gamma_i(t_0) = \alpha (i = 0, 1)$. If $\gamma_0(t_0) = \alpha$, plug it into (1.3.5) to get $|\gamma_1(t_0) - \alpha| < |\alpha - \gamma_1(t_0)|$, a contradiction. If $\gamma_1(t_0) = \alpha$, plug it into (1.3.5) to get $|\alpha - \gamma_0(t_0)| < |\alpha - \gamma_0(t_0)|$, which too is a contradiction. Let

$$\gamma(t) = \frac{\gamma_1(t) - \alpha}{\gamma_0(t) - \alpha}.$$

Then $\gamma : z = \gamma(t)$ is a closed chain in $\mathbb{C}$. By (1.3.5),

$$|\gamma(t) - 1| = \left|\frac{\gamma_1(t) - \alpha}{\gamma_0(t) - \alpha} - 1\right| = \frac{|\gamma_1(t) - \gamma_0(t)|}{|\alpha - \gamma_0(t)|} < 1 + |\gamma(t)|.$$

Then the closed chain γ lies entirely in the simply connected domain

$$D = \mathbb{C} \setminus (-\infty, 0] = \{w : w \notin (-\infty, 0]\},$$

where $1/\zeta$ is analytic on D. From Lemma 1.2.25, $\int_\gamma d\zeta/\zeta = 0$. Note that $(\gamma_0(t) - \alpha)\gamma(t) = \gamma_1(t) - \alpha$. Differentiating both sides with respect to t gives $\gamma'(\gamma_0 - \alpha) + \gamma \gamma_0' = \gamma_1'$, so that

$$\frac{\gamma'}{\gamma} = \frac{\gamma_1'}{\gamma(\gamma_0 - \alpha)} - \frac{\gamma_0'}{\gamma_0 - \alpha} = \frac{\gamma_1'}{\gamma_1 - \alpha} - \frac{\gamma_0'}{\gamma_0 - \alpha}.$$

Then

$$\frac{1}{2\pi i}\int_{\gamma_1}\frac{d\zeta}{\zeta-\alpha}-\frac{1}{2\pi i}\int_{\gamma_0}\frac{d\zeta}{\zeta-\alpha}$$
$$=\frac{1}{2\pi i}\left(\int_0^1\frac{\gamma_1'(t)dt}{\gamma_1(t)-\alpha}-\int_0^1\frac{\gamma_0'(t)dt}{\gamma_0(t)-\alpha}\right)$$
$$=\frac{1}{2\pi i}\int_0^1\left(\frac{\gamma_1'(t)}{\gamma_1(t)-\alpha}-\frac{\gamma_0'(t)}{\gamma_0(t)-\alpha}\right)dt$$
$$=\frac{1}{2\pi i}\int_0^1\frac{\gamma'(t)}{\gamma(t)}dt=\frac{1}{2\pi i}\int_\gamma\frac{d\zeta}{\zeta}=0,$$

i.e., $Ind_{\gamma_0}(\alpha)=Ind_{\gamma_1}(\alpha)$. □

Theorem 1.3.7 (Rouché's Theorem). *Suppose that γ is a closed chain in Ω, such that $Ind_\gamma(\alpha)\in\{0,1\}$ for any $\alpha\notin\gamma^*$ and $Ind_\gamma(\alpha)=0$ for any $\alpha\notin\Omega$. Let $f\in H(\Omega)$ and N_f be the number of zeros f in $\Omega_1=\{\alpha\in\Omega:Ind_\gamma(\alpha)=1\}$ (a zero of multiplicity k is counted k times).*

(a) *If f has no zeros on γ^*, then*

$$N_f=\frac{1}{2\pi i}\int_\gamma\frac{f'(z)}{f(z)}dz=Ind_\Gamma(0), \tag{1.3.6}$$

where $\Gamma=f\circ\gamma$.
(b) *If $g\in H(\Omega)$ and for all $z\in\gamma^*$, we have*

$$|f(z)-g(z)|<|f(z)|+|g(z)|,$$

then $N_g=N_f$, where N_g is the number of zeros of g in Ω_1 (a zero of multiplicity k is counted k times).

Proof. (a) Let $\varphi(z)=\frac{f'(z)}{f(z)}$. Then $\varphi(z)$ is a meromorphic function in Ω. If $a\in\Omega$ is a zero of order $m=m(a)$ of f, then $f(z)=(z-a)^m h(z)$, where $h\in H(\Omega)$ and $h(a)\neq 0$. Hence, a is a pole of order 1 of φ and $\text{Res}(\varphi,a)=m(a)$. Let $A=\{a\in\Omega_1:f(a)=0\}$. By the residue theorem (Theorem 1.3.5), we have

$$\frac{1}{2\pi i}\int_\gamma\frac{f'(z)}{f(z)}dz=\sum_{a\in A}\text{Res}(\varphi,a)=\sum_{a\in A}m(a)=N_f.$$

On the other hand,

$$Ind_\Gamma(0) = \frac{1}{2\pi i} \int_\Gamma \frac{dw}{w} = \frac{1}{2\pi i} \int_\gamma \frac{f'(z)}{f(z)} dz.$$

Therefore (1.3.6) holds.

To prove (b), we note that the assumption in (b) shows that g has no zeros on γ^*. Then the conclusion of (a) applies to g, and consequently, from Lemma 1.3.6, we have $N_g = Ind_{\Gamma_0}(0) = Ind_\Gamma(0) = N_f$, where $\Gamma = f \circ \gamma$, $\Gamma_0 = g \circ \gamma$. □

Definition 1.3.8 (Homotopy). Suppose γ_0 and γ_1 are two continuous closed curves in an open set Ω, i.e., there are continuous mappings $\gamma_i(t)$ from $[0,1]$ to Ω such that $\gamma_i(0) = \gamma_i(1)$, $i = 0, 1$. If there is a continuous mapping $H : [0,1] \times [0,1] \to \Omega$, such that for any $s, t \in [0,1]$, we have

$$H(s,0) = \gamma_0(s), \quad H(s,1) = \gamma_1(s), \quad H(0,t) = H(1,t),$$

then γ_0 and γ_1 are said to be homotopic in Ω. In particular, if γ_1 is a point, i.e., γ_1 is constant, γ_0 is said to be homotopic to zero in Ω.

Theorem 1.3.9 (Homotopy Version of Cauchy's Theorem). *If closed paths γ_0 and γ_1 are homotopic in Ω and $f \in H(\Omega)$, then*

$$\int_{\gamma_0} f(z)dz = \int_{\gamma_1} f(z)dz.$$

In particular, if a closed path γ is homotopic to zero in Ω, then

$$\int_\gamma f(z)dz = 0.$$

To prove this theorem, we only need to prove the following proposition.

Proposition 1.3.10. *If closed paths γ_0 and γ_1 are homotopic in Ω, then they are homologous in Ω. If a closed path γ is homotopic to zero in Ω, then it is homologous to zero in Ω.*

Proof. Suppose that closed paths γ_0 and γ_1 are homotopic in Ω. By definition, there is a continuous mapping $H : [0,1] \times [0,1] \to \Omega$, such that

$$H(s,0) = \gamma_0(s), \quad H(s,1) = \gamma_1(s), \quad H(0,t) = H(1,t)$$

for $s,t \in [0,1]$. Obviously, $\alpha \notin H([0,1] \times [0,1])$ for any $\alpha \notin \Omega$. Since $[0,1] \times [0,1]$ is compact, the continuity of H ensures that $H([0,1] \times [0,1])$ is also compact. Then there is $\varepsilon > 0$, such that

$$|\alpha - H(s,t)| > 2\varepsilon, \tag{1.3.7}$$

for any $s,t \in [0,1]$. Since $H(s,t)$ is uniformly continuous on $[0,1] \times [0,1]$, for the above $\varepsilon > 0$, there is a positive integer n, such that

$$|H(s,t) - H(s',t')| < \varepsilon \tag{1.3.8}$$

for $|s - s'| + |t - t'| < \frac{1}{n}$. In Ω, choose $n+1$ closed polygonal lines $P_k : z = \varphi_k(s), s \in [0,1](k = 0,1,\ldots,n)$, where φ_k is linear on the interval $\left[\frac{m-1}{n}, \frac{m}{n}\right]$ $(m = 1,2,\ldots,n)$:

$$\varphi_k(s) = H\left(\frac{m}{n}, \frac{k}{n}\right)(ns - m + 1) + H\left(\frac{m-1}{n}, \frac{k}{n}\right)(m - ns). \tag{1.3.9}$$

From (1.3.8) and (1.3.9),

$$\left|\varphi_k(s) - H\left(s, \frac{k}{n}\right)\right| < \varepsilon \quad (k = 0,1,\ldots,n; 0 \leqslant s \leqslant 1), \tag{1.3.10}$$

In particular, $H(s,0) = \gamma_0(s)$ for $k = 0$ and $H(s,1) = \gamma_1(s)$ for $k = 1$. Hence,

$$|\varphi_0(s) - \gamma_0(s)| < \varepsilon, \quad |\varphi_n(s) - \gamma_1(s)| < \varepsilon \ (0 \leqslant s \leqslant 1). \tag{1.3.11}$$

By (1.3.7) and (1.3.10),

$$|\alpha - \varphi_k(s)| \geqslant \left|\alpha - H\left(s, \frac{k}{n}\right)\right| - \left|H\left(s, \frac{k}{n}\right) - \varphi_k(s)\right| > \varepsilon. \tag{1.3.12}$$

Using (1.3.8) and (1.3.9), we have

$$|\varphi_k(s) - \varphi_{k-1}(s)| < \varepsilon \quad (k = 1,\ldots,n; 0 \leqslant s \leqslant 1). \tag{1.3.13}$$

From (1.3.11) and (1.3.12),

$$|\varphi_0(s) - \gamma_0(s)| < \varepsilon < |\alpha - \varphi_0(s)|, \quad |\varphi_n(s) - \gamma_1(s)| < \varepsilon < |\alpha - \varphi_n(s)|.$$

By (1.3.12) and (1.3.13),

$$|\varphi_k(s) - \varphi_{k-1}(s)| < \varepsilon < |\alpha - \varphi_{k-1}(s)| \quad (k = 1,2,\ldots,n; 0 \leqslant s \leqslant 1).$$

Since φ_k is a closed path, Lemma 1.3.6 gives

$$\mathrm{Ind}_{\gamma_0}(\alpha) = \mathrm{Ind}_{\varphi_0}(\alpha), \quad \mathrm{Ind}_{\varphi_n}(\alpha) = \mathrm{Ind}_{\gamma_1}(\alpha),$$
$$\mathrm{Ind}_{\varphi_{k-1}}(\alpha) = \mathrm{Ind}_{\varphi_k}(\alpha) \quad (k = 1, 2, \ldots, n),$$

so $\mathrm{Ind}_{\gamma_0}(\alpha) = \mathrm{Ind}_{\gamma_1}(\alpha)$ for any $\alpha \notin \Omega$. By definition, γ_0 and γ_1 are homologous in Ω. This completes the proof. □

Exercises I

1. Prove the following:
   (a) $\overline{\partial f/\partial z} = \partial \overline{f}/\partial \overline{z}$.
   (b) If $f \in H(\Omega)$, then the function $g(z) = \overline{f(\overline{z})}$ is analytic on $\{z : \overline{z} \in \Omega\}$.
   (c) If Ω is a domain, $f \in H(\Omega)$, and f is a real function, then f is constant.
   (d) If Ω is a domain, $f \in H(\Omega)$, and $|f|$ is constant, then f is constant.

2. Suppose that $f(z)$ is an entire function and satisfies
$$\liminf_{|z|\to\infty}(\log|f(z)| - a\log|z|) < +\infty,$$
where a is a positive constant. Prove that $f(z)$ is a polynomial of degree a or lower.

3. If $f \in C^m(D(0,R))$, show that
$$f(z) = f(0) + \sum_{k=1}^{m}\sum_{j=0}^{k} \frac{z^{k-j}\overline{z}^j}{(k-j)!j!} \left.\frac{\partial^k f}{\partial z^{k-j}\partial \overline{z}^j}\right|_0 + o(|z|^m).$$

4. Suppose that f is analytic inside and on a positively oriented simple closed curve γ and has no zeros on γ^*. Prove that if f has n zeros $z_k (k = 1, 2, \ldots, n)$ in γ, where the multiplicity of z_k is m_k, then
$$\int_\gamma \frac{zf'(z)}{f(z)}\,dz = 2\pi i \sum_{k=1}^{n} m_k z_k.$$

5. **Schwarz Integral Formula.** Suppose $f \in H(D(0,R)) \cap C^1(\overline{D(0,R)})$, $f = u + iv$. Prove that f can be represented by the real part u:

$$f(z) = \frac{1}{2\pi} \int_0^{2\pi} \frac{Re^{i\theta} + z}{Re^{i\theta} - z} u(Re^{i\theta}) d\theta + iv(0) \quad (z \in D(0,R)).$$

6. Suppose that $\{f_n(z)\}$ is analytic and uniformly bounded in the unit disk U, $|f_n(z)| \leq M$ ($z \in U, n = 1, 2, \ldots$), and for $\zeta \in \sigma = \{e^{i\theta} : |\theta| \leq \frac{\pi}{6}\}$, it satisfies

$$\varlimsup_{\substack{z \to \zeta \\ z \in U}} |f_n(z)| \leq \varepsilon_n, \quad \text{with} \lim_{n \to \infty} \varepsilon_n = 0.$$

Then $f_n(z)$ is uniformly convergent to zero on compact subset of U. In particular, if $f(z)$ is bounded and analytic in U and

$$\varlimsup_{z \in U, z \to \zeta} |f(z)| = 0$$

for $\zeta \in \sigma = \{e^{i\theta} : |\theta| \leq \frac{\pi}{6}\}$, then $f(z)$ is identically zero in U.

7. Suppose C_N is the boundary of a square bounded by the following lines: $x = \pm(N + \frac{1}{2})\pi$, $y = \pm(N + \frac{1}{2})\pi$, where N is a positive integer. Prove that

$$\int_{C_N} \frac{dz}{z^2 \sin z} = 2\pi i \left[\frac{1}{6} + 2 \sum_{n=1}^{N} \frac{(-1)^n}{n^2 \pi^2} \right].$$

Then use the fact that this integral approaches zero as $N \to \infty$ to prove that

$$\sum_{n=1}^{\infty} \frac{(-1)^{n+1}}{n^2} = \frac{\pi^2}{12}.$$

8. Use the residue theorem and the function

$$f(z) = \frac{z^{1/3}}{(z+a)(z+b)} = \frac{e^{(1/3)\log z}}{(z+a)(z+b)}, \quad (|z| > 0, 0 < \arg z < 2\pi)$$

to evaluate the integral

$$\int_0^{\infty} \frac{x^{1/3}}{(x+a)(x+b)} dx \quad (a > b > 0).$$

9. Find the number of roots in the annulus $1 \leq |z| < 2$ of the equation $2z^5 - 6z^2 + z + 1 = 0$.
10. Let $P_n(z) = z^n + a_{n-1}z^{n-1} + \cdots + a_1 z + a_0$ be a polynomial of degree n with a leading coefficient of 1. Show that there is a $z_0, |z_0| \leq 1$ so that
$$|P_n(z_0)| = \max\{|P_n(z)| : |z| \leq 1\} \geq 1.$$
11. Suppose Ω is a domain on the plane, $\varphi(z)$ is analytic in Ω, the derivative $\varphi'(z)$ of $\varphi(z)$ is not zero everywhere in Ω, $f(w)$ is analytic in $\varphi(\Omega)$, and $g(z) = f(\varphi(z))$, $z_0 \in \Omega$, $w_0 = \varphi(z_0)$. Prove that if f has a zero of order m at w_0, then g has a zero of order m at z_0, How will you modify the above conclusion if $\varphi'(z)$ has a zero of order k at z_0 and f has a zero of order m at w_0? Justify your conclusion.
12. If a non-constant function $f(z)$ is analytic on $0 < |z| < 1$, and there is a sequence $\{z_n\}, 0 < |z_n| < 1$, such that $f(z_n) = 0$ ($n = 1, 2, \ldots$) and $z_n \to 0$ ($n \to \infty$), show that 0 is an essential singularity of $f(z)$.
13. Find the number of roots in the annulus $2 < |z| < 3$ of the equation $z^5 + z + 35 = 0$.
14. Suppose that $f(z) = u(z) + iv(z)$ is an entire function and there are two positive numbers R and M, such that for $|z| \geq R$, the real part $u(z)$ of $f(z)$ satisfies $u(z) \leq M|z|$. Prove that $f(z)$ is constant or a polynomial of degree at most 1.
15. Let C be a positively oriented simple closed curve on the z-plane and
$$g(w) = \int_C \frac{z^3 + 2z}{(z-w)^3} dz.$$
Show $g(w) = 6\pi i w$ for w inside C and $g(w) = 0$ for w outside C.
16. Let $f(z) = u(x,y) + iv(x,y)$ be continuous on the bounded closed domain R and non-constant analytic inside the domain. Prove that $u(x,y)$ achieves maximum and minimum values on and only on the boundary of R.
17. Suppose $f(z)$ is analytic on the unit disk U, $|f(z)| \leq 1$ for $z \in U$, $a \in U$, $f(a) = 0$. Show

(a)
$$|f(z)| \leq \left|\frac{z-a}{1-\bar{a}z}\right|$$

for all $z \in U$, and

$$|f'(a)| \leq \frac{1}{1-|a|^2}.$$

(b) If

$$|f'(a)| = \frac{1}{1-|a|^2},$$

then $w = f(z)$ is a fractional linear transformation.

18. Show that the equation $z^4 + 2z^3 - 2z + 10 = 0$ has exactly one root in each quadrant.

19. Suppose $f(z)$ is analytic on the unit disk U and $f(0) = 0$.
    (a) If Re $f(z) \leq A$ $(A > 0)$ holds for all $z \in U$, then

    $$|f(z)| \leq \frac{2A|z|}{1-|z|}$$

    holds for all $z \in U$.
    (b) If $|\text{Re} f(z)| \leq A$ $(A > 0)$ holds for all $z \in U$, then the following inequalities hold for all $z \in U$ ($w = \arctan z$ is the inverse function of $z = \tan w = \frac{\sin z}{\cos z}$):

    (i) $|f'(0)| \leq \frac{4A}{\pi}$; (ii) $|\text{Im} f(z)| \leq \frac{2A}{\pi} \log \frac{1+|z|}{1-|z|}$;

    (iii) $|\text{Re} f(z)| \leq \frac{4A}{\pi} \arctan |z|$; (iv) $|f(z)| \leq \frac{2A}{\pi} \log \frac{1+|z|}{1-|z|}$.

20. Assume that ω is a domain surrounded by a finite number of piecewise smooth curves and $z_1, z_2, \ldots, z_n$ are n distinct points in ω. If $f \in H(\omega) \cap C(\bar{\omega})$, show that

    $$P(z) = \frac{1}{2\pi i} \int_{\partial \omega} \frac{f(\zeta)}{Q_n(\zeta)} \frac{Q_n(\zeta) - Q_n(z)}{\zeta - z} d\zeta$$

    is a polynomial of degree $n - 1$ or lower and

    $$P(z_k) = f(z_k), \quad k = 1, 2, \cdots, n,$$

    where $Q_n(z) = (z - z_1)(z - z_2) \ldots (z - z_n)$.

21. Suppose Ω is a domain, $a \in \Omega$, $D(a, R) \subset \Omega$, and $f \in H(D(a, R) \setminus \{a\})$. Show that there is $F \in H(\Omega \setminus \{a\})$, such that $\lim_{z \to a}(F(z) - f(z)) = 0$, and thus, $F - f \in H(D(a, R))$.

22. (a) Suppose μ is a finite complex measure on the measure space X, Ω is a domain in the plane, $\varphi(z, t)$ is a bounded measurable function on $\Omega \times X$, such that $\varphi(z, t)$ is a measurable function of t when $z \in \Omega$ is fixed, and $\varphi(z, t)$ is an analytic function of z when $t \in X$ is fixed. Define $f(z) = \int_X \varphi(z, t) d\mu(t)$. Show that $f(z)$ is analytic in Ω.
   (b) Suppose $F(t)$ is a measurable function on $X = (0, \infty)$ and $\int_0^\infty |F(t)|^2 dt < \infty$. Show that
   $$f(z) = \int_0^\infty F(t) e^{itz} dt$$
   is analytic in the upper-half plane $\{z = x + iy : y > 0\}$.

23. Use the residue theorem to evaluate the integrals:
   (a) $\displaystyle\int_{|z|=2} \frac{z}{(z-i)(z-1)^2} dz$;   (b) $\displaystyle\int_{-\infty}^{+\infty} \frac{\cos x dx}{4 + x^2}$.

24. Suppose the radius of convergence of the power series $f(z) = \sum_{n=0}^\infty a_n z^n$ is 1, $a_n \geq 0$, $n \geq 0$. Prove that f has a singularity at point 1.

25. Suppose $R(z)$ is a rational function, such that $|R(z)| = 1$ for $|z| = 1$. Show that $R(z)$ may be written as
$$R(z) = cz^m \prod_{n=1}^k \frac{z - \alpha_n}{1 - \overline{\alpha_n} z},$$
where c is a constant, m is an integer, $\alpha_1, \alpha_2, \ldots \alpha_k$ are complex numbers such that $0 < |\alpha_n| \neq 1$, $n = 1, 2, \ldots, k$.

26. Suppose Γ is the boundary of the unbounded region $\Omega \neq \mathbb{C}$, $f \in H(\Omega)$ is continuous on $\Omega \cup \Gamma$, there are constants B and $M \leq B$ such that $|f(z)| \leq M$ on Γ and $|f| \leq B$ in Ω. Show that $|f| \leq M$ in Ω.

27. Suppose Ω is a region, $f \in H(\Omega)$, $1 \leqslant p \leqslant \infty$, $C > 0$, $f_n \in H(\Omega)$, $|f_n| \leqslant C$, and $f_n \to f$ in L^p_{loc}, i.e., for any compact set $K \subset \Omega$,

$$\lim_{n \to \infty} \int_K |f_n(z) - f(z)|^p d\lambda(z) = 0.$$

Show that there is $g \in H(\Omega)$, such that $f = g$ for almost all $z \in \Omega$ and $f_n \to g$ u.c. on Ω, i.e., for any compact set $K \subset \Omega$, f_n uniformly converges to g on K.

28. Let E be a compact set in the plane and $E^c = \mathbb{C} \setminus E$ be the complement of E. If there exists a non-constant bounded analytic function $f(z) \in H(E^c)$, E is said to be non-removable; otherwise, it is said to be removable.

   (a) Prove that each countable compact set is removable.
   (b) If E is a compact set in a line whose one-dimensional Lebesgue measure is zero, then E is removable.
   (c) Suppose E is a removable compact set, Ω is a domain, $E \subset \Omega$, and $f(z)$ is a bounded analytic function on $\Omega \setminus E$. Prove that $f(z)$ can be extended to become bounded and analytic on Ω.
   (d) Suppose E is a connected compact set in the plane that contains at least two points. Prove that E is non-removable.

29. Determine the sets on which the following functions are defined and the domains on which they are analytic:

$$f(z) = \int_0^1 \frac{dt}{1+tz}, \quad g(z) = \int_0^\infty \frac{e^{tz} dt}{1+t^2}, \quad h(z) = \int_{-1}^1 \frac{e^{tz} dt}{1+t^2}.$$

Chapter 2

Normal Family and Conformal Mapping

2.1 Normal Family

Definition 2.1.1. Let $f_n(z)$ be a function defined on the open set Ω_n $(n = 1, 2, 3, \ldots)$. $\{f_n(z)\}$ is considered to be uniformly bounded on compact subsets of $\Omega = \bigcup_{N=1}^{\infty} \text{int} \left(\bigcap_{n=N}^{\infty} \Omega_n \right)$ if $\{f_n\}$ is uniformly bounded on each compact subset of Ω, i.e., if corresponding to each compact subset $K \subset \Omega$, there exist N and $M = M(K, N)$ (dependent on K, N), such that $K \subset \text{int}\left(\bigcap_{n=N}^{\infty} \Omega_n \right)$ and $|f_n(z)| \leqslant M$ for any $n \geqslant N$ and $z \in K$, where $\text{int} E$ denotes the set of interior points of E.

Definition 2.1.2. Let $f_n(z)$ be continuous function defined on the open set Ω_n $(n = 1, 2, 3, \ldots)$. $\{f_n(z)\}$ is called to be equicontinuous on compact subset of $\Omega = \bigcup_{N=1}^{\infty} \text{int}\left(\bigcap_{n=N}^{\infty} \Omega_n \right)$ if $\{f_n\}$ is equicontinuous on each compact subset of Ω, i.e., if corresponding to each compact subset $K \subset \Omega$ and any $\varepsilon > 0$, there exist N and $\delta > 0$ (dependent on K, ε), such that $K \subset \text{int}\left(\bigcap_{n=N}^{\infty} \Omega_n \right)$ and $|f_n(z) - f_n(z')| < \varepsilon$ for every $n \geqslant N$ and for all pairs of $z, z' \in K$, with $|z - z'| < \delta$.

Theorem 2.1.3. Let $f_n(z)$ be a holomorphic function defined on the open set Ω_n $(n = 1, 2, 3, \ldots)$.

(a) If $\{f_n(z)\}$ is uniformly bounded on compact subsets of $\Omega = \bigcup_{N=1}^{\infty} \text{int}\left(\bigcap_{n=N}^{\infty} \Omega_n \right)$, then $\{f_n\}$ is equicontinuous on compact subsets of Ω.

(b) If $\{f_n(z)\}$ is uniformly bounded on compact subsets of $\Omega = \bigcup_{N=1}^{\infty} int(\bigcap_{n=N}^{\infty} \Omega_n)$, and there exists a dense subset B of Ω, such that $\{f_n(z)\}$ converges pointwise in B, then there exists $f \in H(\Omega)$, such that $f_n(z) \to f$ u.c. on Ω.

Proof. (a) Suppose that K is a bounded closed set in Ω and ρ_0 is the distance of K to the boundary of Ω, and ρ is a positive constant such that $0 < \rho < \rho_0$. Let $K_j = \{z : d(z, K) \leqslant \frac{\rho}{1+j}\}$ ($j = 0, 2$), then K_j ($j = 0, 2$) also denote bounded closed sets in Ω, satisfying $K \subset K_2 \subset K_0$. By this assumption, there exists N_1, such that $K_0 \subset \Omega_n$ ($n \geqslant N_1$) and $\{f_n(z) : n \geqslant N_1\}$ is uniformly bounded on compact subset K_0, i.e., there exists a positive constant M, such that $|f_n(z)| \leqslant M$ for all $n \geqslant N_1$ and all $z \in K_0$. Let z_1 and z_2 be in K_2, such that $|z_1 - z_2| < \rho/12$, then for all $n \geqslant N_1$, it follows from the Cauchy integral formula that

$$|f_n(z_1) - f_n(z_2)|$$
$$= \left| \frac{1}{2\pi i} \int_{\partial D(z_1, \rho/6)} \frac{f_n(\zeta)}{\zeta - z_1} d\zeta - \frac{1}{2\pi i} \int_{\partial D(z_1, \rho/6)} \frac{f_n(\zeta)}{\zeta - z_2} d\zeta \right|$$
$$\leqslant \frac{1}{2\pi} \int_{\partial D(z_1, \rho/6)} \frac{|f_n(\zeta)||z_1 - z_2|}{|\zeta - z_1||\zeta - z_2|} |d\zeta|.$$

For $\zeta \in \partial D(z_1, \rho/6)$, $|\zeta - z_2| \geqslant |\zeta - z_1| - |z_1 - z_2| \geqslant \frac{\rho}{6} - \frac{\rho}{12} = \frac{\rho}{12}$, so

$$|f_n(z_1) - f_n(z_2)| \leqslant \frac{1}{2\pi} M |z_1 - z_2| \frac{1}{\frac{\rho}{6} \cdot \frac{\rho}{12}} 2\pi \frac{\rho}{6} = \frac{12M}{\rho} |z_1 - z_2|.$$

Therefore, for all $\varepsilon > 0$, there is $\delta = \min\{\frac{\rho}{6}, \frac{\rho \varepsilon}{36M}\}$ such that, when $z_1, z_2 \in K_2$ and $|z_1 - z_2| < \delta$,

$$|f_n(z_1) - f_n(z_2)| < \frac{\varepsilon}{3}, \quad n \geqslant N_1. \tag{2.1.1}$$

(b) Since the family of open sets $\{D(z, \frac{\delta}{3}) : z \in K\}$ is the open cover of the compact set K, there are the finite open disk $D(z_k, \frac{\delta}{3})$ ($k = 1, 2, 3, \ldots, l$) covering K. Since B is dense in Ω, there are points $z'_k \in D(z_k, \frac{\delta}{3}) \cap B$ ($k = 1, 2, 3, \ldots, l$). Since $\{f_n(z)\}$ converges at z'_k

as $n \to \infty$, hence there is an integer $N \geqslant N_1$, such that

$$|f_m(z'_k) - f_n(z'_k)| < \frac{\varepsilon}{3} \qquad (2.1.2)$$

for $m > N, n > N$ and all $z'_k (1 \leqslant k \leqslant l)$.

To finish, pick $z \in K$. $z \in D(z_k, \frac{\delta}{3})$ for some k. Our choice of δ and N shows that $|z'_k - z| < \delta$ and

$$|f_n(z) - f_n(z'_k)| < \frac{\varepsilon}{3}, \quad |f_m(z) - f_m(z'_k)| < \frac{\varepsilon}{3}$$

if $m > N, n > N$. Thus, if $m, n > N$,

$$|f_n(z) - f_m(z)| < \varepsilon.$$

Therefore, $\{f_n(z)\}$ converges compactly on Ω. □

Theorem 2.1.4 (The Montel Theorem). Let $f_n(z)$ be a holomorphic function defined on the open set Ω_n ($n = 1, 2, 3, \ldots$). If $\{f_n(z)\}$ is uniformly bounded on compact subsets of $\Omega = \bigcup_{N=1}^{\infty} int(\bigcap_{n=N}^{\infty} \Omega_n)$. Then there exist a subsequence $\{f_{n_k}(z)\}$ of $\{f_n(z) : n = 1, 2, 3, \ldots\}$ and $f \in H(\Omega)$, such that $f_{n_k}(z) \to f$ u.c. on Ω, where $\Omega \subset \bigcup_{N=1}^{\infty} int(\bigcap_{k=N}^{\infty} \Omega_{n_k})$.

According to the Weierstrass theorem, the limit function of the subsequence $\{f_{n_k}(z)\}$ of $\{f_n(z) : n = 1, 2, 3, \ldots\}$ is analytic in Ω.

Proof. We recall that the set $B_0 = \{z \in \Omega : \text{Re} z \text{ and } \text{Im} z \text{ are rational}\}$ is dense in Ω and it is possible to list the element of B_0 in a sequence. Let $z_1, z_2, z_3, \ldots$ be such a listing. We begin a construction by considering the sequence of complex numbers $\{f_n(z_1)\}$. Since there is N_1 such that $z_1 \in \Omega_n (n \geqslant N_1)$ and the pointwise boundedness of $\{f_n(z_1), n \geqslant N_1\}$, it is possible to select a sequence of indices $m_1^{(1)} < m_2^{(1)} < m_3^{(1)} < \cdots$, such that $\{f_{m_k^{(1)}}(z_1)\}$ converges to w_1. The superscript in $m_k^{(1)}$ is there to serve notice that this sequence of integers is associated with the point z_1 of B_0. In the interest of preserving some notational sanity, we shall abbreviate $\{f_{m_k^{(1)}}\}$ to $\{f_{1,k} : n = 1, 2, 3, \ldots\}$ and do likewise in similar situation throughout the course of this argument. Now, there is N_2 such that $z_2 \in \Omega_n (n \geqslant N_2)$, $\{f_{1,k}(z_2) : k = 1, 2, 3, \ldots\}$ is another bounded sequence of complex numbers. We can, therefore, select a sequence of indices $m_1^{(2)} < m_2^{(2)} < m_3^{(2)} < \cdots$ from $m_1^{(1)} < m_2^{(1)} < m_3^{(1)} < \cdots$, such that $\{f_{m_k^{(2)}}(z_2)\}$ converges to w_2. Contining this process inductively, we construct corresponding to each positive integer l a complex number w_l and a strictly increasing sequence of positive integers $m_1^{(l)} < m_2^{(l)} < m_3^{(l)} < \cdots$, such that $\{f_{m_k^{(l)}}(z_l)\}$ converges to w_l and $m_1^{(l+1)} < m_2^{(l+1)} < m_3^{(l+1)} < \cdots$ is a subsequence of $m_1^{(l)} < m_2^{(l)} < m_3^{(l)} < \cdots$. For $k \geq 1$, we set $n_k = m_k^{(k)} < m_2^{(2)}$. By construction, $n_< n_2 < n_3 < \cdots$. As a result, $\{f_{n_k}(z) : k = 1, 2, 3, \ldots\}$ is a legitimate subsequence of $\{f_n(z) : n = 1, 2, 3, \ldots\}$. Moreover, for fixed $l \geq 1$, the sequence $\{f_{n_k}(z) : k = 1, 2, 3, \ldots\}(k = 1, 2, 3, \ldots)$ is, with the possible exception of its first $l-1$ terms, also a subsequence of $\{f_{l,k}(z) : k = 1, 2, 3, \ldots\}$. This observation has the consequence that

$$\lim_{k \to \infty} f_{n_k}(z_l) = \lim_{k \to \infty} f_{l,k}(z_l) = w_l$$

for each l, which is to say that

$$\lim_{k \to \infty} f_{n_k}(z_l) = w_l$$

possesses a limit for every point z of the set B_0. Theorem 2.1.3 implies that there exists $f \in H(\Omega)$, such that $f_{n_k}(z) \to f$ u.c. on Ω. $\square$

Theorem 2.1.5 (Vitali's Theorem). *Let $f_n(z)$ be a holomorphic function defined on the open set Ω_n ($n = 1, 2, 3, \ldots$) and $\{f_n(z)\}$ is uniformly bounded on compact subset of $\Omega = \bigcup_{N=1}^{\infty} int(\bigcap_{n=N}^{\infty} \Omega_n)$. If the set $B \subset \Omega$ has at least one accumulation in every component of Ω and $\{f_n(z)\}$ converges pointwise in B, then there exists $f \in H(\Omega)$ such that $f_n(z) \to f$ u.c. on Ω.*

Proof. First, we prove that for any $z_0 \in \Omega$, there is N_0 such that $z_0 \in \Omega_n$ for $n \geq N_0$, and $\{f_n(z_0) : n \geq N_0\}$ converges. There is $z_0 \in \Omega$, and N_0 such that $z_0 \in \Omega_n$ for $n \geq N_0$. If $\{f_n(z_0) : n \geq N_0\}$ does not converge, then due to its boundedness, it must have two subsequences $\{f_{n_k}(z_0) : k = 1, 2, 3, \ldots\}$ and $\{f_{m_k}(z_0) : k = 1, 2, 3, \ldots\}$ converge to different numbers a and b, respectively. By Montel's theorem 2.1.4, we can choose from $\{f_{n_k}(z) : k = 1, 2, 3, \ldots\}$ and $\{f_{m_k}(z) : k = 1, 2, 3, \ldots\}$ subsequences $\{f_{n_{k_j}} : j = 1, 2, 3, \ldots\}$, and $\{f_{m_{k_j}} : j = 1, 2, 3, \ldots\}$ converges compactly on Ω to the limit functions $f(z)$ and $g(z)$. By Weierstrass theorem, $f(z)$ and $g(z)$ are holomorphic in Ω, and $f(z_0) = a \neq b = g(z_0)$. But on the other hand, since $\{f_n(z)\}$ converge for any point at B, then $\{f_{n_{k_j}}(z) : j = 1, 2, 3, \ldots\}$ and $\{f_{m_{k_j}}(z) : j = 1, 2, 3, \ldots\}$ also shows convergence at any point at B, therefore $f(z)$ and $g(z)$ are equal for any point at B. Let Ω_0 be the component of Ω containing z_0 and B has at least one accumulation in Ω_0. The uniqueness principle, $f(z) = g(z)$ for all $z \in \Omega_0$, in particular, $f(z_0) = g(z_0)$, contradicts with $a \neq b$. □

Vitali's Theorem 2.1.5 is an improvement of Theorem 2.1.3.

Choosing a subsequence of a sequence of functions that is uniformly convergent on each compact subset of a region is an effective method for solving many problems in function theory. Therefore, we introduce the important concept of normal families, which was first introduced by Montel in 1927.

Definition 2.1.6. Let $f_n(z)$ be a holomorphic function defined on the open set Ω_n ($n = 1, 2, 3, \ldots$). $\{f_n(z)\}$ is considered to be uniformly convergent to ∞ on compact subsets of $\Omega = \bigcup_{N=1}^{\infty} \text{int} \left(\bigcap_{n=N}^{\infty} \Omega_n \right)$ (which will be abbreviated to $f_n \to \infty$ u.c. on Ω) if $\{f_n\}$ uniformly converges to ∞ on each compact subset of Ω, i.e., if corresponding to each compact subset $K \subset \Omega$ and $M > 0$, there exist N (dependent on K, N), such that $K \subset \text{int} \left(\bigcap_{n=N}^{\infty} \Omega_n \right)$ and $|f_n(z)| \geq M$ for any $n \geq N$.

Definition 2.1.7 (Normal Family). Let $\mathfrak{F}$ be a family of functions, in which each $f_n(z)$ is a holomorphic function defined on the open set Ω_n ($n = 1, 2, 3, \ldots$), $\Omega = \bigcup_{N=1}^{\infty} \text{int} \left(\bigcap_{n=N}^{\infty} \Omega_n \right)$, $\mathfrak{F} = \{f_n(z)\}$ or let $\mathfrak{F} = \{f(z)\}$ be the family of analytic functions on open set Ω (if $\mathfrak{F} = \{f(z)\}$ is uncountable), the family $\mathfrak{F}$ is considered as a normal family on Ω if every sequence $\{f_{n_k}(z) : n = 1, 2, 3, \ldots\}$ in $\mathfrak{F}$ has either a subsequence $\{f_{n_{k_j}}(z) : j = 1, 2, 3, \ldots\}$ that converges uniformly on compact subsets of Ω to an analytic function (which will be abbreviated to $f_{n_{k_j}}$ u.c. on Ω) or a subsequence $\{f_{n_{k_j}}(z) : j = 1, 2, 3, \ldots\}$, such that $f_{n_{k_j}} \to \infty$ u.c. on Ω.

Definition 2.1.8. Let $\mathfrak{F}$ be a family of functions, in which each $f_n(z)$ is a holomorphic function defined on the open set Ω_n ($n = 1, 2, 3, \ldots$),

$\Omega = \bigcup_{N=1}^{\infty} \text{int} \left(\bigcap_{n=N}^{\infty} \Omega_n \right)$, $\mathfrak{F} = \{f_n(z)\}$ or let $\mathfrak{F} = \{f(z)\}$ be the family of analytic functions on open set Ω (if $\mathfrak{F} = \{f(z)\}$ is uncountable), the family $\mathfrak{F}$ is considered as a normal family at the point $z_0 \in \Omega$ if there is z_0 and $r > 0$, such that $D(z_0, r) = \{z : |z - z_0| < r\} \subset \Omega$, and $\mathfrak{F}$ is a normal family on $D(z_0, r)$.

The concept of normal family leads to the formulation of Theorem 2.1.4 as follows: If the analytic function family $\mathfrak{F}$ is uniformly bounded on an open set Ω (i.e., for any compact subset $K \subset \Omega$, there exists $M > 0$ such that $|f(z)| \leq M$ for any function $f(z) \in \mathfrak{F}$ and any point $z \in K$, then $\mathfrak{F}$ is normal in Ω. Vitali's Theorem 2.1.5 can be extended to this case.

Theorem 2.1.9 (Vitali's Theorem). *Let $f_n(z)$ be a sequence of analytic functions defined on open sets Ω_n, $(n = 1, 2, 3, \ldots)$, $\Omega = \bigcup_{N=1}^{\infty} \text{int} \left(\bigcap_{n=N}^{\infty} \Omega_n \right)$ and the family $\{f_n(z)\}$ is normal on Ω. If the set $B \subset \Omega$ has at least one accumulation in every component of Ω and $\{f_n(z)\}$ converges pointwise in B, then there exists $f \in H(\Omega)$, such that $f_n(z) \to f$ u.c. on Ω.*

Proof. If the sequence $\{f_n(z)\}$ does not converge uniformly on compact subsets of Ω, there would exist a compact set K and a positive number ε_0 and two subsequences $\{f_{n_k}(z) : k = 1, 2, 3, \ldots\}$ and $\{f_{m_k}(z) : k = 1, 2, 3, \ldots\}$ of the sequence $\{f_n(z)\}$ and $z_k \in K (k = 1, 2, 3, \ldots)$, such that $|f_{n_k}(z_k) - f_{m_k}(z_k)| \geq \varepsilon_0$. We can assume that $\{z_k : k = 1, 2, 3, \ldots\}$ converges to $z_0 \in K$. Since the family $\{f_n(z)\}$ is *normal* on Ω, $\{f_{n_k}(z) : k = 1, 2, 3, \ldots\}$ has either a subsequence $\{f_{n_{k_j}}(z) : j = 1, 2, 3, \ldots\}$ that converges uniformly on compact subsets of Ω to an analytic function or a subsequence $\{f_{n_{k_j}}(z) : j = 1, 2, 3, \ldots\}$ that converges uniformly on compact subsets of Ω to ∞, and $\{f_{m_k}(z) : k = 1, 2, 3, \ldots\}$ has either a subsequence $\{f_{m_{k_j}}(z) : j = 1, 2, 3, \ldots\}$ that converges uniformly on compact subsets of Ω to an analytic function or a subsequence $\{f_{m_{k_j}}(z) : j = 1, 2, 3, \ldots\}$ that converges uniformly on compact subsets of Ω to ∞. But the set $B \subset \Omega$ has at least one accumulation in every component of Ω and $\{f_n(z)\}$ converges pointwise in B, the set $B \subset \Omega$ has at least one accumulation in every component of Ω and $\{f_n(z)\}$ converges pointwise in B. We can choose from $\{f_{n_k}(z) : k = 1, 2, 3, \ldots\}$ and $\{f_{m_k}(z) : k = 1, 2, 3, \ldots\}$ subsequence

$\{f_{n_{k_j}} : j = 1, 2, 3, \ldots\}$ and $\{f_{m_{k_j}} : j = 1, 2, 3, \ldots\}$, respectively, both of these converge uniformly on compact subsets of Ω to two analytic function Ω to $f(z)$ and $g(z)$, respectively. Weierstrass theorem shows that $f(z)$ and $g(z)$ are both analytic in Ω, and $|f(z_0) - g(z_0)| \geqslant \varepsilon_0$. But on the other hand, since $\{f_n(z)\}$ converges pointwise in B, their subsequences $\{f_{n_{k_j}}(z) : j = 1, 2, 3, \ldots\}$ and $\{f_{m_{k_j}}(z) : j = 1, 2, 3, \ldots\}$ converge to the same numbers at each point at B, so $f(z)$ and $g(z)$ are equal at each point of B. $f(z)$ and $g(z)$ are equal for any point at B. Let Ω_0 be the component of Ω containing z_0, and B has at least one accumulation in Ω_0, The uniqueness principle, $f(z) = g(z)$ for all $z \in \Omega_0$, in particular, $f(z_0) = g(z_0)$, contradicts with $|f(z_0) - g(z_0)| \geqslant \varepsilon_0$. □

Theorem 2.1.10 (Vitali's Theorem). *Let $f_n(z)$ be a sequence of analytic functions defined on open sets Ω_n, $(n = 1, 2, 3, \ldots)$, $\Omega = \bigcup_{N=1}^{\infty} int \left(\bigcap_{n=N}^{\infty} \Omega_n \right)$ and $\{f_n(z)\}$ a normal family at each point of Ω. If there is a set $B \subset \Omega$ which has at least one accumulation in every component of Ω and $\{f_n(z)\}$ converges pointwise in B, then $\{f_n(z)\}$ converges uniformly on every compact subsets of Ω.*

Proof. For every component Ω_0 of Ω, let $D_0 = \{a \in \Omega_0 :$ there is $\delta > 0, \{f_n\}$ converges uniformly on compact subsets of $D(a, \delta) \subset \Omega_0\}$. First, we prove that D_0 is non-empty. Since there exists $a \in B' \cap \Omega$ (where B' is the set of all limit points of B) with $\delta > 0$ such that $D(a, \delta) \subset \Omega$ and $\{f_n(z)\}$ is normal in $D(a, \delta)$, Vitali's theorem 2.1.9 implies that $\{f_n\}$ converges uniformly on compact subsets of $D(a, \delta)$, which is closed in $D(a, \delta)$, so $a \in D_0$. Next, we prove that D_0 is an open set. For any $a \in D_0$, there exists $\delta > 0$ such that $\{f_n\}$ converges uniformly on compact subsets of $D(a, \delta)$, which implies that, for $b \in D(a, \delta)$, there is $\delta_1 > 0$ such that $D(b, \delta_1) \subset D(a, \delta)$. So, $\{f_n\}$ converges uniformly on compact subsets of $D(b, \delta_1)$, thus $b \in D_0$, this proves that D_0 is open. Finally, we prove that D_0 satisfies $\overline{D_0} \cap \Omega_0 \subset D_0$. In fact, if $a \in \overline{D_0} \cup \Omega_0$, then there exists $\delta > 0$, such that $D(a, \delta) \subset \Omega_0$ and $\{f_n(z)\}$ is normal in $D(a, \delta)$. Since a is a limit point of $D(a, \delta) \cap D_0$, Vitali's theorem 2.1.9 implies that $\{f_n\}$ converges uniformly on compact subsets of $D(a, \delta)$, so $a \in D_0$. Thus, $\overline{D_0} \cap \Omega_0 \subset D_0$. Theorem 1.1.12 implies that $D_0 = \Omega_0$. Therefore, for any compact $K \subset \Omega$ and $b \in K$, there is $\delta_b > 0$ such that $D(b, \delta_b) \subset \Omega$, $\{f_n\}$ converges uniformly on each compact of $D(b, \delta_b)$.

Therefore, by the finite cover theorem, there exist $b_1, b_2, \ldots, b_l \in K$ such that $\bigcup_{j=1}^{l} D(b_j, \delta_{b_j}) \supset K$. Thus, $\{f_n\}$ converges uniformly on compact subsets of Ω. □

Theorem 2.1.11. *The necessary and sufficient condition for a family of analytic functions $\mathfrak{F}$ to be normal in a domain Ω is that each point in Ω has a neighborhood, on which the family $\mathfrak{F}$ is normal.*

Proof. The necessity of Theorem 2.1.11 is evident. □

Proof of sufficiency. Consider an arbitrary sequence $\{f_n\}$ in $\mathfrak{F}$. If there exists $a \in \Omega$, such that

$$\liminf_{n \to \infty} |f_n(a)| < \infty,$$

then there exists a subsequence $\{f_{m_k}\}$ of $\{f_n\}$, such that $\lim_{k \to \infty} f_{m_k}(a)$ exists and is finite. Therefore, there exists $\delta > 0$, such that $D(a, \delta) \subset \Omega$ and $\{f_{m_k}\}$ is normal in $D(a, \delta)$, so there exists subsequence $\{f_{m_{k_j}}\}$ of $\{f_{m_k}\}$, $\{f_{m_{k_j}}\}$ converges uniformly on compact subsets of $D(a, \delta)$. $\{f_{m_{k_j}}\}$ converges uniformly on compact subsets of Ω by Theorem 2.1.10 (Vitali theorem).

Next, if $\lim_{n \to \infty} |f_n(a)| = \infty$ for any $a \in \Omega$, we will claim that $\{f_n\}$ converges uniformly to ∞ on compact subsets of Ω. If $\{f_n\}$ does not converge uniformly to ∞ on compact subsets of Ω, then there exists a compact $K \subset \Omega$, a subsequence $\{f_{n_k}\}$ of $\{f_n\}$, and points $z_k \in K, k = 1, 2, 3, \ldots$, such that $z_k \to a$ as $k \to \infty$ for some $a \in K$ $\{f_{n_k}(z_k)\}$ is bounded. Since there is $\delta > 0$, such that $D(a, \delta) \subset \Omega$, $\{f_{n_k}\}$ is normal in $D(a, \delta)$, there exists a subsequence $\{f_{n_{k_j}}\}$ of $\{f_{n_k}\}$ such that $\{f_{n_{k_j}}\}$ converges uniformly on compact subsets of $D(a, \delta)$. Thus, $\lim_{j \to \infty} f_{n_{k_j}}(a)$ exists and is finite, the existence and finiteness contradict that for all $a \in \Omega$, $\lim_{n \to \infty} |f_n(a)| = \infty$. This proves that $\{f_n\}$ converges uniformly to ∞ on compact subsets of Ω. □

Lemma 2.1.12 (Hurwitz Theorem). *Suppose that f_n is analytic in open set Ω_n ($n = 1, 2, 3, \ldots$) and $\{f_n\}$ converges uniformly to f on compact subsets of*

$$\Omega = \bigcup_{N=1}^{\infty} \text{int}\left(\bigcap_{n=N}^{\infty} \Omega_n\right).$$

Let Ω_0 be a component of Ω.

(a) If f_n does take the value b_n in $\Omega_0 \cap \Omega_n$ and f is not constant b on Ω_0, where $b_n \in \mathbb{C}$, and $b_n \to b (n \to \infty)$, then $f(z) \neq b$ for all $z \in \Omega_0$.

(b) If all f_n are injective in $\Omega_0 \cap \Omega_n$ and f is not constant in Ω_0, then f is also injective in Ω_0.

Proof. (a) By Weierstrass Theorem 1.2.9, we know that f is analytic in Ω_0. If f takes the value b on Ω_0, then there exists $a \in \Omega_0$ such that $f(a) = b$. Observe that $g_n(z) = f_n(z) - b_n$ are non-vanishing in $\Omega_0 \cap \Omega_n$, since $\{f_n\}$ converges uniformly to f on compact subsets of Ω, so $\{g_n\}$ converges uniformly to $f(z) - b = g(z)$ on compact subsets of Ω. Since g is non-constant in Ω_0, a is an isolated zero of g. Therefore, there exists $R > 0$, such that the closed disk $\overline{D(a,R)} \subset \Omega_0$, $\overline{D(a,R)} \setminus \{a\}$ does not contain any zeros of g. Thus, $\inf\{|g(z)| : z \in \partial D(a,R)\} = 2\delta > 0$ and there is N such that $\overline{D(a,R)} \subset \Omega_n$ for all $n \geqslant N$. Since $\{g_n(z)\}$ converges uniformly to $g(z)$ on the compact $\partial D(a,R)$, there is $N_1 \geqslant N$, such that $|g_n(z) - g(z)| < \delta < |g(z)|$ for all $z \in \partial D(a,R)$ and for $n > N_1$. Using the Rouché theorem, we can conclude that $g(z)$ and $g_n(z)$ have the same number of zeros in $D(a,R)$. Since $g_n(z)$ has no zeros in $D(a,R)$, it follows that $g(z)$ also has no zeros in $D(a,R)$, which contradicts with $g(a) = 0$. Hence, $g(z) \neq 0$ for $z \in \Omega_0$, i.e., $f(z) \neq b$ for $z \in \Omega_0$. This proves that $f(z) \neq b$ for all $z \in \Omega_0$.

(b) Since f_n is injective on $\Omega_0 \cap \Omega_n$, $f_n(z)$ does take the value $f_n(a)$ in $\Omega_0 \cap \Omega_n \setminus \{a\}$ for all $a \in \Omega_0$ and $f_n(a) \to f(a)$ $(n \to \infty)$, f is not constant in Ω_0, so f does not take the value $f(a)$ in Ω_0, so by (a), we have $f(a') \neq f(a)$ for all $a' \in \Omega_0 \setminus \{a\}$. This shows that f is injective in Ω_0. $\square$

Lemma 2.1.13. *Let $\mathfrak{F}$ be a family of analytic functions in a region Ω. If there exists a positive constant M such that for each $f(z) \in \mathfrak{F}$, we have $|f(z)| \geqslant M$ for all $z \in \Omega$, then $\mathfrak{F}$ is normal in Ω.*

Proof. Let $L_f(z) = \frac{1}{f(z)}, f \in \mathfrak{F}$, then $L_f(z)$ is analytic in Ω and $\mathfrak{F}_1 = \{L_f : f \in \mathfrak{F}\}$ is uniformly bounded in Ω. Montel Theorem 2.1.4 show that $\mathfrak{F}_1$ is normal, i.e., every sequence $\{L_{f_n}\}$ in $\mathfrak{F}_1$ has either a subsequence $\{L_{f_{n_k}}\}$ that converges uniformly on compact subsets of Ω to an analytic function $g(z)$. If $g(z) \equiv 0$, then $\{f_{n_k}\}$ converges uniformly to ∞ on compact subsets of Ω; if $g \not\equiv 0$, since $L_f(z)$ does not take the value 0 in Ω, applying Lemma 2.1.12 (Hurwitz theorem) to g, we have that $g(z) \neq 0$ all $z \in \Omega$. Therefore, for each compact $K \subset \Omega$, we have $\min\{|g(z)| : z \in K\} > 0$; thus, $\{f_{n_k}\}$ converges uniformly to $\frac{1}{g(z)}$ on the compact K. $\square$

Theorem 2.1.14 (Marty's Theorem). *A family of analytic functions $\mathfrak{F}$ is normal in a region Ω if and only if for any compact subset K of Ω, there exists a positive number M (depending on K), such that for any function $f(z)$ in $\mathfrak{F}$, we have*

$$\frac{|f'(z)|}{1+|f(z)|^2} \leqslant M. \tag{2.1.3}$$

Proof. **Necessity:** If not, then there exists a compact set K of Ω, a sequence $\{f_n(z)\}$ with points $z_n \in K$, such that

$$\frac{|f'_n(z_n)|}{1+|f_n(z_n)|^2} \to \infty \quad (n \to \infty). \tag{2.1.4}$$

We can assume, without loss of generality, that $z_n \to z_0$ $(n \to \infty)$. $\mathfrak{F}$ is normal in a region Ω, there is $r_0 > 0$, such that $D(z_0, 2r_0) \subset \Omega$, and has either a subsequence $\{f_{n_k}(z)\}$ of $\{f_n(z)\}$ that converges uniformly on compact subsets of $D(z_0, 2r_0)$ to an analytic function $\varphi(z)$ or a subsequence $\{f_{n_k}(z) : j = 1, 2, 3, \ldots\}$ that converges uniformly on compact subsets of Ω to ∞. Therefore, by the Weierstrass theorem, the sequence $\left\{ \frac{|f'_{n_k}(z)|}{1+|f_{n_k}(z)|^2} \right\}$ converges uniformly to $\frac{|\varphi'(z)|}{1+|\varphi(z)|^2}$ on $D(z_0, r_0)$. In the first case, $\varphi(z)$ is analytic in the disk $D(z_0, 2r_0)$. Therefore, the set $\left\{ \frac{|f'_{n_k}(z_k)|}{1+|f_{n_k}(z_k)|^2} \right\}$ is bounded, which contradicts (2.1.4). In the second case, $\{f_{n_{k_j}}(z) : j = 1, 2, 3, \ldots\}$ converges uniformly on compact subsets of $D(z_0, 2r_0)$ to ∞, so $\left\{ \frac{1}{f_{n_k}(z)} \right\}$ converges uniformly on compact subsets of $D(z_0, 2r_0)$ to 0. Note that

$$\frac{|f'_{n_k}(z)|}{1+|f_{n_k}(z)|^2} = \frac{\left|\left(\frac{1}{f_{n_k}(z)}\right)'\right|}{1+\left|\frac{1}{f_{n_k}(z)}\right|^2} \to 0 \quad (k \to \infty).$$

This also contradicts with (2.1.4).

Sufficiency: Let z_0 be an arbitrary point in Ω and consider the disk $D(z_0, 2r_0) \subset \Omega$ for some $r_0 > 0$. By the given condition, there exists $M > 0$ such that for every function $f(z)$ in $\mathfrak{F}$, we have

$$\frac{|f'(z)|}{1+|f(z)|^2} \leqslant M, \quad |z - z_0| \leqslant r_0.$$

Let $h(t) = \arctan |f(z_0 + te^{i\alpha})|$ ($0 \leqslant t \leqslant r_0$) for fixed real number α, note that

$$\left|\frac{d}{ds}|f(z_0 + se^{i\alpha})|\right| \leqslant |f'(z_0 + se^{i\alpha})|.$$

Then we have

$$|h(t) - h(0)| \leqslant \left|\int_0^t h'(s)ds\right| \leqslant \int_0^t \frac{|f'(z_0 + se^{i\alpha})|ds}{1+|f(z_0 + se^{i\alpha})|^2} \leqslant Mt.$$

Take $r_1 > 0$ such that $r_1 \leqslant \min\{\frac{\pi}{12M}, r_0\}$, then

$$\arctan |f(z_0)| - \frac{\pi}{12} \leqslant \arctan |f(z)| \leqslant \arctan |f(z_0)| + \frac{\pi}{12} \quad (2.1.5)$$

for all $z : |z - z_0| \leqslant r_1$. If $|f(z_0)| \geqslant 1$, then

$$\arctan |f(z)| \geqslant \frac{\pi}{4} - \frac{\pi}{12} = \frac{\pi}{6}, \quad z \in D(z_0, r_1).$$

If $|f(z_0)| < 1$, then

$$\arctan |f(z)| \leqslant \frac{\pi}{4} + \frac{\pi}{12} = \frac{\pi}{3}, \quad z \in D(z_0, r_1).$$

In conclusion, for any point z_0 in Ω, there exists a neighborhood $D(z_0, r_1)$ such that in this neighborhood, every function $f(z)$ in $\mathfrak{F}$ satisfies $|f(z)| \leqslant \sqrt{3}$ or has $|f(z)| \geqslant \frac{1}{\sqrt{3}}$. Using Lemma 2.1.12 and Theorem 2.1.11, it follows that $\mathfrak{F}$ is normal in Ω. □

2.2 Conformal Mapping for Simply Connected Regions

Definition 2.2.1 (Conformal Mapping). Suppose f is a one-to-one continuous mapping of an open set Ω into the plane, and $a \in \Omega$.

We say that f preserves angles at a if, for any real number θ, there exists $e^{i\varphi}$ and is independent of θ, such that

$$\lim_{R \to 0} e^{-i\theta} \frac{f(a + Re^{i\theta}) - f(a)}{|f(a + Re^{i\theta}) - f(a)|} = e^{i\varphi}.$$

f is called the conformal mapping in Ω if f preserves angles at each point of Ω.

We call two regions Ω_1 and Ω_2 conformally equivalent if there exists a $\psi \in H(\Omega_1)$, such that $\psi(\Omega_1) = \Omega_2$ and ψ is one-to-one in Ω, i.e., if there exists a conformal one-to-one mapping of Ω_1 onto Ω_2.

It follows that conformally equivalent regions are homeomorphic. Due to the topological invariance of the connectedness of a region, it is also invariant under conformal mappings. Therefore, a necessary condition for two regions to be conformally equivalent is that they have the same number of connected components.

In the class of simply connected regions, the extended complex plane $\overline{\mathbb{C}}$ cannot be conformally mapped to any other region. All regions obtained by removing a point from $\overline{\mathbb{C}}$ form a conformal equivalence class, implying that any two such regions are conformally equivalent, while these regions are not conformally equivalent to any other region. The complex plane $\mathbb{C}$ can be regarded as the representative of this class of regions. What about other simply connected regions?

Riemann's mapping theorem asserts that, apart from the above two special classes of simply connected regions, any other simply connected region with more than one boundary point is conformally equivalent to the unit disk. Thus, there are three conformal equivalence classes of simply connected regions, represented, respectively, by $\overline{\mathbb{C}}$, $\mathbb{C}$, and the unit disk. In other words, from the perspective of conformal equivalence, there are only these three distinct simply connected regions. This theorem is one of the fundamental and important theorems in complex analysis.

Now, let's take a look at the relationship between conformal mappings and one-to-one analytic functions.

Theorem 2.2.2. *Let f be a one-to-one holomorphic mapping of an open set Ω, then f is a conformal mapping in Ω. Conversely, if $f \in C^1(\Omega)$ is a conformal mapping in Ω, then f is a one-to-one holomorphic mapping in Ω.*

Proof. If f is a one-to-one holomorphic mapping of an open set Ω, Theorem 1.2.32 implies that $f'(a) \neq 0$ for each $a \in \Omega$. Let $f'(a) = |f'(a)|e^{i\varphi}$, then

$$\lim_{R \to 0} e^{-i\theta} \frac{f(a+Re^{i\theta}) - f(a)}{|f(a+Re^{i\theta}) - f(a)|} = \frac{f'(a)}{|f'(a)|} = e^{i\varphi}$$

for each θ, so f preserves angles at a. Therefore, f is a conformal mapping in Ω.

Conversely, $f \in C^1(\Omega)$ is a conformal mapping in Ω, then

$$f(a+Re^{i\theta}) - f(a) = \left.\frac{\partial f}{\partial z}\right|_a Re^{i\theta} + \left.\frac{\partial f}{\partial \bar{z}}\right|_a Re^{-i\theta} + o(R)$$

for each $a \in \Omega$. Since f preserves angles at a, there exists $e^{i\varphi}$, such that

$$e^{i\varphi} = \lim_{r \to 0} e^{-i\theta} \frac{f(a+Re^{i\theta}) - f(a)}{|f(a+Re^{i\theta}) - f(a)|} = \frac{\left.\frac{\partial f}{\partial z}\right|_a + \left.\frac{\partial f}{\partial \bar{z}}\right|_a e^{-2i\theta}}{\left|\left.\frac{\partial f}{\partial z}\right|_a + \left.\frac{\partial f}{\partial \bar{z}}\right|_a e^{-2i\theta}\right|}$$

exists and is independent of θ, and it is possible only when $\left.\frac{\partial f}{\partial \bar{z}}\right|_a = 0$. Therefore, $f \in H(\Omega)$. □

Lemma 2.2.3 (Schwarz' Lemma). *Suppose $f \in H(U)$, $|f(z)| \leq 1$, and $f(0) = 0$. Then (a) $|f(z)| \leq |z|$ $(z \in U)$; (b) $|f'(0)| \leq 1$; (c) if equality $|f(z_0)| = |z_0|$ for one $z_0 \in U$, $z_0 \neq 0$, or equality $|f'(0)| = 1$ holds, then $f(z) = \lambda z$, where $\lambda \in \partial U$ is a constant.*

Proof. Since $f(0) = 0$, $g(z) = \frac{f(z)}{z}$ has a removable singularity at $z = 0$, and $g(0) = \lim_{z \to 0} g(z) = f'(0)$. Hence, $g(z)$ is holomorphic in U. If $z \in D(0, 1-\varepsilon) = \{z : |z| < 1-\varepsilon\}$ and $0 < \varepsilon < 1$, then

$$|g(z)| \leq \frac{\max\{|f(z)| : |z| = 1-\varepsilon\}}{1-\varepsilon} \leq \frac{1}{1-\varepsilon}.$$

Letting $\varepsilon \to 0$, we see that $|g(z)| \leq 1$ at every $z \in U$. So, if $z \neq 0$, $|f(z)| \leq |z|$. Since $|g(0)| = |f'(0)| \leq 1$, (a) and (b) follow.

If $|f(z_0)| = |z_0|$ for some $z_0 \neq 0, z_0 \in U$ or $f'(0)| = 1$, then $|g(z)| = 1$ for some $z \in U$, then g is constant $\lambda \in \partial U$ by another application of the maximum modulus theorem. □

Definition 2.2.4 (Analytic Automorphism Group of the Unit Disk). The analytic automorphism group of the unit disk is defined as follows: If $f(z)$ is an analytic function that maps the unit disk U onto itself in a one-to-one manner, and $f(U) = U$, then $f(z)$ is called an analytic automorphism of U. The set of all analytic automorphisms on U forms a group, which is noted by $\text{Aut}(U)$.

Analytic automorphism of U refers to an analytic function that maps the unit disk onto itself in a one-to-one manner. It is also known as an analytic automorphism.

The structure of the group $\text{Aut}(U)$ of the unit disk can be immediately obtained from the Schwarz lemma.

Let

$$\phi_\alpha(z) = \frac{z - \alpha}{1 - \bar{\alpha}z}$$

for $\alpha \in U$, then $\phi_\alpha(z) \in \text{Aut}(U)$. In fact, $\phi'_\alpha(z) = 1 - |\alpha|^2/(1-\bar{\alpha}z)^2 \neq 0$ for $z \in U$, so $\phi_\alpha(z)$ is a conformal mapping on U:

$$|\phi_\alpha(z)| = \left|\frac{z - \alpha}{1 - \bar{\alpha}z}\right| = \frac{|z - \alpha|}{|z||\bar{z} - \bar{\alpha}|} = 1$$

for $|z| = 1$, and by the maximum modulus principle, we conclude that

$$|\phi_\alpha(z)| \leqslant \max_{|z|=1} |\phi_\alpha(z)| = 1$$

for all $z \in U$. Since $\phi_\alpha(z)$ is an open mapping, $|\phi_\alpha(z)| < 1$ for $z \in U$, $\phi_\alpha(z)$ is a mapping from U into U. Finally, we make the following very simple observation: $\phi_\alpha(\phi_{-\alpha}(z)) = \phi_{-\alpha}(\phi_\alpha(z)) = z$, from which we conclude that the inverse of $\phi_\alpha(z)$ is $\phi_{-\alpha}(z)$, so the function $\phi_\alpha(z)$ is an analytic automorphism mapping on U. Clearly, $\phi_\alpha(z)$ has the following two important properties: (1) $\phi_\alpha(\alpha) = 0$; (2) $\phi_\alpha^{-1} = \phi_{-\alpha}$. For the rotation transformation $\xi = \rho_\tau(\zeta) = e^{i\tau}\zeta, \tau \in \mathbb{R}$, it is obvious that all rotation transformations also belong to $\text{Aut}(U)$, and the group consisting of all rotation transformations is called the rotation group, which is a subgroup of $\text{Aut}(U)$.

The following theorem says that the rotations combined with the maps $\phi_\alpha(z)$ exhaust all the automorphisms of the disk.

Theorem 2.2.5 (Automorphisms of the Disk). *If $f \in \mathrm{Aut}(U)$ is an automorphism of the disk, then there exist $a, |a| < 1$ and $\tau \in \mathbb{R}$, such that*

$$f(z) = \varphi_a(\rho_\tau(z)). \tag{2.2.1}$$

Proof. If $f \in \mathrm{Aut}(U)$ is an automorphism of the disk, we consider the automorphism g defined by $g = \phi_b \circ f$, where $f(0) = b$. Then $g(0) = \phi_b \circ f(0) = \phi_b(b) = 0$, and the Schwarz lemma gives $|g'(0)| \leqslant 1$. Moreover, $g^{-1}(0) = 0$, so applying the Schwarz lemma to g^{-1}, we find that $\left|\frac{1}{g'(0)}\right| = |(g^{-1})'(0)| \leqslant 1$, and so, $|g'(0)| = 1$. So, applying the Schwarz lemma again, we conclude that $g(\xi) = e^{i\tau}\xi = \rho_\tau(\xi)$ for some $\tau \in \mathbb{R}$, i.e., $\phi_b \circ f = \rho_\tau$, we deduce that $f = \phi_{-b} \circ \rho_\tau$, as claimed by replacing $-b$ by a. □

The important Schwarz–Pick theorem can be derived from Theorem 2.2.5.

Lemma 2.2.6 (Schwarz–Pick Lemma). *If f is an analytic function from U into U, and suppose that $z_1, z_2 \in U$ and $w_1 = f(z_1), w_2 = f(z_2)$, then*

$$\left|\frac{w_1 - w_2}{1 - w_1 \bar{w}_2}\right| \leqslant \left|\frac{z_1 - z_2}{1 - z_1 \bar{z}_2}\right| \tag{2.2.2}$$

and

$$\frac{|dw|}{1 - |w|^2} \leqslant \frac{|dz|}{1 - |z|^2}. \tag{2.2.3}$$

Equality in (2.2.2) or (2.2.3) will hold for $f \in \mathrm{Aut}(U)$.

Proof. Put

$$\varphi(z) = \frac{z + z_1}{1 + \bar{z}_1 z}, \quad \phi(z) = \frac{z - w_1}{1 - \bar{w}_1 z},$$

then $\varphi, \phi \in \mathrm{Aut}(U)$, and

$$\phi \circ f \circ \varphi(0) = \phi \circ f(z_1) = \phi(w_1) = 0.$$

Applying the Schwarz lemma to $\phi \circ f \circ \varphi$, $|(\phi \circ f \circ \varphi)(z)| \leqslant |z|$ for $z \in U, z \neq 0$. Put $z = \varphi^{-1}(z_2)$, then $|\phi \circ f(z_2)| \leqslant |\varphi^{-1}(z_2)|$, so $|\phi(w_2)| \leqslant |\varphi^{-1}(z_2)|$, (2.2.2) has been proved.

Applying the Schwarz lemma again, we find that $|(\phi \circ f \circ \varphi)'(0)| \leqslant 1$ for $z = 0$, i.e., $|\phi'(w_1)f'(z_1)\varphi'(0)| \leqslant 1$. Since

$$\varphi'(z) = \frac{1 - z_1\bar{z}_1}{(1 + \bar{z}_1 z)^2}, \quad \varphi'(0) = 1 - |z_1|^2,$$

$$\phi'(z) = \frac{1 - w_1\bar{w}_1}{(1 - \bar{w}_1 z)^2}, \quad \phi'(w_1) = \frac{1}{1 - |w_1|^2}.$$

So, $|f'(z_1)| \leqslant \frac{1-|w_1|^2}{1-|z_1|^2}$, and (2.2.3) holds. The Schwarz lemma implies that the equality holds if and only if $\phi \circ f \circ \varphi(z) = e^{i\tau}z = \rho_\tau(z)$, so $f = \phi^{-1} \circ \rho_\tau \circ \varphi^{-1} \in \text{Aut}(U)$. □

Now, let $\alpha \in U$, we consider the family $\Sigma(\alpha)$ of all analytic functions from U into U. How large can $|g'(\alpha)|$ be if $g \in \Sigma$?

To solve this, put $f(w) = \phi_{g(\alpha)} \circ g \circ \phi_{-\alpha}(w)$ ($w = \phi_\alpha(z)$), f is an analytic function from U to U; also, $f(0) = 0$. The passage from g to f has reduced our problem to the Schwarz lemma, which gives $|f'(0)| \leqslant 1$, the equality can occur if and only if $f(w) = \lambda w$ ($\lambda \in \partial U$), i.e., $g(z) = \phi_{-g(\alpha)}(\lambda\phi_\alpha(z))$. The chain rule gives

$$f'(0) = \phi'_{g(\alpha)}(g(\alpha))g'(\alpha)\phi'_{-\alpha}(0) = g'(\alpha)\frac{1 - |\alpha|^2}{1 - |g(\alpha)|^2}.$$

The inequality $|f'(0)| \leqslant 1$ implies that

$$|g'(\alpha)| \leqslant \frac{1 - |g(\alpha)|^2}{1 - |\alpha|^2},$$

so

$$|g'(\alpha)| \leqslant \frac{1}{1 - |\alpha|^2}. \qquad (2.2.4)$$

This solves our problem, since equality can occur in (2.2.4). This happens if and only if $g(\alpha) = 0$ and $|f'(0)| = 1$, in which case $g(z) = \phi_{-g(\alpha)}(\lambda\phi_\alpha(z)) = \lambda\phi_\alpha(z)$ is a rotation for some constant λ with $|\lambda| = 1$, so that $\phi_{-g(\alpha)}(z) = (z + g(\alpha))/(1 - \overline{g(\alpha)}z) = z$.

This observation may serve as the motivation for the Riemann mapping theorem.

Theorem 2.2.7 (The Riemann Mapping Theorem). *Every simply connected region $\Omega \subset \overline{\mathbb{C}}$ with the number of its boundary points* more than 2, $z_0 \in \Omega$ and $z_o \in \mathbb{C}$, then there is a unique conformal map f of Ω onto the open unit disk U, such that $f(z_0) = 0, f'(z_0) > 0$.

Proof. **Existence:** (i) Let Σ be the class of all holomorphic functions ψ which are one-to-one, which maps Ω to U and is 0 at the point z_0.

We first prove that Σ is not empty. Let $w_0 \in \mathbb{C} \setminus \Omega$, since Ω is simply connected, $z - w_0$ is holomorphic in Ω, Lemma 1.2.25 implies that there exists a $\phi \in H(\Omega)$ so that $(\phi(z))^2 = z - w_0$ in Ω. If $z_1, z_2 \in \Omega$, and $\phi(z_1) = \pm\phi(z_2)$, then also $\phi(z_1)^2 = \phi(z_2)^2$, hence $z_1 = z_2$; thus, ϕ is one-to-one, and ϕ is an open mapping, $\phi(\Omega)$ is an open set, $\phi(\Omega)$ contains a disk $D(a, r)$ with $0 < r < |a|$. The same argument shows that there are no two distinct z_1 and z_2 in Ω, such that $\phi(z_1) = -\phi(z_2)$, and the open set $\phi(\Omega)$ there fails to intersect $D(-a, r)$. We define $\psi_0(z) = r/(\phi(z) + a)$, since $\phi(\Omega) \cap D(-a, r)$ is empty, for all $z \in \Omega$, we have $|\phi(z) + a| \geq r > 0$, ψ_0 is holomorphic in Ω, and $|\psi_0(z)| = \left|\frac{r}{\phi(z)+a}\right| \leq 1$. ψ_0 is open mapping, so $|\psi_0(z)| < 1$, thus ψ_0 is a holomorphic function from Ω into U. ϕ_0 is also one-to-one in $\Omega\psi_0$. Let $\psi(z) = \phi_{\psi_0(z_0)}(\psi_0(z))$, then ψ is one-to-one from Ω to U with $\psi(z_0) = 0$, we see that $\psi \in \Sigma$, this proves that Σ is not empty.

(ii) The next step consists in showing that if $\psi \in \Sigma$, if $\psi(\Omega)$ does not cover all of U, then there exists a $\tilde{\psi} \in \Sigma$ with $|\tilde{\psi}'(z_0)| > |\psi'(z_0)|$. Since ψ does not cover all of U and $\psi(z_0) = 0$, there is $\beta \in U \setminus \psi(\Omega)$, $\beta \neq 0$, then $\phi_\beta \circ \psi \in \Sigma$, and $\phi_\beta \circ \psi$ 在 has no zero in Ω; Lemma 1.2.25 implies that there exists a $g \in H(\Omega)$, such that $g^2 = \phi_\beta \circ \psi$. We see that g is one-to-one, in fact, if $g(z_1) = g(z_2)$, then $\phi_\beta(\psi(z_1)) = \phi_\beta(\psi(z_2))$, so $z_1 = z_2$. Let $\tilde{\psi} = \phi_{g(z_0)} \circ g$, then $\tilde{\psi} \in \Sigma$. Since

$$(g(z_0))^2 = \phi_\beta(0) = -\beta, \quad \tilde{\psi}'(z_0) = \phi'_{g(z_0)}(g(z_0))g'(z_0)$$
$$= \frac{g'(z_0)}{1 - |g(z_0)|^2} = \frac{g'(z_0)}{1 - |\beta|}$$

and

$$2g(z_0)g'(z_0) = \phi'_\beta(0)\psi'(z_0) = (1 - |\beta|^2)\psi'(z_0),$$

the chain rule gives

$$|\tilde{\psi}'(z_0)| = \frac{1+|\beta|}{2|g(z_0)|}\psi'(z_0) = \frac{1+|\beta|}{2\sqrt{|\beta|}}|\psi'(z_0)|.$$

Since $0 < |\beta| < 1$, then $1+|\beta| > 2\sqrt{|\beta|}$, thus $|\tilde{\psi}'(\alpha)| > |\psi'(\alpha)|$.

(iii) Put $\eta = \sup\{|\psi'(z_0)| : \psi \in \Sigma\}$. (i) shows that there is $\psi_0 \in \Sigma$. Theorem 1.2.32 implies that $\eta \geqslant |\psi_0'(z_0)| > 0$. Since there is $r_0 > 0$, such that $D(\alpha, r_0) \subset \Omega$, and for all $\psi \in \Sigma$, $\psi_{z_0}(z) = \psi(z_0 + r_0 z) \in H(U)$, (2.2.4) implies that $|\psi'(z_0)| \leqslant 1/r_0$, so $\eta \leqslant 1/r_0$.

Since $|h(z)| < 1$ for all $h \in \Sigma$ and $z \in \Omega$, the Montel Theorem 2.1.4 shows that Σ is a normal family. The definition of η shows that there is a sequence $\{\psi_n\}$ in Σ, such that

$$\lim_{n \to \infty} |\psi_n'(z_0)| = \eta,$$

and by normality of Σ, we can extract a subsequence $\{\psi_{n_k}\}$, from $\{\psi_n\}$, which converges uniformly on compact subsets of Ω to a limit ψ. The Weierstrass Theorem 1.2.9 implies that $\psi \in H(\Omega)$ and

$$|\psi'(z_0)| = \lim_{k \to \infty} |\psi_{n_k}'(z_0)| = \eta.$$

Since $|\psi'(z_0)| > 0$, then ψ is not constant. Lemma 2.1.12 implies that ψ is one-to-one. Since $|\psi_n(z)| < 1$ for all $z \in \Omega$ and $n = 1, 2, 3, \ldots$, we have $|\psi(z)| \leqslant 1$, but the open mapping theorem shows that actually $\psi(\Omega) \subset U$, thus $\psi \in \Sigma$. (ii) implies that ψ map Ω onto U. In fact, if there were $\tilde{\psi} \in \Sigma$ such that $|\tilde{\psi}'(z_0)| > |\psi'(z_0)| = \eta$, which is a contradiction with the definition of η.

Uniqueness: Assume that $w = f(z)$ an $\zeta = g(z)$ are two conformal mapping of Ω onto U which have $f(z_0) = g(z_0) = 0$ and $f'(z_0) > 0$, $g'(z_0) > 0$. Consider the function $w = h(\zeta) = f(g^{-1}(\zeta))$. This function provides a conformal self-mapping from $U = \{\zeta : |\zeta| < 1\}$ onto $U = \{w : |w| < 1\}$ that fixes the origin and satisfies $h'(0) > 0$. Theorem 2.2.5 implies that $h(\zeta) = \varphi_a(\rho_\tau(\zeta))$, where $a \in U, \tau$ is a real number. $h(0) = 0$ implies $a = 0$, and $h'(0) > 0$ implies $e^{i\tau} = 1$, so $h(\zeta) = \zeta$, thus $f(z) = g(z)$. This proves the uniqueness. $\square$

Note: Let (P_Ω) be all $f \in H(\Omega)$ such that, if f has no zero in Ω, then there is a $\varphi \in H(\Omega)$, such that $\varphi^2 = f$. We see that the property (P_Ω) implies the Riemann mapping.

2.3 Boundary Correspondence Theorem

In the previous section, the Riemann mapping theorem states that there exists a one-to-one correspondence between simply connected regions and points inside the unit disk. It does not involve any correspondence between any two regions. Due to the complexity of the boundary of a region, the study of boundary correspondence is more challenging.

Let Ω be a bounded simply connected region, $\overline{\Omega}$ the closure of Ω, $\partial\Omega$ boundary of Ω, U the unit disk. ∂U denotes the unit circle, $H(U)$ denotes the set of all analytic functions on U, $C([0,1))$ denotes the set of all continuous functions on $[0,1)$ and $\mathbb{N}$ the set of all natural numbers.

Definition 2.3.1. Let $z_0 \in \partial\Omega$, $\ell : [0,1) \mapsto \Omega$ continuous and $\lim_{t \to 1} \ell(t) = z_0$, the point (z_0, ℓ) is called an accessible boundary point of Ω. The boundary point of $z_0 \in \partial\Omega$ will be called a simply boundary point of Ω if z_0 has the following property: To every sequence $\{z_n\}$ in Ω such that $z_n \to z_0$ as $n \to \infty$), there corresponds a curve $\ell : [0,1) \mapsto \Omega$, with parameter interval $[0,1]$ and a sequence $\{t_n\}, 0 < t_1 < t_2 < t_3 \cdots, t_n \to 1$, such that $\ell(t_n) = z_n$ $(n = 1, 2, 3, \ldots)$ and $\ell(t) \in \Omega$ for $t \in [0,1)$.

Every boundary point of a region Ω bounded by a piecewise smooth closed Jordan curve is a simple boundary point (in this case, the Jordan curve theorem holds).

Let $\mathcal{D} = \{(z_0, \ell) : (z_0, \ell)$ be an accessible boundary point of Ω. Introduce the following relation in $\mathcal{D}$: $\sim$:$(z_1, \ell_1) \sim (z_2, \ell_2)$ if and only if $z_1 = z_2 = z_0$ and there are $z'_n \in \ell_1 \cap \Omega, z''_n \in \ell_2 \cap \Omega$, such that $z'_n \to z_0, z''_n \to z_0$, and also there is $\ell \in C([0,1))$, such that $(z_0, \ell) \in \mathcal{D}$ and $z'_n, z''_n \in \ell \cap \Omega (n = 1, 2, \ldots)$. Clearly, $\sim$ is an equivalence relation on $\mathcal{D}$. When (z_1, ℓ_1) and (z_2, ℓ_2) are not equivalent, we denote this as $(z_1, \ell_1) \nsim (z_2, \ell_2)$ and denote $[(z_0, \ell)] = \{(z_0, \ell_1) : (z_0, \ell) \sim (z_0, \ell_1)\}$ the equivalence class of (z_0, ℓ).

Example: Let

$$\Omega_0 = \{(x, y) : |x| < 2; |y| < 1\}, \quad E = \{(x, y) : x = 0; y \in [0, 1)\},$$

and $\Omega_1 = \Omega_0 \setminus E$. For any chosen $z_0 \in E$, we can select a point $z_1 \in \Omega$ from the left side of E. We can then connect z_0 and z_1 with a continuous curve ℓ_1. Similarly, we can select a point $z_2 \in \Omega_1$ from the right side of E, and we can then connect z_0 and z_2 with another continuous curve ℓ_2. As a result, (z_0, ℓ_1) and (z_0, ℓ_2) determine two distinct accessible boundary points of Ω_1. Let

$$G_1 = \bigcup_{n=1}^{\infty} \left\{ (x,y) : x = \frac{1}{2^n}; 0 \leqslant y \leqslant \frac{3}{4} \right\},$$

$$G_2 = \bigcup_{n=1}^{\infty} \left\{ (x,y) : x = \frac{3}{2^{n+1}}; \frac{1}{4} \leqslant y \leqslant 1 \right\},$$

$$E_1 = \{(x,y) : x \in [0,1]; y = 0\},$$

Let $G = G_1 \cup G_2 \cup E_1$. and $\Omega = \Omega_1 \setminus G$. For any chosen $z_0 \in E$, we can select a point $z_3 \in \Omega$ from the left side of E. We can then connect z_0 and z_3 with a continuous curve ℓ_3, (z_0, ℓ_3) determine an accessible boundary point of Ω. For any chosen $z_0 \in E$, a sequence $\{z_n : n \in \mathbb{N}\} \subset \Omega$ from the right side of E such that $z_n \to z_0$ ($n \to \infty$), and $\operatorname{Re} z_n > 0$, $\operatorname{Im} z_n > 0$, then for any curve $\ell : [0,1) \mapsto \Omega$, with parameter interval $[0,1]$ and a sequence $\{t_n\}, 0 < t_1 < t_2 < t_3 \cdots, t_n \to 1$, such that $\ell(t_n) = z_n$ ($n = 1, 2, 3, \ldots$) and $\ell(t) \in \Omega$ for $t \in [0,1)$ $z_n (n \in \mathbb{N})$, $\lim_{t \to 1} \ell(t)$ does not exist. Since one of the endpoints is missing in the middle, we can infer that using only then point z_0 is not sufficient to fully express the complete connotation of accessible boundary point.

Lemma 2.3.2. *Let $C_n (n = 1, 2, \ldots)$ be Jordan curves and $C_n \subset U$ ($n = 1, 2, \ldots$), and there is $\delta_0 \in (0,1)$, such that $C_n \subset U \setminus D(0, \delta_0)$, let's denote the two endpoints of C_n as z_n', z_n'', and $z_n' \to a$, $z_n'' \to b$, where $a \neq b$. Suppose that $f \in H(U)$ is bounded in U. If $\lim_{n \to \infty} \max\{|f(z)| : z \in C_n\} = 0$, then $f(z) \equiv 0$.*

Proof. Let $M > 0$ such that $|f(z)| \leqslant M$ for all $z \in U$. We can assume, without loss of generality, that $\operatorname{Im} a > 0, \operatorname{Im} b < 0$, then there is positive integer $m \geqslant 2$, such that $\operatorname{Im} a > \operatorname{Im} e^{\frac{\pi}{m}i} = \cos \frac{\pi}{m}$, $\operatorname{Im} b < -\operatorname{Im} e^{\frac{\pi}{m}i} = -\cos \frac{\pi}{m}$. We can also assume that all C_n pass through open intervals $(0,1)$ on the real axis and also pass through the segment $\{z = \operatorname{Re}^{\frac{\pi}{m}i} : 0 < r < 1\}$. Then there exists an endpoint of C_n on the interval $(0,1)$ and the other endpoint is on the arc $(0, e^{\frac{\pi}{m}i})$. There exist arc C_n' such that C_n' lies entirely on

$\{z = Re^{i\theta} : 0 < \theta < \frac{\pi}{m}, \delta_0 < r < 1\}$ excluding the two endpoints. Put $\overline{C_n'} = \{z : \bar{z} \in C_n'\}, \Gamma_n = C_n \cup \overline{C_n'}$.

Since $f \in H(U)$, there exists a Taylor series expansion

$$f(z) = \sum_{n=0}^{\infty} a_n z^n$$

for $z \in U$, so

$$\overline{f(\bar{z})} = \sum_{n=0}^{\infty} \bar{a}_n z^n \in H(U).$$

Let $\varphi(z) = f(z)\overline{f(\bar{z})}$, then $\varphi(z) \in H(U)$. Put $\eta = e^{\frac{2}{\pi}mi}$,

$$\psi(z) = \varphi(z)\varphi(\eta z) \cdots \varphi(\eta^{m-1} z)$$

and

$$\tilde{\Gamma}_n = \Gamma_n \cup \eta\Gamma_n \cup \cdots \cup \eta^{n-1}\Gamma_n,$$

and $\psi \in H(U)$. $\tilde{\Gamma}_n$ is a Jordan closed curve contained in U, and the interior of $\tilde{\Gamma}_n$ contains $D(0, \delta_0)$. Let $\varepsilon_n = \max\{|f(z)| : z \in C_n\}$, then

$$|\psi(z)| \leq M^{2m-1}\varepsilon_n \quad (z \in \tilde{\Gamma}_n).$$

The maximum modulus principle implies that

$$|\psi(z)| \leq M^{2m-1}\varepsilon_n$$

for all $z \in D(0, \delta_0)$. $\psi(z) = 0$ by letting $n \to \infty$. These functions $\varphi(z), \ldots, \varphi(\eta^{m-1}z)$ at least one contains infinitely many zeros in $D(0, \delta_0)$, $\varphi(z)$ is identically zero in $D(0, \delta_0)$ by the uniqueness theorem, so $f(z) \equiv 0$. □

Theorem 2.3.3 (Koebe's Boundary Correspondence Theorem). *Suppose that Ω is a bounded simply connected region, $f : \Omega \mapsto U$ is a conformal mapping, and $g = f^{-1} : U \mapsto \Omega$ is an inverse mapping of $w = f(z)$.*

(a) If (z_0, ℓ) is an accessible boundary point of Ω, then the limit $\lim_{t \to 1} f(\ell(t))$ exists, and this limit is denoted by $f(z_0, \ell)$, then $f(z_0, \ell) \in \partial U$.
(b) If $(z_1, \ell_1) \sim (z_2, \ell_2)$, then $f(z_1, \ell_1) = f(z_2, \ell_2)$.
(c) If $(z_1, \ell_1) \nsim (z_2, \ell_2)$, then $f(z_1, \ell_1) \neq f(z_2, \ell_2)$.
(d) Let $\ell_\theta(t) = g(te^{i\theta})$, $t \in [0, 1)$, then $g(e^{i\theta}) \in \partial \Omega$ exists for almost everywhere all $\theta \in [0, 2\pi]$, such that $\lim_{t \to 1} \ell_\theta(t) = g(e^{i\theta})$. $(g(e^{i\theta}), \ell_\theta)$ is accessible boundary point of Ω and $e^{i\theta} = f(g(e^{i\theta}), \ell_\theta) \in F$ for almost everywhere all $\theta \in [0, 2\pi]$, where $F = \{f(z_0, \ell) : (z_0, \ell)$ is the accessible boundary point of $\Omega\}$.

Proof. (a) Let $\ell : [0, 1] \mapsto \overline{\Omega}$ be a continuous curve and $\ell[0, 1) \subset \Omega$, $\ell(1) = z_0 \in \partial \Omega$.

We shall first prove that

$$\lim_{t \to 1} |f(\ell(t))| = 1$$

exists and is equal to 1. If this is false, then

$$\liminf_{t \to 1} |f(\ell(t))| < 1.$$

Let $g = f^{-1}$ be the converse function of $w = f(z)$, then $g \in H(U)$, $\alpha \in U$ and there is a sequence $\{t_n\}$ such that $t_n \to 1$, and $f(\ell(t_n)) \to \alpha$, so $\ell(t_n) = g(f(\ell(t_n))) \to g(\alpha) \in \Omega$. This contradicts with $\ell(t_n) \to z_0 \in \partial \Omega$. Next, we prove that the limit $\lim_{t \to 1} f(\ell(t))$ exists. If not, suppose that the limit $\lim_{t \to 1} f(\ell(t))$ does not exist, then there are $\varepsilon_0 > 0$ and two sequences $\{t'_n\}$ and $\{t''_n\}$ with $t'_n \to 1, t''_n \to 1$, and

$$|f(\ell(t'_n)) - f(\ell(t''_n))| \geqslant \varepsilon_0.$$

Since two sequences $\{f(\ell(t'_n))\}$ and $\{f(\ell(t''_n))\}$ are bounded, we can assume, without loss of generality, that

$$f(\ell(t'_n)) \to a, \quad f(\ell(t''_n)) \to b,$$

and $t'_n < t''_n, n = 1, 2, \ldots$, this implies that $|a - b| \geqslant \varepsilon_0$, and let

$$\lambda(t) = f(\ell(t)), C_n = \{\lambda(t) : t'_n \leqslant t \leqslant t''_n\},$$

then

$$g(\lambda(t)) = g(f(\ell(t))) = \ell(t) \to z_0 \quad (t \to 1).$$

so

$$\max\{|g(w) - z_0| : w \in C_n\} \to 0 \ (n \to \infty).$$

By Lemma 2.3.2, applied to $g(w)$, $-z_0$ shows that g is constant. But g is one-to-one in U, and we have a contradiction. Thus, $f(z_1, \ell_1) = f(z_2, \ell_2)$, and (a) is proved.

(b) If $(z_1, \ell_1) \sim (z_2, \ell_2)$, then $z_1 = z_2$ (noted as z_0), and by the definition of equivalence class, there are $z'_n \in \ell_1 \cap \Omega$, $z''_n \in \ell_2 \cap \Omega$, such that $z'_n \to z_0$, $z''_n \to z_0$ and a continuous curve $\ell : [0, 1] \to \overline{\Omega}$ ℓ which connects z'_n, z''_n and $\lim_{t \to 1} \ell(t) = z_0$, so

$$f(\ell, z_0) = f(\ell_1, z_0) = f(\ell_2, z_0).$$

This proves (b).

(c) Let $(z_1, \ell_1) \nsim (z_2, \ell_2)$, we want to prove that $f(z_1, \ell_1) \neq f(z_2, \ell_2)$. Suppose (c) is false. If we multiply f by a suitable constant of an absolute value 1, we then have $w_0 = -1$, but

$$f(z_1, \ell_1) = f(z_2, \ell_2) = w_0 = -1 \in \partial U.$$

Put

$$\lambda_1(t) = f(\ell_1(t)), \ t \in [0, 1), \ \lambda_1(1) = f(z_1, \ell_1),$$
$$\lambda_2(t) = f(\ell_2(t)), \ t \in [0, 1), \ \lambda_2(1) = f(z_2, \ell_2),$$
$$\delta_0 = \min\{|\lambda_1(0) + 1|, |\lambda_2(0) + 1|\} > 0.$$

Given an $r \in (0, \delta_0)$, there is $\varphi(r) = \arccos \frac{r}{2} \leqslant \frac{\pi}{2}$, such that $-\varphi(r) < \theta < \varphi(r)$ describes the corresponding arc C_r in $\partial D(-1, r)$ and the two endpoints of C_r with $w'_r \in \lambda_1, w''_r \in \lambda_2$, $g(C_r) \subseteq \Omega$ is a smooth curve segment with two endpoints $z'_r \in \ell_1, z''_r \in \ell_2$ and $z'_r \to z_1, z''_r \to z_2, \ (r \to 0)$. Let L_r be the length of $g(C_r)$. Note that $L_r = \int_{C_r} |g'(z)||dz|$.

We prove a contradiction (c) by considering the following two cases:

Case (i): $z_1 \neq z_2$, since $z'_r \to z_1, z''_r \to z_2, \ (r \to 0)$, there are $r_0 > 0, r_0 \leqslant \delta_0$, such that the length $L_r \geqslant \frac{1}{2}|z_1 - z_2|$ (write as $\frac{1}{2}|z_1 - z_2| = \varepsilon_1$) for all $r \in (0, r_0)$.

Case (ii): $z_1 = z_2(= z_0)$, then $z'_r \to z_0$, $z''_r \to z_0$, $(r \to 0)$. Since $(z_0, \ell_1) \sim (z_0, \ell_2)$, there are $\varepsilon_2 > 0$ and $1 > r_0 > 0$, such that there is $w_r \in g(C_r) \setminus D(z_0, \varepsilon_2)$ for $0 < r < r_0$, the length $L_r \geqslant \varepsilon_2$.

In both cases, we have proved that there are $\varepsilon_0 > 0$ and $r_0 > 0$ such that the length $L_r \geqslant \varepsilon_0$ for all $r \in (0, r_0)$. Schwarz's inequality implies that

$$\varepsilon_0^2 \leqslant L_r^2 \leqslant \left(\int_{-\varphi(r)}^{\varphi(r)} |g'(-1 + Re^{i\theta})| r d\theta \right)^2$$

$$\leqslant \pi r^2 \int_{-\varphi(r)}^{\varphi(r)} |g'(-1 + Re^{i\theta})|^2 d\theta,$$

where $\varphi(r) = \arccos \frac{r}{2} \leqslant \frac{\pi}{2}$, so

$$\infty = \int_0^{r_0} \frac{\varepsilon_0^2}{\pi r} dr \leqslant \int_0^{r_0} \int_{-\varphi(r)}^{\varphi(r)} |g'(-1 + Re^{i\theta})|^2 r d\theta dr \leqslant \iint_\Omega du dv < \infty.$$

This is impossible. This proves (c).

(d) Suppose that $h(\theta)$ is the length of the curve $l_\theta(t) : z = g(te^{i\theta})$, $t \in [\frac{1}{2}, 1)$, the Schwarz's inequality implies that

$$h^2(\theta) = \left(\int_{\frac{1}{2}}^1 |g'(te^{i\theta})| dt \right)^2 \leqslant \int_{\frac{1}{2}}^1 |g'(te^{i\theta})|^2 t dt,$$

since Ω is a bounded region,

$$\int_0^{2\pi} h^2(\theta) d\theta \leqslant \int_0^{2\pi} \int_{\frac{1}{2}}^1 |g'(te^{i\theta})|^2 t dt d\theta \leqslant \iint_\Omega du dv < \infty,$$

therefore $h(\theta) < \infty$ for almost everywhere $\theta \in [0, 2\pi]$. Thus, there is $g(e^{i\theta}) \in \partial\Omega$ for some $\theta \in [0, 2\pi]$, such that

$$\lim_{t \to 1} l_\theta(t) = g(e^{i\theta}),$$

so $(g(e^{i\theta}), l_\theta)$ is accessible boundary point of Ω, and $e^{i\theta} = f(g(e^{i\theta}), l_\theta) \in F$, this proves (d). □

Theorem 2.3.4 (Continuity to Boundary of Conformal Mappings). *Let Ω be a bounded simply connected region in the plane, every boundary point of $\partial\Omega$ of Ω is a simple boundary point of Ω. If f is a conformal mapping from Ω onto U, then f extends continuously and one-to-one to $\partial\Omega$, i.e., there is a continuous, one-to-one function $\tilde{f}$ from $\overline{\Omega}$ onto $\overline{U}$, such that $\tilde{f}|_{\Omega} = f$.*

Proof. Suppose that f is a conformal mapping from Ω onto U. Theorem 2.3.3 implies that f can be extended a mapping $\tilde{f}$ from $\overline{\Omega}$ into $\overline{U}$, which satisfies that $\{f(\alpha_n)\}$ converges to $\tilde{f}(\alpha)$ for all $\alpha \in \overline{\Omega}$ and all sequences $\{\alpha_n\}$ in Ω which converges to α, if a sequence $\{z_n\}$ in $\overline{\Omega}$ converges to z, there is $\alpha_n \in \Omega$ for each n such that $|\alpha_n - z_n| < 1/n$ and $|f(\alpha_n) - f(z_n)| < 1/n$. So, $\alpha_n \to z$, thus $f(\alpha_n) \to f(z)$, so $f(z_n) \to \tilde{f}(z)$. This proves that $\tilde{f}$ is continuous on $\overline{\Omega}$ and $U \subset \tilde{f}(\overline{\Omega}) \subset \overline{U}$. Since $\overline{\Omega}$ is compact, $\tilde{f}(\overline{\Omega})$ is compact, so $\tilde{f}(\overline{\Omega})$ is a closed set, thus $\tilde{f}(\overline{\Omega}) = \overline{U}$. Theorem 2.3.3,(b) implies that $\tilde{f}$ is one-to-one on $\overline{\Omega}$, $f(K)$ for any compact $K \subset \overline{\Omega}$, therefore $\tilde{f}^{-1}$ is a continuous, one-to-one function from $\overline{U}$ onto $\overline{\Omega}$. $\square$

2.4 Univalent Function

A function $w = f(z)$ is called univalent on a region Ω if it is analytic and injective (=one-to-one) in Ω, i.e., $f \in H(\Omega)$ and $f(z_1) \neq f(z_2)$ for all $z_1 \in \Omega, z_2 \in \Omega$ with $z_1 \neq z_2$.

If $f(z) \in H(\Omega)$ is univalent, then f is called a univalent function (univalent function is also called schlicht function). Ω is called univalent region of $f(z)$.

According to the Riemann mapping theorem, any simply connected region Ω in the complex plane that is not a finite complex plane can be conformally mapped to the unit disk U. Therefore, univalent analytic functions (=schlicht function) in Ω can be reduced to univalent analytic functions in the unit disk. Let $f(z)$ be a univalent analytic function in the unit disk U. Without loss of generality, we assume that it satisfies the normalization conditions: $f(0) = 0, f'(0) = 1$. Otherwise, we can consider the function

$$\frac{f(z) - f(0)}{f'(0)}.$$

Let S be the class of all $f \in U$ which is one-to-one in U and satisfy $f(0) = 0, f'(0) = 1$. Thus, every $f \in S$ has a power series expansion

$$f(z) = z + a_2 z^2 + a_3 z^3 + \cdots \quad (z \in U) \qquad (2.4.1)$$

The family of functions closely related to the family S is characterized by the following conditions:

Let $\sum$ be the class of all $f(z)$ which is one-to-one in $\{z : |z| > 1\}$ and has a Laurent series expansion

$$f(z) = z + b_0 + b_1 z^{-1} + b_2 z^{-2} + \cdots. \qquad (2.4.2)$$

If $f \in \sum$, the infinity is a pole of order 1 of $f(z)$.

From the perspective of conformal mapping, univalent functions play a significant role in a given region.

First, we prove the area theorem, which is a fundamental theorem in the theory of single-valued functions and was first given by Gronwall in 1914.

Theorem 2.4.1 (Area Theorem). *If $f(z) \in \sum$ and f has the expansion* (2.4.2)*, then*

$$\sum_{n=1}^{\infty} n |b_n|^2 \leqslant 1. \qquad (2.4.3)$$

Proof. For $\rho > 1$, Let $f(C_\rho) : w = f(\rho e^{i\theta}), 0 \leqslant \theta \leqslant 2\pi$ be the curve that is the image under f of the circle $C_\rho = \{z : |z| = \rho\}$. Since f is univalent, $f(C_\rho)$ is a smooth Jordon curve; let $E(\rho)$ be the inside of $f(C_\rho)$. Denote the area of $E(\rho)$ by $I_\rho(f)$. Applying the analytic Green's theorem to the function $w = f(z)$, we get

$$I_\rho(f) = \frac{1}{2i} \int_{f(C_\rho)} \overline{f} df = \frac{1}{2i} \int_{C_\rho} \overline{f(z)} f'(z) dz.$$

Using the expansion (2.4.2) of $f(z)$, this means that

$$I_\rho(f) = \frac{1}{2} \int_0^{2\pi} \left(\rho e^{-i\theta} + \sum_{n=0}^{\infty} \overline{b_n} \rho^{-n} e^{in\theta} \right)$$

$$\times \left(\rho e^{i\theta} - \sum_{m=1}^{\infty} m b_m \rho^{-m} e^{-im\theta} \right) d\theta$$

$$= \pi \left(\rho^2 - \sum_{m=1}^{\infty} m |b_m|^2 \rho^{-2m} \right).$$

We can integrate term by term in view of the uniform convergence. Since the area $I(f)$ is non-negative, it follows that

$$0 \leq I_\rho(f) = \pi \rho^2 - \pi \sum_{m=1}^{\infty} m |b_m|^2 \rho^{-2m}.$$

Therefore,

$$\sum_{m=1}^{\infty} m |b_m|^2 \rho^{-2m-2} \leq 1$$

for all $\rho > 1$. If we let $\rho \to 1$, we obtain (2.4.3). $\square$

Corollary 2.4.2. *Let $f(z) \in \Sigma$ and f has the expansion (2.4.2), then $|b_1| \leq 1$. Moreover, $|b_1| = 1$ if and only if the set E which is the complement of the image domain $\{f(z) : |z| > 1\}$ is a line segment of length 4. In this case,*

$$f(z) = z + b_0 + b_1 z^{-1}, \tag{2.4.4}$$

where $|b_1| = 1$ and

$$E = [b_0 - 2e^{\frac{i\beta}{2}}, b_0 + 2e^{\frac{i\beta}{2}}],$$

where $\beta = \arg b_1$.

Proof. Since $|b_1|$ is one of the terms in the sum appearing in the area theorem, it is clear that $|b_1| \leq 1$. If $|b_1| = 1$, then $b_n = 0$ for $n \geq 2$. Thus, let $b_1 = e^{i\beta}$, then

$$f(z) = b_0 + 2e^{\frac{i\beta}{2}} \frac{1}{2} \left(\zeta + \frac{1}{\zeta} \right),$$

where $\zeta = e^{-\frac{i\beta}{2}} z$. Since Joukowski's function maps the set $\{z : |z| > 1\}$ onto the complex plane slit along the interval $[-1, 1]$, the function

f defined by (2.4.4) maps $\{z : |z| > 1\}$ onto the complex plane slit along the interval $[b_0 - 2e^{\frac{i\beta}{2}}, b_0 + 2e^{\frac{i\beta}{2}}]$. In particular, E is a straight line of length 4.

Conversely, assume that E is a straight line of length 4; so, E has the form $E = [b_0 - 2e^{\frac{i\beta}{2}}, b_0 + 2e^{\frac{i\beta}{2}}]$, where b_0 is a complex number and β is a real number. If $g(z) = z + b_0 + e^{i\beta}z^{-1}$, then $g \in \Sigma$ and the complement of the image domain $\{g(z) : |z| > 1\}$ is equal to the complement of the image domain $\{f(z) : |z| > 1\}$. Therefore, $f \circ g^{-1} \in \Sigma$ and maps $\{z : |z| > 1\}$ onto itself. Therefore, $f = g$ and $b_1 = e^{i\beta}$, so that $|b_1| = 1$. $\square$

Theorem 2.4.3 (Bieberbach's Theorem). *If $f(z) \in S$, then $|a_2| = |\frac{f''(0)}{2}| \leqslant 2$.*

Proof. Since the function $f(z)z^{-1}$ is holomorphic in the disk $U = \{z : |z| < 1\}$ ($z = 0$ is a removable singularity) and $f(z)z^{-1} \neq 0$ for all $z \in U$, there is a holomorphic function $h(z)$ such that $h(0) \neq 0$ and $h(z) = \{f(z)z^{-1}\}^2$. The function $g(z) = zh(z^2)$ is univalent in U. In fact, if $g(z_1) = g(z_2)$, then $f(z_1^2) = f(z_2^2)$, and since $f(z)$ is univalent, it follows that $z_1^2 = z_2^2$, i.e., $z_1 = z_2$ or $z_1 = -z_2$. The minus sign is impossible $z_1 = -z_2$ because the function $g(z) = zh(z^2)$ is an odd function, $g(z_1) = -g(z_1) \neq g(z_1)$. So, $g(z)$ is univalent. Let

$$g(z) = z + a'_3 z^3 + a'_5 z^5 + \cdots.$$

Since $(g(z))^2 = f(z^2)$,

$$(z + a'_3 z^3 + a'_5 z^5 + \cdots)^2 = z^2 + a_2 z^4 + a_3 z^6 + \cdots,$$

so $a'_3 = \frac{a_2}{2}$. Let

$$F(z) = \left(g\left(\frac{1}{z}\right)\right)^{-1} = z - \frac{a_2}{2}\frac{1}{z} + \cdots,$$

then $F(z) \in \Sigma$. The area theorem implies that $|\frac{a_2}{2}| \leqslant 1$, i.e., $|a_2| \leqslant 2$. $\square$

By Corollary 2.4.2 of the area theorem, the equality holds in Theorem 2.4.3 if and only if

$$F(z) = \left(g\left(\frac{1}{z}\right)\right)^{-1} = z + \frac{e^{i\beta}}{z},$$

i.e.,

$$g(z) = \frac{z}{1 + e^{i\beta}z^2}.$$

So,

$$f(z) = \frac{z}{(1 + e^{i\beta}z)^2}.$$

This function maps U onto the complex plane slit along the half line $\{w = te^{-i\beta} : 4^{-1} \leqslant t < \infty\}$. For $\beta = \pi$, let

$$k(z) = \frac{z}{(1-z)^2} = z + 2z^2 + 3z^3 + \cdots, \qquad (2.4.5)$$

this function maps U onto the complex plane slit along the negative real axis from $-\frac{1}{4}$ to $-\infty$. The function $k(z)$ is called the Koebe function. The function

$$f(z) = \frac{z}{(1 + e^{i\beta}z)^2}$$

can be rewritten as $f(z) = -e^{i\beta}k(-e^{i\beta}z)$.

From the estimate of a_2, we can easily obtain the following Köebe's one-quarter theorem.

Theorem 2.4.4 (Koebe's One-Quarter Theorem).

(a) If $f(z) \in S$, then $f(U) \supset D(0, \frac{1}{4}) = \{w : |w| < \frac{1}{4}\}$.
(b) If $F \in \Sigma$ with the Laurent expansion

$$F(z) = z + b_0 + b_1 z^{-1} + b_2 z^{-2} + \cdots, \qquad |z| > 1,$$

then

$$F(\mathbb{C} \setminus \overline{U}) \supset \mathbb{C} \setminus \overline{D(b_0, 2)}.$$

In particular, if w_1 and w_2 are not in $F(\mathbb{C} \setminus \overline{U})$, then $|w_1 - w_2| \leqslant 4$.

Proof. (a) Fix f in $\mathcal{S}$ and let w be a complex number that does not belong to $f(U)$; it must be shown that $|w| \geq 1/4$. Since $0 \in f(U)$, $w \neq 0$, and so,

$$g(z) = \frac{wf(z)}{w - f(z)}$$

is an analytic function on U. In fact, $g \in \mathcal{S}$. To see this, first observe that $g(0) = 0$ and

$$g'(0) = \lim_{z \to 0} \frac{wf(z)}{z(w - f(z))} = f'(0) = 1.$$

Finally, g is the composition of f and a fractional linear transformation and hence must be univalent.

Since $f(0) = 0$, there is a small value of r, such that $|f(z)| < |w|$ for $|z| < r$. In this neighborhood of 0, we get

$$g(z) = \frac{wf(z)}{w - f(z)} = z + \left(a_2 + \frac{1}{w}\right)z^2 + \cdots \in \mathcal{S}.$$

By Theorem 2.4.3, this implies that $|a_2 + w^{-1}| \leqslant 2$. But $|a_2| \leqslant 2$ so that $\frac{1}{|w|} \leqslant |a_2| + |a_2 + \frac{1}{w}| \leqslant 4$, or $|w| \geqslant \frac{1}{4}$.

(b) If $F \in \sum$, w_1 and w_2 are not in the $F(\mathbb{C} \setminus U)$, then

$$f(z) = \left(F_1\left(\frac{1}{z}\right)\right)^{-1} = z - b_1 z^3 + \cdots \in \mathcal{S},$$

where $F_1 = F - b_0$. If $w \notin f(U)$, by the proof of (a) above, we have $|w^{-1}| \leq 2$, so $|w| \geqslant \frac{1}{2}$. Equivalently, the values that $f(z)$ do not take lie in the exterior of the disk $|w| < \frac{1}{2}$, thus $f(U) \supset D(0, \frac{1}{2})$, so

$$F(\mathbb{C} \setminus \overline{U}) = F_1(\mathbb{C} \setminus \overline{U}) + b_0 \supset \mathbb{C} \setminus \overline{D(0,2)} + b_0 = \mathbb{C} \setminus \overline{D(b_0, 2)}.$$

Especially, if w_1 and w_2 are not in $F(\mathbb{C} \setminus \overline{U})$, then $|w_1 - b_0| \leqslant 2$, $|w_2 - b_0| \leqslant 2$, so $|w_2 - w_1| \leqslant 4$. $\square$

The function (2.4.5) indicates that the constant $\frac{1}{4}$ is exact in this case.

We continue by deriving some further implications from Theorem 2.4.3 regarding the estimation of a_2. These estimations on $|f(z)|$ and $|f'(z)|$ are known as the distortion theorems.

First, we prove the following lemma. $\square$

Lemma 2.4.5. If $f(z) \in \mathcal{S}$, then

$$\left| \frac{zf''(z)}{f'(z)} - \frac{2|z|^2}{1-|z|^2} \right| \leq \frac{4|z|}{1-|z|^2}, \quad |z| < 1. \tag{2.4.6}$$

Proof. For each complex number a in U, define the function

$$g_a(z) = \frac{f\left(\frac{z+a}{1+\bar{a}z}\right) - f(a)}{(1-|a|^2)f'(a)}.$$

(f_a is called the Koebe's transform of f with respect to a). g_a is univalent on U since it is the composition function of univalent functions. Therefore, $g_a \in \mathcal{S}$. Let $g_a(z) = z + b_2 z^2 + \cdots$ in U.

Another computation reveals that

$$g''(0) = (1-|a|^2)\frac{f''(a)}{f'(a)} - 2\bar{a}.$$

Since $b_2 = g_a''(0)/2$ and $b_2| \leq 2$ by Theorem 2.4.3, then

$$|g_a''(0)| = \left|(1-|a|^2)\frac{f''(a)}{f'(a)} - 2\bar{a}\right| \leq 4.$$

Multiply both sides of this inequality by $\frac{|a|}{1-|a|^2}$ and substitute $z = a$ to get (2.4.6). $\square$

Theorem 2.4.6. If $f \in \mathcal{S}$, then

$$\frac{1-|z|}{(1+|z|)^3} \leq |f'(z)| \leq \frac{1+|z|}{(1-|z|)^3}, \tag{2.4.7}$$

$$|\arg f'(z)| \leq 2\log\frac{1+|z|}{1-|z|} \tag{2.4.8}$$

for all $z \in U$.

Proof. By Lemma 2.4.5,

$$\frac{2|z|^2 - 4|z|}{1-|z|^2} \leq \operatorname{Re}\frac{zf''(z)}{f'(z)} \leq \frac{2|z|^2 + 4|z|}{1-|z|^2},$$

$$\left|\operatorname{Im}\frac{zf''(z)}{f'(z)}\right| \leq \frac{4|z|}{1-|z|^2}.$$

Now, f' does not vanish on U, so there is an analytic branch of $\log f'(z)$ with $\log f'(0) = 0$. Note that $\log f'(z) = \log |f'(z)| + i \arg f'(z)$, and using the chain rule,

$$\operatorname{Re} \frac{zf''(z)}{f'(z)} = r \frac{\partial}{\partial r} \log |f'(z)|, \quad \operatorname{Im} \frac{zf''(z)}{f'(z)} = r \frac{\partial}{\partial r} \arg f'(z).$$

This implies that

$$\frac{2r-4}{1-r^2} \leqslant \frac{\partial}{\partial r} \log |f'(\operatorname{Re}^{i\theta})| \leqslant \frac{2r+4}{1-r^2},$$

$$-\frac{4}{1-r^2} \leqslant \frac{\partial}{\partial r} \arg f'(\operatorname{Re}^{i\theta}) \leqslant \frac{4}{1-r^2}.$$

Integrating both sides of the expressions with respect to r from 0 to r yields these inequalities (2.4.7) and (2.4.8). $\square$

Theorem 2.4.7. *If $f \in \mathcal{S}$, then*

$$\frac{|z|}{(1+|z|)^2} \leqslant |f(z)| \leqslant \frac{|z|}{(1-|z|)^2}, \qquad (2.4.9)$$

$$\frac{1-|z|}{1+|z|} \leqslant \left|\frac{zf'(z)}{f(z)}\right| \leqslant \frac{1+|z|}{1-|z|} \qquad (2.4.10)$$

for all $z \in U$.

Proof. Note that

$$|f(z)| = \left| \int_{[0,z]} f'(\zeta) d\zeta \right| \leqslant \int_{[0,z]} |f'(\zeta)| |d\zeta|.$$

Parameterize the line segment by $\zeta = tz, 0 \leq t \leq 1$, and use (2.4.7) to get an upper estimate for $|f'(\zeta)|$. After performing the required calculations, we have

$$|f(z)| \leqslant \int_0^{|z|} \frac{1+t}{(1-t)^3} dr = \frac{|z|}{(1-|z|)^2},$$

the right-hand side of (2.4.9).

To get the left-hand side of (2.4.9), first note that an elementary argument using calculus shows that $t(1+t)^2 \leq \frac{1}{4}$ for $0 \leq t \leq 1$;

so it is sufficient to establish the inequality under the assumption that $|f(z)| < \frac{1}{4}$. But, here, Koebe's $\frac{1}{4}$ theorem implies that $\{\zeta : |\zeta| > \frac{1}{4}\} \subset f(U)$. So, fix z in U with $|f(z)| < \frac{1}{4}$ and let γ be the path in U such that $f \circ \gamma$ is the straight line segment $[0, f(z)]$. i.e., $f(\gamma(t)) = tf(z)$ for $0 \le t \le 1$. Thus, $|f(z)| = |\int_\gamma f'(w)dw| = |\int_0^1 f'(\gamma(t))\gamma'(t)dt|$. Now, $f'(\gamma(t))\gamma'(t) = [tf(z)]' = f(z)$ for all t. Thus, $|f(z)| = \int_\gamma |f'(w)||dw|$. Using the left-hand side of the inequalities (2.4.7), we get that $|f'(w)| \ge \frac{1-|w|}{(1+|w|)^3}$. On the other hand, if we take $0 \ge s < t \ge 1, |\gamma(t) - \gamma(s)| \ge ||\gamma(t)| - |\gamma(s)||$, and so, (symbolically), $|dw| \ge d|w|$. Combining these inequalities gives

$$|f(z)| \ge \int_0^{|z|} \frac{1-t}{(1+t)^3} dt = \frac{|z|}{(1+|z|)^2}.$$

This proves the left-hand side of the inequalities (2.4.9).

Proof of (2.4.10). For each complex number a in U, the Koebe's transform of f

$$g_a(z) = \frac{f\left(\frac{z+a}{1+\bar{a}z}\right) - f(a)}{(1-|a|^2)f'(a)} \in \mathcal{S},$$

and taking $z = -a$, we see that

$$g_a(-a) = \frac{-f(a)}{(1-|a|^2)f'(a)}, \quad \text{i.e.,} \quad \frac{af'(a)}{f(a)} = \frac{-a}{(1-|a|^2)g(-a)}$$

because $g_a \in \mathcal{S}, g_a(0) = 0$, hence, replaced a by z, we obtain (2.4.10) from (2.4.9). □

Lemma 2.4.8. *If $f(z) \in \mathcal{S}$, then*

$$\frac{1}{2\pi} \int_0^{2\pi} |f(Re^{i\theta})| d\theta \le \frac{r}{1-r}, \quad 0 \le r < 1. \tag{2.4.11}$$

Proof. Write $f(z) = zf_1(z)$, $f_1(z)$ holomorphic in U. Note that $f_1(0) = 1$ and f_1 is non-vanishing on U. Thus, f_1 has a holomorphic square root $h(z)$ on U satisfying $h(0) = 1$. Define $g(z) = zh(z^2) \in \mathcal{S}$. By (2.4.9),

$$|g(z)| = \sqrt{|f(z^2)|} \leqslant \sqrt{\frac{|z|^2}{(1-|z|^2)^2}} = \frac{|z|}{1-|z|^2}.$$

On the other hand, we need to check that $g(z)$ is one-to-one. If $g(a) = g(b)$, then $f(a^2) = f(b^2)$ so $a^2 = b^2$. Thus, $a = \pm b$, then the definition of g shows that $g(a) = -g(b)$, whence $g(a) = g(b) = 0$, Since h is non-vanishing, this forces $a = b = 0$. If $g(z) = \sum_{n=1}^{\infty} b_n z^n$, D_r is the image of the disk $|z| < r$ under the mapping $g(z)$. The area of D_r is denoted as A_r, then

$$A_r = \iint_{|z|<r} |g'(z)|^2 dxdy = \pi \sum_{n=1}^{\infty} n|b_n|^2 r^{2n} \leqslant \frac{\pi r^2}{(1-r^2)^2},$$

$$\sum_{n=1}^{\infty} n|b_n|^2 r^{2n-1} \leqslant \frac{r}{(1-r^2)^2}.$$

By integrating both sides from 0 to r, we get

$$\sum_{n=1}^{\infty} |b_n|^2 r^{2n} \leqslant \frac{r^2}{1-r^2}.$$

Note that

$$\frac{1}{2\pi} \int_0^{2\pi} |g(Re^{i\theta})|^2 d\theta = \sum_{n=1}^{\infty} |b_n|^2 r^{2n}.$$

Thus,

$$\frac{1}{2\pi} \int_0^{2\pi} |g(Re^{i\theta})|^2 d\theta \leqslant \frac{r^2}{1-r^2},$$

i.e.,

$$\frac{1}{2\pi} \int_0^{2\pi} |f(r^2 e^{2i\theta})| d\theta \leqslant \frac{r^2}{1-r^2}.$$

Replacing r^2 by r, 2θ by θ, we obtain (2.4.11). □

Theorem 2.4.9 (Littlewood Theorem). *If $f(z) \in \mathcal{S}$ with*

$$f(z) = z + \sum_{n=2}^{\infty} a_n z^n,$$

then $|a_n| < en$ for all $n \geq 2$.

Proof. By Cauchy formula,

$$a_n = \frac{1}{2\pi i} \int_{|z|=r<1} \frac{f(z)dz}{z^{n+1}}.$$

By (2.4.11),

$$|a_n| \leq \frac{1}{2\pi} \int_0^{2\pi r^n} |f(Re^{i\theta})| d\theta \leq \frac{1}{r^{n-1}(1-r)}.$$

It is easy to verify that the function $r^{n-1}(1-r)$ attains its maximum value at $r = 1 - \frac{1}{n}$. Thus,

$$|a_n| \leq n\left(1 + \frac{1}{n-1}\right)^{n-1} < en.$$

□

Definition 2.4.10 (Convergence of Regions). Let $\{\Omega_n\}$ be sequence of regions in $\mathbb{C}$ and $z_0 \in \Omega_0 = \bigcup_{N=1}^{\infty} \text{int}(\bigcap_{n=N}^{\infty} \Omega_n)$. The kernel $\ker(\Omega_n, z_0)$ with respect to z_0 is defined as the connected component of Ω_0, which contains z_0, i.e., the set $\ker(\Omega_n, z_0)$ is defined as the set consisting of z_0 together with all points $w \in \mathbb{C}$ with the following property: there exists a region H with $z_0 \in H$, $w \in H$, such that $H \subset \Omega_n$ for all sufficiently large n. If $z_0 \notin \bigcup_{N=1}^{\infty} \text{int}(\bigcap_{n=N}^{\infty} \Omega_n)$, but $z_0 \in \bigcup_{N=1}^{\infty} \bigcap_{n=N}^{\infty} \Omega_n$, then the kernel $\ker(\Omega_n, z_0)$ with respect to z_0 is defined as the point $\{z_0\}$, i.e., $\ker(\Omega_n, z_0) = \{z_0\}$. When $\{\Omega_n\}$ converges to Ω, if Ω is the kernel of every subsequence of $\{\Omega_n\}$, this is denoted by $\Omega_n \to \Omega$.

Example 2.4.11. Consider Ω_n ($n = 1, 2, \ldots$) $= \mathbb{C}\setminus(E_n \cap F_n)$, the plane minus $E_n = \{x + iy : 1 \leq x < \infty, y = 0\}$ and $F_n = \{z : |z| = 1, 0 \leq \arg z \leq 2\pi - \frac{1}{n}\}$, then the kernel $\ker(\Omega_n, 0)$ (with respect to 0) is U and $\Omega_n \to U$.

We will now present and prove the following significant theorem.

Theorem 2.4.12 (Carathéodory's Convergence Theorem).
Let $\{\Omega_n\}$ be a sequence of simple connected regions in $\mathbb{C}$. Suppose that $0 \in \bigcap_{n=1}^{\infty} \Omega_n$, $f_n(z)$ is conformal mapping from U onto Ω_n, with $f_n(0) = 0, f'_n(0) > 0$, and that $\Omega = \ker(\Omega_n, 0)$ is the kernel with respect to 0 of $\{\Omega_n\}$. Then $\{f_n(z)\}$ converges uniformly on compact subsets of U if and only if $\{\Omega_n\}$ converges uniformly on compact subsets of its kernel $\Omega = \ker(\Omega_n, 0)$ with respect to 0. When this happens, if $\Omega = \{0\}$, then $\{f_n(z)\}$ converges uniformly to 0 on compact subsets of U; and if $\Omega \neq \{0\}$, then Ω is simply connected domain, and the limit function $f(z)$ is univalent on Ω and maps from U onto Ω, and the sequence $\{f_n^{-1}\}$ of converse functions converges uniformly on U to the converse function $f^{-1}(w)$ of $f(z)$.

Proof. Necessity: Let $\{f_n(z)\}$ converge uniformly to $f(z)$ on compact subsets of U. $f(z)$ is holomorphic in U by Weierstrass's theorem. By Lemma 2.1.12 (Hurwitz's theorem), the limit function $f(z)$ is univalent or constant in U. In the case of the limit function, $f(z)$ is constant, then $f(z) \equiv 0$. We prove the necessity in the following two cases:

Case (i): $f(z) \equiv 0$. We must prove that $\Omega = \{0\}$. If not, there is $N, \rho > 0$, such that, if $n \geq N$, then Ω_n contains $|w| < \rho$ for $n \geq N$, the converse function $\varphi_n(w) = f_n^{-1}(w)$ holomorphic in the disk $|w| < \rho$ with $\varphi_n(0) = 0$, $|\varphi_n(w)| < 1$. Schwarz's lemma implies that $|\varphi'_n(0)| \leq \frac{1}{\rho}$. Thus, $f'_n(0) \geq \rho$. But this is a contradiction with the point that $\{f_n(z)\}$ converges to f uniformly to 0 on compact subsets of U because every subsequence $\{\Omega_{n_k}\}$ of $\{\Omega_n\}$ also converges to $f(z)$, so $\{\Omega_n\}$ converges to $\Omega = \{0\}$.

Case (ii): Let $f(z)$ be analytic and univalent in U. Suppose that $f(z)$ maps U onto its image $G = f(U)$, we want to prove that $G = \Omega$ and $\Omega_n \to \Omega$.

We first prove that $G \subset \Omega$. Let

$$E = f(\overline{D(0, r_{N_0})}) \subset G = \bigcup_{n=1}^{\infty} f(D(0, r_n)) \quad \left(r_n = 1 - \frac{1}{n+1}\right),$$

and choose $N_1 > N_0$, such that $E \subset f(D(0, r_{N_1}))$, and a Joran curve

$$\Gamma = \partial f(D(0, r_{N_1})) = f(\partial D(0, r_{N_1})).$$

Note that δ is the distance from E to Γ ($\delta > 0$), $\gamma = \partial D(0, r_{N_1})$. We prove that $E \subset \Omega_n$ for large n. For every point $w_0 = f(z_0) \in E$, $z_0 \in \overline{D(0, r_{N_0})}$ and $z \in \gamma$, $|f(z) - w_0| \geq \delta$, since $\{f_n(z)\}$ converges uniformly on compact subsets of U, there is an integer $N > N_1$ such that $|f_n(z) - f(z)| < \delta$ for all $n \geq N$ and all $z \in \gamma$. The Rouché theorem implies the function

$$f_n(z) - w_0 = (f_n(z) - f(z)) + (f(z) - w_0)$$

and the function $f(z) - w_0$ have the same number of zeros in the inner domain $D(0, r_{N_1})$ of γ for $n \geq N$. This implies that $w_0 = f(z_0) \in \Omega_n = f_n(U)$ $n \geq N$ here. N is dependent only on E and independent of w_0, i.e., $E = f(\overline{D(0, r_{N_0})}) \subset \Omega_n$ for $n \geq N$. Since $E = f(\overline{D(0, r_{N_0})})$ is connected, the definition of the kernel and

$$E = f(\overline{D(0, r_{N_0})}) \subset \bigcup_{N=1}^{\infty} \text{int} \bigcap_{n=N}^{\infty} \Omega_n,$$

we have that $G \subset \Omega$. Therefore, the converse function $\varphi_n(w) = f_n^{-1}(w)$ of $w = f_n(z)$ is well defined on $E = f(\overline{D(0, r_{N_0})})$ ($n \geq N$), satisfying $|\varphi_n(w)| \leq 1$. By the Montel theorem, a subsequence $\{\varphi_{n_k}(w)\}$ of the sequence $\{\varphi_n(w)\}$ converges uniformly to a $\varphi(w)$ on compact subsets of G with $\varphi(0) = 0$, $\varphi'(0) > 0$. In fact,

$$\varphi'(0) = \lim_{k \to \infty} \varphi_{n_k}(0) = \lim_{k \to \infty} \frac{1}{f_{n_k}(0)} = \frac{1}{f'(0)} > 0.$$

Therefore, $\varphi(w)$ is univalent in G.

Next, we show that $z = \varphi(w)$ is the inverse function of $w = f(z)$. Let $z_0 \in U$, and put $w_0 = f(z_0)$. $C : |z - z_0| = \varepsilon$ is contained in U for small $\varepsilon > 0$. Let $\Gamma = f(C)$ and δ the distance from w_0 to Γ, then $|f(z) - w_0| \geq \delta$ for all $z \in C$, there is k_0 such that $|f_{n_k}(z) - f(z)| < \delta$ for $z \in C$ and all $k \geq k_0$. The Rouché theorem implies that, for each $k \geq k_0$, there is z_k in the inner of C such that $f_{n_k}(z_k) = w_0$. Thus, $|z_k - z_0| < \varepsilon$, $z_k = \varphi_{n_k}(w_0)$. Therefore, $z_k \to \varphi(w_0)$ as $k \to \infty$. This means that $\varphi(w_0)$ satisfies $|\varphi(w_0) - z_0| \leq \varepsilon$, since $\varepsilon > 0$ is arbitrarily small, so $z_0 = \varphi(w_0) \in U$. Thus, $\varphi = f^{-1}$. Note that the preceding argument can be applied to any subsequence of $\{\varphi_n(w)\}$, the sequence $\{\varphi_n(w)\}$ converges uniformly to $f^{-1}(w)$ on compact subsets of G. The preceding argument can also be applied to the sequence $\{\varphi_n(w)\}$, which converges uniformly to a univalent function $\psi(w)$ on compact subsets of G with $|\psi(w)| < 1$. $\psi(w)$ is an analytic extension of f^{-1} from G to Ω. Since f^{-1} is a conformal mapping from G to U, there must be the equality $G = \Omega$.

Finally, we want to prove that $\Omega_n \to \Omega$. With the same reasoning to the subsequence $\{\Omega_{n_k}\}$ of $\{\Omega_n\}$, we can show that the mapping $w = f(z)$ is a conformal mapping from U to the kernel of $\{\Omega_{n_k}\}$, i.e., the kernels are the same for any subsequence $\{\Omega_{n_k}\}$. Therefore, $\Omega_n \to \Omega$.

Sufficiency: Let $\Omega_n \to \Omega \neq \mathbb{C}$. We prove the sufficiency in the following two cases:

Case (iii): $\Omega = \{0\}$. We want to prove that $\{f_n(z)\}$ converges uniformly to 0 on compact subsets of $U \not\subseteq U$. First, prove that $f_n'(0) \to 0$ as $n \to \infty$. If there is $\varepsilon_0 > 0$ and a subsequence $\{f_{n_k}(z)\}$, such that $f_{n_k}'(0) \geq \varepsilon_0$, the Koebe $\frac{1}{4}$ theorem, applied to $f_{n_k}(z)/f_{n_k}'$ shows that Ω_{n_k} contains $|w| < \frac{1}{4}\varepsilon_0$. This contradicts the point that the kernel of Ω_{n_k} is $\{0\}$. Thus, $f_n'(0) \to 0$ as $k \to \infty$. Theorem 2.4.7 implies that

$$f_n'(0)\frac{|z|}{(1+|z|)^2} \leq |f_n(z)| \leq f_n'(0)\frac{|z|}{(1-|z|)^2}, \quad |z| < 1.$$

So, $\{f_n(z)\}$ converges uniformly to a 0 on compact subsets of U.

Case (iv): $\Omega \neq \{0\}, \Omega \neq \mathbb{C}$. We want to prove that $\{f_n'(0)\}$ is bounded. If not, there is a subsequence $\{f_{n_k}'(0)\}$ of $\{f_n'(0)\}$ such that $f_{n_k}'(0) \to \infty$ $(k \to \infty)$. Theorem 2.4.7 implies that the kernel of $\{\Omega_{n_k}\}$ is $\mathbb{C}$, giving a contradiction. Therefore, $\{f_n'(0)\}$ is bounded. Next, Theorem 2.4.7 implies that $\{f_n(z)\}$ is uniformly bounded on compact subset of U. According to the Vitali Theorem 2.1.9, case (iv) will prove to be true if we can prove that $\{f_n(z)\}$ converges for each point of U. If $\{f_n(z)\}$ does not converge at some point $z_0 \in U$, once again Montel theorem implies that there are two subsequence $\{f_{n_k}(z)\}$ and $\{f_{m_k}(z)\}$ of $\{f_n(z)\}$, such that $\{f_{n_k}(z)\}$ $\{f_{m_k}(z)\}$ converges uniformly to $f(z)$ and $\widetilde{f}(z)0$ on compact subsets of U, respectively. The proof of the necessity shows that $\{\Omega_{n_k}\}$ and $\{\Omega_{m_k}\}$ converge different kernels $f(U)$ and $\widetilde{f}(U)$, respectively. This contradicts that $\Omega_n \to \Omega$. Therefore, $\{f_n(z)\}$ converges uniformly to $f(z)$ on compact subsets of U. □

Remark. According to the proof of the theorem, if $\{f_n(z)\}$ is a sequence of univalent functions in U, satisfying $f_n(0) = 0, f_n'(0) > 0$ for each n, then $\{f_n(z)\}$ converges to $f(z)$ on compact subsets of U if and only if $\{f_n(z)\}$ converges $f(z)$ for all $z \in U$.

Lemma 2.4.13. Let $f(z)$ be holomorphic in Ω, $M > 0$ and $|f(z)| \leqslant M$ for all $z \in \Omega$. If suppose that $z_0 \in \Omega$, the subarc γ of $|z - z_0| = r$ with the length more than $\frac{2\pi r}{m}$ (m is a positive integer), the subarc γ is contained in the complement of Ω, $|f(z)| \leqslant \varepsilon$ ($\varepsilon > 0$) for z in the boundary of $\Omega \cup D(z_0, r)$, then

$$|f(z_0)| \leqslant \sqrt[m]{\varepsilon M^{m-1}}.$$

Proof. We can assume, without loss of generality, that $z_0 = 0$. The regions obtained by rotating Ω around $z = 0$ the angles of $\frac{2\pi}{m}, \frac{4\pi}{m}, \ldots, \frac{2(m-1)\pi}{m}$ are denoted as $\Omega_1, \Omega_2, \ldots, \Omega_{m-1}$, respectively. Let B be a connected component which contains 0 of the open set $\Omega \cap \Omega_1 \cap \Omega_2 \cap \cdots \cap \Omega_{m-1}$, then B does not contain any points on $|z| = r$. Therefore, B is completely contained within the disk $|z| < r$, and its boundary ∂B is the set formed by rotating by angles $\frac{2\pi}{m}, \frac{4\pi}{m}, \ldots, \frac{2(m-1)\pi}{m}$ of the boundary of Ω, which are fully contained the disk $|z| < r$. The function

$$F(z) = f(z)f(\eta z)f(\eta^2 z)\cdots f(\eta^{m-1}z) \quad (\eta = e^{\frac{2\pi}{m}})$$

is analytic in B. When z tends to the boundary of B from its interior, there exists an integer $0 < k < m$ such that $\eta^k z$ to the boundary of Ω within $|z| < r$. By assumption, $|f(\eta^k z)| \leqslant \varepsilon$. Consequently, we have $|F(z)| \leqslant \varepsilon M^{m-1}$. According to the maximum modulus theorem, this inequality holds within B. In particular,

$$|F(0)| = |f(0)|^m \leqslant \varepsilon M^{m-1}.$$

This proves Lemma 2.4.13. □

The following is the relationship between the convergence of a sequence of single-valued functions on the closed unit disk and the convergence of the corresponding sequence of regions.

Theorem 2.4.14 (Rado's Theorem). Let $\{\Omega_n\}$ be a sequence of simply connected regions, $0 \in \text{int}\Omega_n (n \geqslant 1)$, $\{\Omega_n\}$ converge to a region $\Omega = \ker(\Omega_n, 0)$, the boundaries of Ω_n and Ω are both simple curves. Let $w = f_n(z)$ be a sequence of conformal mappings from U to the domain Ω_n, satisfying $f_n(0) = 0$, $f_n'(0) > 0$. The sequence of functions $\{f_n(z)\}$ converges uniformly on the closed unit disk $\overline{U} = \{z : |z| \leqslant 1\}$ to $w = f(z)$, and this function maps U onto Ω if and only if for any $\varepsilon > 0$, there exists an $N > 0$ such that for all $n > N$, there exists a one-to-one continuous correspondence between C_n and C such that the distance between corresponding points is less than ε.

Proof. The necessity is evident. The correspondence between the two curves C_n and C can be defined as follows: For a point z on $|z| = 1$, let $w_n = f_n(z)$ correspond to $w = f(z)$. Since $\{f_n(z)\}$ converges uniformly to $f(z)$ on $|z| = 1$, for any $\varepsilon > 0$, there exists an $N > 0$ such that for all $n > N$, we have

$$|w_n - w| = |f_n(z) - f(z)| < \varepsilon.$$

To prove the sufficiency, it is sufficient to show that $\{f_n(z)\}$ is equicontinuous on $|z| = 1$. This is because of the following:

(i) By the Ascoli–Arzelà theorem, there must exist a subsequence $\{f_{n_k}(z)\}$ that converges uniformly on $|z| = 1$. Then, by the maximum modulus principle, $\{f_{n_k}(z)\}$ converges uniformly on $|z| \leqslant 1$.

(ii) We can prove that $\{f_n(z)\}$ converges on each point $z : |z| = 1$. Otherwise, suppose there exists a point $z_0, |z_0| = 1$, such that $\{f_n(z_0)\}$ does not converge. Then, there exist two subsequences $\{f_{n_1}(z)\}$ and $\{f_{n_2}(z)\}$ that converge to different limits. From (i), we have that both $\{f_{n_1}(z)\}$ and $\{f_{n_2}(z)\}$ have a subsequence (let's assume it is still denoted by $\{f_{n_1}(z)\}$ and $\{f_{n_2}(z)\}$) that uniformly converges on $|z| \leqslant 1$ to two distinct functions, $g_1(z)$ and $g_2(z)$. Since the kernel Ω of $\{\Omega_n\}$ is a domain, both $g_1(z)$ and $g_2(z)$ are non-constant and non-zero. Moreover, they are both one-to-one functions with $g_1(0) = g_2(0) = 0$ and $g_1'(0) > 0$, $g_2'(0) > 0$. By Theorem 2.4.12, $g_1(z)$ and $g_2(z)$ both map $|z| < 1$ to Ω. Therefore, $g_1(z) \equiv g_2(z)$, which leads to a contradiction. This proves that $\{f_n(z)\}$ converges on each point $z : |z| = 1$.

(iii) From the equicontinuity of $\{f_n(z)\}$ on $|z| = 1$ as shown in (ii), we can conclude that $\{f_n(z)\}$ converges uniformly on $|z| = 1$.

Therefore, $\{f_n(z)\}$ converges uniformly on the closure of the unit disk $\overline{U}$ to a function $f(z)$. By Theorem 2.4.12, $f(z)$ maps U to Ω.

The following proves that $\{f_n(z)\}$ is equicontinuous on $|z|=1$.

Suppose not, then for some $\varepsilon_0 > 0$, there exists a sequence $\{\eta_k\}$ with $\eta_k \to 0$, and two sequences of points $\{z'_k\}, \{z''_k\}$ with $|z'_k| = |z''_k| = 1$, such that for sufficiently large k, we have $|z'_k - z''_k| < \eta_k$, and

$$|f_{n_k}(z'_k) - f_{n_k}(z''_k)| > \varepsilon_0.$$

Without loss of generality, we can assume $z'_k \to \alpha, z''_k \to \alpha, |\alpha| = 1$. Otherwise, we can replace them with their respective subsequences. Similarly, we can assume without loss of generality that $w'_k = f_{n_k}(z'_k)$ and $w''_k = f_{n_k}(z''_k)$ converge to w' and w'', respectively. Here, w'_k and w''_k both belong to C_{n_k}, so w' and w'' belong to C. The arc $\widehat{w'_k w''_k}$ on the curve C_{n_k} corresponds to circular arc $\widehat{z'_k z''_k}$ on $|z|=1$. As k tends to infinity, $\widehat{z'_k z''_k}$ converges to a point α. However, for sufficiently large k, $|w'_k - w''_k| > \varepsilon_0$, therefore, on $\widehat{w'_k w''_k}$, there must be a point w_k such that

$$|w'_k - w_k| > \frac{\varepsilon_0}{2}, \quad |w''_k - w_k| > \frac{\varepsilon_0}{2}.$$

Let's assume $w_k \to w_0$ ($k \to \infty$). Then, on C, we have

$$|w' - w_0| \geqslant \frac{\varepsilon_0}{2}, \quad |w'' - w_0| \geqslant \frac{\varepsilon_0}{2}.$$

Within the open arc $\widehat{w'w''}$, we can take a subarc σ that contains w_0.

Let δ be a positive number smaller than the distance between w_0 and $C \setminus \sigma$ and also smaller than the distance between σ and $C \setminus \widehat{w'w''}$. Additionally, when $k > K$, in the one-to-one correspondence of points between C and C_{n_k}, the distance between corresponding points is less than $\frac{\delta}{2}$, i.e.,

$$|w'_k - w'| < \frac{\delta}{2}, \quad |w''_k - w''| < \frac{\delta}{2}, \quad |w_k - w_0| < \frac{\delta}{2}.$$

In this case, we can prove that when $k > K$, there are no points in the disk $|w - w_0| < \frac{\delta}{2}$, which belong to $C_{n_k} \setminus \widehat{w'_k w''_k}$. Otherwise, for some $k > K$, suppose there exists a point p_k in $C_{n_k} \setminus \widehat{w'_k w''_k}$ such that $|p_k - w_0| < \frac{\delta}{2}$. Let's suppose that the points w_k and p_k on C_{n_k} correspond to the points w_k^* and p_k^* on C, then

$$|w_k^* - w_0| \leqslant |w_k^* - w_k| + |w_k - w_0| < \delta,$$
$$|p_k^* - w_0| \leqslant |p_k^* - p_k| + |p_k - w_0| < \delta.$$

Hence, both w_k^* and p_k^* belong to the arc σ. Since $w_k \in \widehat{w'_k w''_k}$, and $p_k \in C_{n_k} \setminus \widehat{w'_k w''_k}$, it follows that w_k and p_k are separated by w'_k and w''_k. Therefore, the subarc $\widehat{w_k^* p_k^*}$ of the arc σ corresponds to a subarc on C_{n_k}, which must contain one of the points w'_k, or w''_k, let's say it contains w'_k. Then, w'_k corresponds to a point on C that lies on σ, and so, its distance to w' is greater than δ. However, by the choice of k, this point must also be within a distance of $\frac{\delta}{2}$ from w'_k, which contradicts the inequality $|w'_k - w'| < \frac{\delta}{2}$. Therefore, when $k > K$, only the points in $\widehat{w'_k w''_k}$ on C_{n_k} will fall within the disk $|w - w_0| < \frac{\delta}{2}$.

For $k > K$, let's define

$$\varphi_{n_k}(w) = f_{n_k}^{-1}(w), \quad \psi_k(w) = \varphi_{n_k}(w) - \alpha,$$

then $\psi_k(w)$ is analytic in Ω_{n_k}. Let

$$\varepsilon_k = \max\{|\psi_k(w)| : w \in \widehat{w'_k w''_k}\},$$

then $\varepsilon_k \to 0$ $(k \to \infty)$. Let w^* be any point in Ω, such that $|w^* - w_0| < \frac{\delta}{4}$. Construct the circle $|w - w^*| = \frac{\delta}{2}$, which contains w_0 in its interior. Therefore, there must be a point outside of Ω within this circle, implying that there is an arc S on the circle $|w - w^*| = \frac{\delta}{2}$ that lies outside of Ω.

Due to the correspondence property between C_{n_k} and C, when k is sufficiently large, w^* will be inside each Ω_{n_k}, and there exists a positive integer m such that the central angle corresponding to the arc on $|w - w^*| = \frac{\delta}{2}$ outside of Ω_{n_k} is greater than $\frac{2\pi}{m}$. Since the portion of C_{n_k} inside $|w - w^*| = \frac{\delta}{2}$ is contained in $\widehat{w'_k w''_k}$, when w approaches the boundary C_{n_k} of Ω_{n_k} within $|w - w^*| < \frac{\delta}{2}$, we have $|\psi_k(w)| < \varepsilon_k$. This follows from the proof of Lemma 2.3.2, and thus, we have

$$|\psi_k(w^*)| \leqslant \sqrt[m]{\varepsilon_k M_k^{m-1}}, \quad M_k = \max\{|\psi_k(w)| : w \in \Omega_{n_k}\}.$$

Since $|\psi_k(w)| \leqslant |\varphi_{n_k}(w)| + |\alpha| \leqslant 2$, when k is sufficiently large,

$$|\psi_k(w^*)| \leqslant \sqrt[m]{\varepsilon_k 2^{m-1}}.$$

As $k \to \infty$, we have $\psi(w^*) = 0$, which implies $\varphi(w^*) = 0$. According to Theorem 2.4.12, $\varphi(w)$ is a one-to-one mapping from the domain Ω onto the region $|z| < 1$. Since w^* is any point in the intersection of $|w^* - w_0| < \frac{\delta}{4}$ and Ω, we have $\varphi(w) \equiv \alpha$. This leads to a contradiction. Therefore, $f_n(z)$ is uniformly continuous on the circle $|z| = 1$. $\square$

2.5 Picard's Theorem

In the final part of this chapter, we will prove a profound theorem in the theory of complex analysis: Picard's theorem. Picard's theorem states that for an entire function f, if there exist two distinct complex numbers α and β such that $f(z)$ does not assume the values α and β, then f must be constant. More generally, suppose f is analytic in a punctured neighborhood V of z_0, where z_0 is an isolated singularity of f. If there exist two distinct complex numbers α and β such that f omits two values α and β in V, then z_0 is a removable singularity or a pole of f. This so-called "Big Picard's theorem" is a remarkable strengthening of the so-called "Little Picard's theorem" which merely asserts that every non-constant entire function attains each value, with one possible exception. In fact, Big Picard's theorem implies Little Picard's theorem. This is because we consider a function $f(z)$ that satisfies the conditions stated in Little Picard's theorem. Let $w = 1/z$, $g(w) = f(z) = f(\frac{1}{w})$, then $g(w)$ is analytic in

$$V = (\mathbb{C} \cup \{\infty\}) \setminus \{0\},$$

$g(w)$ does not assume the values α and β on V, Big Picard's theorem implies that 0 is a removable singularity or a pole of $g(w)$. If 0 is a pole of $g(w)$ of order m, it can be easily deduced that $f(z)$ is a polynomial of degree m. Let us denote it as

$$f(z) = c_0 + c_1 z + c_2 z^2 + \cdots + c_m z^m \quad (c_m \neq 0)$$

for any $\gamma \in \mathbb{C}$, according to the fundamental theorem of algebra, an mth degree algebraic equation

$$f(z) - \gamma = (c_0 - \gamma) + c_1 z + c_2 z^2 + \cdots + c_m z^m = 0$$

has exactly m roots in the complex plane. Therefore, it can be concluded that for any given $\gamma \in \mathbb{C}$, there exists $z_\gamma \in \mathbb{C}$, such that $f(z_\gamma) = \gamma$. This contradicts the fact that $f(z)$ does not take on the values α and β in V. Hence, 0 can only be a removable singularity of $g(w)$. Therefore, the limit

$$\lim_{z \to \infty} f(z) = \lim_{w \to 0} g(w)$$

exists, thus, $f(z)$ is bounded, and by Liouville's theorem, $f(z)$ is a constant. In this section, our objective is to prove Picard's theorem using the primary tools of Bloch's theorem and Schottky's inequality.

Lemma 2.5.1. *If $f \in H(U), M > 0, f(0) = 0, |f'(0)| = 1$, and $|f(z)| \leqslant M$ for all $z \in U$, then*

$$f(U) \supset D\left(0, \frac{1}{4M}\right).$$

Proof. If $w \notin f(U)$, then we construct the auxiliary function $1 - f(z)/w$, which is non-vanishing on U. Thus, by Lemma 1.2.25, there is $h \in H(U)$, such that

$$h^2(z) = 1 - f(z)/w. \tag{2.5.1}$$

Such a function $h(z)$ has a Taylor power series expansion of the form $h(z) = a_0 + a_1 z + \cdots$ on U, $h^2(0) = 1 - f(0)/w = 1$, we can take $a_0 = h(0) = 1$. Taking the derivative of (2.5.1) on both sides, we obtain that $2h(z)h'(z) = -f'(z)/w$, so $h'(0) = -f'(0)/2w$, thus

$a_1 = h'(0) = -f'(0)/2w$. Since $|f(z)| \leq M$ for all $z \in U$. Applying (2.5.1) again for all $z \in U$, we have

$$|h(z)|^2 = \left|1 - \frac{f(z)}{w}\right| \leq 1 + \frac{|f(z)|}{|w|} \leq 1 + \frac{M}{|w|}$$

for all $z = Re^{i\theta} \in U$,

$$|h(Re^{i\theta})|^2 = h(Re^{i\theta})\overline{h(Re^{i\theta})} = \sum_{n=0}^{\infty}\sum_{m=0}^{\infty} a_n \bar{a}_m r^{n+m} e^{i(n-m)\theta}.$$

Note that the series converges uniformly for all $\theta \in [0, 2\pi]$, and this holds for any integer $k \neq 0$, $\int_0^{2\pi} e^{ik\theta} d\theta = 0$, thus

$$\frac{1}{2\pi}\int_0^{2\pi} |h(Re^{i\theta})|^2 d\theta = \sum_{n=0}^{\infty} |a_n|^2 r^{2n} \geq |a_0|^2 + |a_1|^2 r^2.$$

On the other hand,

$$\frac{1}{2\pi}\int_0^{2\pi} |h(Re^{i\theta})|^2 d\theta \leq \frac{1}{2\pi}\int_0^{2\pi} \left(1 + \frac{M}{|w|}\right) d\theta = 1 + \frac{M}{|w|}.$$

Hence,

$$|a_0|^2 + |a_1|^2 r^2 \leq 1 + \frac{M}{|w|},$$

i.e.,

$$1 + \frac{r^2}{4|w|^2} \leq 1 + \frac{M}{|w|},$$

whence

$$|w| \geq \frac{r^2}{4M}.$$

Letting $r \to 1$, we have

$$|w| \geq \frac{1}{4M}$$

for all $w \notin f(U)$. From this, we can conclude that for any

$$w \in D\left(0, \frac{1}{4M}\right),$$

we have $w \in f(U)$, thus

$$D\left(0, \frac{1}{4M}\right) \subset f(U).$$

□

Lemma 2.5.1 can be generalized to the case of $f \in H(D(a, R))$, where $fD(a, R)$ is the open disk centered at a with radius R.

Lemma 2.5.2. *Let $f \in H(D(a, R))$, $M > 0$, $f(a) = 0$, $f'(a) \neq 0$, and $|f(z)| \leqslant M$ for all $z \in D(a, R)$. Then*

$$f(D(a, R)) \supset D\left(0, \frac{|f'(a)|^2 R^2}{4M}\right). \tag{2.5.2}$$

Proof. Let $\zeta = (z - a)/R$, so $z = R\zeta + a$, and if $z \in D(a, R)$, then $\zeta \in U$. Let

$$h(\zeta) = \frac{f(z)}{|f'(a)|R} = \frac{f(R\zeta + a)}{|f'(a)|R},$$

then $h \in H(U)$, $h(0) = 0$, $|h'(0)| = 1$, and

$$|h(\zeta)| = \frac{|f(z)|}{|f'(a)|R} \leqslant \frac{M}{|f'(a)|R}.$$

Lemma 2.5.1 implies that

$$h(U) \supset D\left(0, \frac{|f'(a)|R}{4M}\right).$$

For given w with

$$|w| < \frac{|f'(a)|^2 R^2}{4M},$$

we have

$$\left|\frac{w}{|f'(a)|R}\right| < \frac{|f'(a)|R}{4M},$$

so

$$\frac{w}{|f'(a)|R} \in D\left(0, \frac{|f'(a)|R}{4M}\right) \subset h(U),$$

thus there is $\zeta_0 \in U$, such that

$$w = |f'(a)|Rh(\zeta_0) = f(R\zeta_0 + a) \in f(D(a, R)).$$

This proves that (2.5.2) holds. □

Theorem 2.5.3 (Bloch's Theorem). *If $f \in H(U)$, $|f'(0)| = 1$, then $f(U)$ contains a disk of radius $l \geqslant 1/16$, Here, l is an absolute constant.*

Proof. Without loss of generality, let's assume that f is analytic in some neighborhood of the closure of U since otherwise we can consider the function

$$f(rz)/r \in H(D(0, 1/r)) \quad \left(\frac{1}{2} < r < 1\right).$$

Let

$$h(t) = t \sup\{|f'(z)| : |z| \leqslant 1 - t\}.$$

Then $h(t)$ is continuous on $[0, 1]$, $h(0) = 0$ and $h(1) = 1$. Let $t_0 = \min\{t : h(t) = 1\}$, then

$$1 = h(t_0) = t_0 \sup\{|f'(z)| : |z| \leqslant 1 - t_0\}.$$

Since $f'(z)$ is continuous in $|z| \leqslant 1 - t_0$, a can be chosen with $|a| \leqslant 1 - t_0$ and $1/t_0 = |f'(a)|$. Consider the disk $D(a, t_0/2)$. a as the center and $t_0/2$ as the radius, if $z \in D(0, 1 - t_0/2)$, then $|z| < |a| + t_0/2 \leqslant 1 - t_0/2$, and so,

$$|f'(z)| \leqslant \sup\{|f'(z)| : |z| \leqslant 1 - t_0/2\} = h\left(\frac{t_0}{2}\right) \cdot \frac{2}{t_0} \leqslant \frac{2}{t_0}$$

for all $z \in D(a, t_0/2)$. The second inequality in the above equation holds because of $h(t_0/2) \leqslant 1$. In fact, if $h(t_0/2) > 1$, by the intermediate value theorem for continuous functions, we know that for $h(0) < 1 < h(t_0/2)$, there must exist $t_1 \in (0, t_0/2)$ (clearly, $t_1 < t_0$), such that $h(t_1) = 1$. This contradicts the assumption that t_0 is the smallest value, for which $t_0 = \min\{t : h(t) = 1\}$.

Define $g(z)$ for $D(a, \frac{t_0}{2})$ by

$$g(z) = f(z) - f(a) = \int_a^z f'(\zeta)d\zeta,$$

then

$$g(z) \in H\left(D\left(a, \frac{t_0}{2}\right)\right), g(a) = 0, |g'(a)| = |f'(a)| = \frac{1}{t_0},$$

If $z \in D\left(a, \frac{t_0}{2}\right)$, then

$$|g(z)| \leqslant \int_{[a,z]} |f'(\zeta)||d\zeta| \leqslant |z-a|\frac{2}{t_0} \leqslant \frac{t_0}{2} \cdot \frac{2}{t_0} = 1,$$

where, the path of integration is the line segment from a to z. By Lemma 2.5.2, we have $g(D(a, t_0/2)) \supset D(0, 1/16)$. If this is translated into a statement about f, it yields that $f(U) \supset D(f(a), 1/16)$. □

Remark. This theorem can also be generalized to the case of $f \in H(D(a, R))$. Its statement is as follows: If $f \in H(D(a, R))$ and $f'(a) \neq 0$, then $f(D(a, R))$ contains a disk of radius $|f'(a)|Rl$, where $l \geqslant 1/16$ is the constant mentioned in Theorem 2.5.3.

To prove this, we only need to apply Theorem 2.5.3 to $h(\zeta)$ with respect to

$$h(\zeta) = f(z)/|f'(a)|R = f(R\zeta + a)/|f'(a)|,$$

and $\zeta = (z-a)/R$. If $h(U)$ a disk of radius l centered at w_0, denoted by $D(w_0, l)$, then

$$D(w_0|f'(a)|R, |f'(a)|Rl) \subset f(D(a, R)).$$

Using Bloch's theorem, we can derive the Little Picard theorem.

Theorem 2.5.4 (Little Picard's Theorem). *If f is an entire function and $f(z)$ does not assume the values α and β on $\mathbb{C}$ ($\alpha \neq \beta$), then f is a constant.*

Proof. Suppose that $f(z)$ is an entire function. Without loss of generality, we may assume that $f(z)$ does not take the values 0 and 1 in $\mathbb{C}$. Otherwise, we can consider the function $\psi(z) = (f(z) - \alpha)/(\beta - \alpha)$. Since f never vanishes, there is branch of $\log f$ on $\mathbb{C}$ by Lemma 1.2.25, i.e., $\varphi(z) = \frac{1}{2\pi i}\log f(z) \in H(\mathbb{C})$, such that $f = e^{2\pi i \varphi(z)}$; if $\varphi(a) = n$ for some integer n, then $f(a) = \exp(2ni\pi) = 1$, which cannot happen. Since φ cannot assume the values 0 and 1, by Lemma 1.2.25, there are two branches φ_1 and φ_2 of $\sqrt{\varphi}$, $\sqrt{\varphi-1}$ defined on $\mathbb{C}$, respectively. For convenience, let's also denote these branches as $\varphi_1 = \sqrt{\varphi}$, $\varphi_2 = \sqrt{\varphi-1}$, thus $\varphi = \sqrt{\varphi} - \sqrt{\varphi-1} \in H(\mathbb{C})$ and never vanishes on $\mathbb{C}$. Again, by Lemma 1.2.25, there exists $g \in H(\mathbb{C})$, such that

$$e^g = \sqrt{\varphi} - \sqrt{\varphi-1},$$

hence

$$g(z) = \log\left(\sqrt{\frac{\log f(z)}{2\pi i}} - \sqrt{\frac{\log f(z)}{2\pi i} - 1}\right), \quad (2.5.3)$$

since $e^g = \sqrt{\varphi} - \sqrt{\varphi-1}$, and $e^{-g} = \frac{1}{\sqrt{\varphi}-\sqrt{\varphi-1}} = \sqrt{\varphi} + \sqrt{\varphi-1}$, $(e^g + e^{-g})/2 = \sqrt{\varphi}$, $(\frac{e^g+e^{-g}}{2})^2 = \varphi = h/2\pi i$, therefore

$$f = e^h = \exp\left\{2\pi i\left(\frac{e^g + e^{-g}}{2}\right)^2\right\}. \quad (2.5.4)$$

Let $E = \{\pm\log(\sqrt{n} + \sqrt{n-1}) + 2\pi i m : n = 1, 2, \ldots, m = 0, \pm 1, \pm 2, \ldots\}$, then $g(\mathbb{C}) \cap E = \varnothing$, these points of E form the vertices of a grid of rectangles in the plane. The height of an arbitrary rectangle is 2π, the width is $\log(\sqrt{n+1} + \sqrt{n}) - \log(\sqrt{n} + \sqrt{n-1}) > 0$. In fact, If there exists a point $a \in E$ with $g(a) = \pm\log(\sqrt{n} + \sqrt{n-1}) + m2\pi i$, substituting the above equation, we can calculate that $f(a) = 1$. Upon observing the set E, it can be noted that there exists $d \in (0, 4)$ such that every disk on the plane with radius d intersects with E. Therefore, we can conclude that for any $z \in \mathbb{C}$, we have $g'(z) = 0$. In fact, if there is $a \in \mathbb{C}$, such that $g'(a) \neq 0$. Consider the disk $D(a, d/|g'(a)|l)$, where l is a constant from Theorem 2.5.3. Since $g \in H(D(a, d/|g'(a)|l))$, according to Bloch's theorem, $g(D(a, d/|g'(a)|l))$ contains a disk D_a of radius

$$|g'(a)| \frac{d}{|g'(a)|l} \cdot l = d,$$

which intersects with E. Therefore, there must be points in $g(D(a, d/|g'(a)|l))$ that lie in E, which contradicts the assumption that $g(\mathbb{C})$ does not intersect with E. Therefore, $g'(z) = 0$ for all $z \in \mathbb{C}$, so g must be constant, thus f must also be constant. □

To prove the Great Picard's theorem, besides utilizing Bloch's theorem, it is also necessary to employ the Schottky inequality as a primary tool.

Proposition 2.5.5 (The Schottky Inequality). *Let $r \in (0,1)$, there are two positive constants $m(\alpha, \beta, r)$ and $M(\alpha, \beta, r)$, such that if $f \in H(U)$ and $f(z)$ omits the values 0 and 1 in U, and $\alpha \leqslant |f(0)| \leqslant \beta$ $(\alpha, \beta > 0)$, then*

$$m(\alpha, \beta, r) \leqslant |f(z)| \leqslant M(\alpha, \beta, r),$$

for all $z \in \overline{D(0, r)}$.

Proof. Assume that $f \in H(U)$, $f(z)$ and omits the values 0 and 1 in U. From the proof process of the Little Picard's theorem, it can be inferred that there exists $g \in H(\Omega)$ such that (2.5.3) and (2.5.4) hold. Since $\alpha \leqslant |f(0)| \leqslant \beta$, we can also choose an appropriate g by first determining the value $g(0)$ so that $|g(0)| \leqslant C(\alpha, \beta)$, here, $C(\alpha, \beta)$ is a positive function that depends on α, β, but is independent of f.

For $z \in U$ with $g'(z) \neq 0$, if $|z| = r < 1$, then the function

$$\varphi(\zeta) = \frac{g(z + (1-r)\zeta)}{(1-r)|g'(z)|}$$

is analytic in U and $\varphi'(\zeta) = g'(z + (1-r)\zeta)/|g'(z)|$, so $|\varphi'(0)| = 1$. According to Bloch's theorem, $\varphi(U)$ contains a disk of radius l, so $g(U)$ contains a disk of radius $(1-r)|g'(z)|l$. In fact, if $\varphi(U) \supset D(w_0, l)$, then

$$g(U) \supset D((1-r)|g'(z)|w_0, \ (1-r)|g'(z)|l).$$

This is because if
$$w \in D((1-r)|g'(z)|w_0, \ (1-r)|g'(z)|l),$$
then
$$|w - (1-r)|g'(z)|w_0| < (1-r)|g'(z)|l,$$
i.e.,
$$\left|\frac{w}{(1-r)|g'(z)|} - w_0\right| < l.$$

Put
$$w' = \frac{w}{(1-r)|g'(z)|},$$
then $|w' - w| < l$, so $w' \in \psi(U)$. There is $\zeta_0 \in U$, such that
$$w' = \varphi(\zeta_0) = \frac{g(z + (1-r)\zeta_0)}{(1-r)|g'(z)|},$$
and
$$w = w'(1-r)|g'(z)| = g(z + (1-r)\zeta_0) \in g(U).$$

Let E be the set defined in the proof of the Little Picard's theorem then $g(U) \cap E = \emptyset$. Since every disk of radius d intersects with E. then $(1-r)|g'(z)|l < d$, otherwise $g(U)$ would contain a disk that intersects E, which contradicts $g(U) \cap E = \emptyset$. Therefore,
$$|g'(z)| < \frac{d}{l(1-r)}.$$

This inequality also holds for z, such that $g'(z) = 0$, so
$$\sup\{|g'(z)| : |z| \leqslant r\} = \sup\{|g'(z)| : |z| = r\} \leqslant \frac{d}{l(1-r)}.$$

Since $g(z) = g(0) + \int_0^z g'(\tau)d\tau$, we thus obtain the estimate
$$|g(z)| \leqslant |g(0)| + \frac{|z|d}{l(1-r)} \leqslant C(\alpha, \beta) + \frac{rd}{l(1-r)}$$

for all z with $|z| \leqslant r$. Furthermore, from (2.5.4), it can be inferred that

$$|f(z)| \leqslant \exp\{2\pi e^{2|g(z)|}\}.$$

Hence, there exists a positive function $M(\alpha, \beta, r)$, such that

$$|f(z)| \leqslant M(\alpha, \beta, r), \quad |z| \leqslant r < 1.$$

On the other hand, note that $\frac{1}{f} \in H(U)$ and for any $z \in U$, $\frac{1}{f(z)} \neq 0$ and $\frac{1}{f(z)} \neq 1$. Applying it, the inequality derived in the previous sentence $\frac{1}{f}$ and setting $m(\alpha, \beta, r) = 1/M(\alpha, \beta, r)$, we get the lower estimate, so

$$m(\alpha, \beta, r) \leqslant |f(z)| \leqslant M(\alpha, \beta, r), \quad |z| \leqslant r < 1.$$

This proves Proposition 2.5.5. □

Remark. The Schottky inequality can also be extended to the case where $f \in H(D(a, R))$. Its statement is as follows: Let $f(z)$ be a function that does not take the values 0 and 1 in $D(a, R)$, $f \in H(D(a, R))$ and $\alpha \leqslant |f(0)| \leqslant \beta (\alpha, \beta > 0)$, then for all $r \in (0, 1)$, there are two positive constants $m(\alpha, \beta, r)$ and $M(\alpha, \beta, r)$, both independent of f, such that

$$m(\alpha, \beta, r) \leqslant |f(z)| \leqslant M(\alpha, \beta, r), \quad |z - a| \leqslant r < R.$$

The proof of the above conclusion only requires letting $\zeta = (z - a)/R$ and applying Proposition 2.5.5 to the function $h(\zeta) = f(z) = f(R\zeta + a)$.

Next, we will prove the Big Picard theorem, which makes use of Theorem 1.2.23 concerning isolated singularities of analytic functions. The statement of the theorem is as follows: (a) If $f(z)$ is analytic in a punctured neighborhood V of an isolated singularity z_0, then z_0 is an essential singularity of $f(z)$ if and only if for any finite or infinite complex number γ, there exists a sequence $\{z_n\}$ in V that converges to z_0 such that $\lim_{n \to \infty} f(z_n) = \gamma$. (b) If $f(z)$ is analytic in the region $R < |z| < \infty$ $(R \geqslant 0)$, then $z = \infty$ is a removable singularity of $f(z)$ if and only if there exists a positive number $\rho_0 \geqslant R$, such that $f(z)$ is bounded in $R < |z| < \infty$.

Theorem 2.5.6 (Big Picard's Theorem). *If $f(z)$ is analytic in a punctured neighborhood V of an isolated singularity a and $f(z)$ omits two distinct values α and β in V, then a is a pole or a removable singularity of $f(z)$.*

Proof. We can assume, without loss of generality, that $\alpha = 0$, $\beta = 1$, $a = \infty$ and $V = \{z : |z| > 1/2\}$ (we can set $\zeta = R/2(z-a)$, and let

$$h(\zeta) = \frac{f(\frac{R}{2\zeta}+a) - \alpha}{\beta - \alpha}$$

for $|\zeta| > 1/2$). We have to prove that if $f(z)$ is analytic for $|z| > 1/2$ and avoid 0 and 1, then either $f(z)$ is bounded or has a pole at infinity. If not, the image of $f(z)$ on each set $|z| > R$ is dense according to Theorem 1.2.22, and therefore, there are $\{\lambda_n\}$, such that $1/2 \leqslant |f(\lambda_n)| \leqslant 1$. Set $f_n(z) = f(\lambda_n z)$. Then f_n is analytic in $|z| > 1/2$ and $1/2 \leqslant |f_n(1)| \leqslant 1$, and therefore, Schottky's inequality yields that there are two positive constants m_0 and M_0, such that

$$m_0 \leqslant |f_n(z)| \leqslant M_0, \quad z \in D(\alpha_0, 1/4).$$

Thus, $m_0 \leqslant |f_n(\alpha_1)| \leqslant M_0$ for some $\alpha_1 \in \partial U \cap D(\alpha_0, 1/4)$, we can repeat the argument and hence get constants m_1 and M_1, such that

$$m_1 \leqslant |f_n(z)| \leqslant M_1, \quad z \in D(\alpha_1, 1/4).$$

After a finite number of steps, we have covered the entire circle, and there are two constants m and M, such that

$$m \leqslant |f_n(z)| \leqslant M, \quad z \in \partial U,$$

i.e., $|f(z)| \leqslant M$ on $|z| = \lambda_n$. By the maximum principle,

$$\sup\{|f(z)| : 1 \leqslant |z| \leqslant |\lambda_n|\} \leqslant M + \sup\{|f(z)| : |z| = 1\}.$$

Hence, $f(z)$ is bounded for $1 \leqslant |z| < \infty$, which is a contradiction. $\square$

Theorem 2.5.7. *If $\mathfrak{F}$ is the family of all analytic functions on a region Ω that do not assume the values 0 and 1, then $\mathfrak{F}$ is normal in Ω.*

Proof. By Theorem 2.1.11, it suffices to prove that the family of analytic functions $\mathfrak{F}$ is normal at every point in Ω. For all $z_0 \in \Omega$, there is $r_0 > 0$, such that $D(z_0, r_0) \subset \Omega$. For any sequence $\{f_n\}$ of functions in $\mathfrak{F}$, we will consider two cases:

(i) If
$$\liminf_{n \to \infty} |f_n(z_0)| < \infty,$$
then there is a subsequence $\{f_{n_k}\}$ of $\{f_n\}$, such that the limit
$$\lim_{k \to \infty} f_{n_k}(z_0)$$
exists and is finite. We can assume, without loss of generality, that the limit is equal to c. When $c \neq 0$, we assume that $0 < \alpha \leqslant |f_n(z_0)| \leqslant \beta$ and when $c = 0$, we assume that $0 < \alpha \leqslant |f_n(z_0) - 1| \leqslant \beta$. Then, by Proposition 2.5.5 (Schottky's inequality), there exists a constant $M(\alpha, \beta)$ that does not depend on f_n, such that when $|z - z_0| \leqslant \frac{r_0}{2}$, $|f_n(z)| \leqslant M(\alpha, \beta)$. (When $c = 0$, similarly, there exists a constant $M(\alpha, \beta)$ does not depend on f_n, such that when $|z - z_0| \leqslant \frac{r_0}{2}$, we have $|f_n(z) - 1| \leqslant M(\alpha, \beta)$.) By Montel's Theorem 2.1.4, there exists a subsequence $\{f_{n_k}\}$ of $\{f_n\}$ that converges uniformly on each compact of the disk $D(z_0, \frac{r_0}{2})$. Therefore, the family of analytic functions $\mathfrak{F}$ is normal at z_0.

(ii) If
$$\lim_{n \to \infty} |f_n(z_0)| = \infty,$$
then there exists $\delta > 0$, such that $D(z_0, \delta) \subset \Omega$, and the analytic functions $g_n(z) = \frac{1}{f_n(z)}$ do not assume the value 0 and 1 on $D(z_0, \delta)$. According to (i), there exists a subsequence $\{g_{n_k}\}$ of $\{g_n\}$ that converges uniformly to an analytic function $g(z)$ on each compact of the disk $D(z_0, \frac{r_0}{2})$. If $g(z) \not\equiv 0$, by Lemma 2.1.12 (Hurwitz's theorem), $g(z)$ is non-vanishing on $D(a, \frac{r_0}{2})$. However, it is clear that $g(a) = 0$, which contradicts our assumption. □

Exercises II

1. Let $\mathfrak{F} = \{f(z)\}$ be a family of normal analytic functions in the domain Ω. If $\mathfrak{F}$ is bounded at a point $z_i n \Omega \in \Omega$, then $\mathfrak{F}$ is uniformly bounded on each compact subset of Ω.

2. Let $\rho(z, f(z)) = \frac{2|f'(z)|}{1+|f(z)|^2}$, the quantity $\rho(z, f(z))$ is called the spherical derivative of $f(z)$. What does its geometric interpretation means?

3. Let Ψ_1 be a family of linear functions in the form of $az + b$, Ψ_2 a family of linear functions in the form of az (a and b are complex constants), and $F(w) = \sin w$.

   (a) Is Ψ_1 a normal family in $\mathbb{C}$?
   (b) Is Ψ_1 a normal family in $\{z : |z| > 1\}$?
   (c) Is Ψ_2 a normal family in $\{z : |z| > 1\}$?
   (d) Is $\Psi_3 = \{F(f(z)) : f \in \Psi_2\}$ a normal family in $\{z : |z| > 1\}$?

4. Is the family formed by the derivative of each function in a normal family also a normal family? Verify the normality of the function family $\{n(z^2 - n^2)\}$ and the family of its derivative.

5. Prove that the analytic function family $\mathfrak{F} = \{z + az^2 : a \in \mathbb{C}\}$ is not a normal family.

6. Prove that the family of analytic functions within the unit disk U, satisfying the normalization conditions $f(0) = 0$, $f'(0) = 1$, is not a normal family.

7. Prove that the family of one-to-one analytic functions of the unit disk U, satisfying the normalization conditions $f(0) = 0$, $f'(0) = 1$, is a normal family.

8. Prove that the function $w = \frac{z}{(1-z)^3}$ is one-to-one on the disk $|z| < \frac{1}{2}$, but not one-to-one within any larger disk centered at the origin.

9. Suppose $w = f(z)$ is a one-to-one analytic function within $|z| < R$ and maps the region $|z| < R$ to a domain D in the w-plane. Prove that if A represents the area of D, then

$$A \geq \pi R^2 |f'(0)|^2.$$

10. Suppose $\Omega = \{x + iy : -1 < y < 1\}$, $f \in H(\Omega)$, $|f(z)| \leq 1$ ($z \in \Omega$) and $f(x) \to 0$ as $x \to \infty$. Prove that, for $y \in (-1, 1)$, $\alpha \in (0, 1)$,

   (a) $\lim\limits_{x \to \infty} f(x + iy) = 0$, (b) $\lim\limits_{x \to \infty} \max\{|f(x + iy)| : |y| \leq \alpha\} = 0$.

   (*Hint*: Consider the sequence $\{f_n\}$, where $f_n(z) = f(z + n)$, in the square $\{z = x + iy : |x| < 1, |y| < 1\}$, $n = 1, 2, 3, \ldots$.

If g is a bounded analytic function in the unit disk U and the radial limit $g(x) \to \alpha$ exists, if the finite, non-tangential limit exists at point 1, what conclusion can be drawn? Please prove your conclusion.

11. (a) Suppose $\Omega = \{z : -1 < \text{Re} z < 1\}$. Find an explicit formula for the one-to-one conformal mapping f of Ω onto the unit disk U, for which $f(0) = 0$ and $f'(0) > 0$. Compute $f'(0)$.

(b) Suppose $g \in H(U)$, $|\text{Re} g(z)| < 1$, and $g(0) = 0$. Prove that
$$|g(z)| \leqslant \frac{2}{\pi} \log\left(\frac{1+|z|}{1-|z|}\right).$$

(c) Let $\Omega = \{x + iy : |y| = |\text{Im } z| < \frac{\pi}{2}\}$. Fix $\alpha + i\beta$ in Ω. Let h be a conformal one-to-one mapping of Ω onto Ω, satisfying $h(\alpha + i\beta) = 0$. Prove that $|h'(\alpha + i\beta)| = 1/\cos\beta$.

12. Let Ψ be the collection of all analytic functions defined on the domain Ω, taking values in the right half-plane. Suppose that there exists a point $z_0 \in \Omega$ such that for every $f \in \Psi$ and $g \in \Psi$, we have $f(z_0) = g(z_0)$. Prove that Ψ is a normal family.

13. Let G and D be two regions containing the point at infinity. Suppose that F and H are two conformal mappings from G to D such that near infinity, they have expansions of the form
$$z + \frac{b_1}{z} + \frac{b_2}{z^2} + \cdots.$$
Prove that F and H are identical.

14. Let Ω be a bounded simply connected region, with $0 \in \Omega$ and let $f(z)$ be a function satisfying the conditions of the Riemann mapping theorem: $f(0) = 0$ and $f'(0) > 0$. Let $g(z)$ be any conformal mapping from Ω to the unit disk U. Express $g(z)$ in terms of $f(z)$.

15. Suppose Ω is a domain, $f \in H(\Omega)$ and $f \not\equiv 0$. Prove that there exists $g \in H(\Omega)$ such that $f(z) = e^{g(z)}$ for all $z \in \Omega$ if and only if for every positive integer n, there exists $g_n \in H(\Omega)$ such that $f(z) = (g_n(z))^n$ for all $z \in \Omega$.

16. If $f(z)$ is analytic in the unit disk U, and $|f(z)| \leqslant 1$ for $z \in U$, then for all $z_0, z \in U$, we have
$$\left|\frac{f(z) - f(z_0)}{1 - \overline{f(z_0)}f(z)}\right| \leqslant \left|\frac{z - z_0}{1 - \overline{z_0}z}\right|, \quad |f'(z_0)| \leqslant \frac{1 - |f(z_0)|^2}{1 - |z_0|^2}.$$

17. Let $\Phi = \{f \in H(U) : \text{Re} f > 0 \text{ and } |f(0)| \leq 1\}$. Prove that the family Φ is normal. Can the condition $|f(0)| \leq 1$ be removed from the family?
18. Assume that $\{f_n(z)\}$ is a sequence of analytic functions in an open set Ω, which is uniformly bounded and pointwise convergent for any $a \in \Omega$. Prove that the sequence $\{f_n(z)\}$ converges uniformly on compact subsets of Ω, i.e., for all compact $K \subset \Omega$, $\{f_n(z)\}$ converges uniformly on compact K.
19. Let $\mathfrak{F}$ be the class of all $f \in H(U)$, for which
$$\int_U |f(z)|^2 d\lambda(z) \leq 1.$$
Is this $\mathfrak{F}$ a normal family?

Chapter 3

Zeros of Holomorphic Functions

3.1 Infinite Products

Definition 3.1.1. Suppose $\{u_n : n = 1, 2, \ldots\}$ is a sequence of complex numbers. If there is $N \geqslant 1$, the partial products

$$p_{N,n} = \prod_{k=N}^{n}(1+u_k) \quad (n = N, N+1, \ldots)$$

converge to a non-zero, number p_N, we shall say that the infinite product

$$\prod_{k=1}^{\infty}(1+u_k) \tag{3.1.1}$$

converges and we write $p = \prod_{k=1}^{\infty}(1+u_k) = p_N(1+u_1)\cdots(1+u_{N-1})$; otherwise, we shall say that the infinite product (3.1.1) is disconvergence.

Therefore, similar to infinite series, changing the values of a finite number of terms in an infinite product does not affect its convergence or divergence. If (3.1.1) converges, then

$$\lim_{n \to \infty} u_n = 0. \tag{3.1.2}$$

Theorem 3.1.2. *The convergence or divergence of the infinite product $\prod_{k=1}^{\infty}(1+|u_k|)$ is the same as that of the infinite series $\sum_{k=1}^{\infty}|u_k|$.*

Proof. For $t \geqslant 0$, the inequality $1+t \leqslant e^t$ is an immediate consequence of the expansion of e^t in powers of t. Replace t by $|u_1|, \ldots, |u_n|$ and multiply the resulting inequalities. This gives

$$\sum_{k=1}^{n} |u_k| \leqslant \prod_{k=1}^{n}(1+|u_k|) \leqslant \exp\left\{\sum_{k=1}^{n} |u_k|\right\}. \qquad (3.1.3)$$

Therefore, the infinite product $\prod_{k=1}^{\infty}(1+|u_k|)$ and the infinite series $\sum_{k=1}^{\infty} |u_k|$ diverge simultaneously. $\square$

If the infinite series $\sum_{k=1}^{\infty} |u_k|$ converges, then the infinite product $\prod_{k=1}^{\infty}(1+u_k)$ is said to be *absolutely convergent*.

Lemma 3.1.3. *If $u_1, u_2, \ldots, u_n$ are complex numbers, and if*

$$p_n = \prod_{k=1}^{n}(1+u_k), \quad \tilde{p}_n = \prod_{k=1}^{n}(1+|u_k|),$$

then

$$|p_n - 1| \leqslant \tilde{p}_n - 1. \qquad (3.1.4)$$

Proof. For $n = 1$, (3.1.4) is trivial. The general case follows by induction: For $n = 1, \ldots, N-1$, so that if (3.1.4) holds with n in place of N, then also

$$|p_{n+1} - 1| = |p_n(1+u_{n+1}) - 1| = |(p_n - 1)(1+u_{n+1}) + u_{n+1}|$$
$$\leqslant (\tilde{p}_n - 1)(1+|u_{n+1}|) + |u_{n+1}| = \tilde{p}_{n+1} - 1. \qquad \square$$

Lemma 3.1.4. *Suppose $\{u_n(z)\}$ is a sequence of bounded complex functions on a set S such that $\sum_{k=1}^{\infty} |u_k(z)|$ converges uniformly on S. Then the product*

$$f(z) = \prod_{k=1}^{\infty}(1+u_k(z)) \qquad (3.1.5)$$

converges uniformly on S, and $f(z_0) = 0$ at some $z_0 \in S$ if and only if $u_{n_0}(z_0) = -1$ for some n_0.

Furthermore, if $\{n_1, n_2, n_3, \ldots\}$ is any permutation of $\{1, 2, 3, \ldots\}$, then we also have

$$f(z) = \prod_{k=1}^{\infty}(1+u_{n_k}(z)) \quad (z \in S). \qquad (3.1.6)$$

Proof. The hypothesis implies that $\sum_{k=1}^{\infty} |u_k(z)|$ is bounded on S, so there is a constant c such that $\sum_{k=1}^{\infty} |u_k(z)| \leqslant c$. If

$$p_n(z) = \prod_{k=1}^{n}(1 + u_k(z))$$

denotes the nth partial product of (3.1.5), we conclude from (3.1.3) that $|p_n(z)| \leqslant e^c = C$ for all $z \in S$ and all $n = 1, 2, 3, \ldots$. Choose $\varepsilon, 0 < \varepsilon < \frac{1}{2}$. There exist an N_0, such that

$$\sum_{k=N_0}^{\infty} |u_k(z)| < \varepsilon \quad (z \in S). \tag{3.1.7}$$

Let $\{n_1, n_2, n_3, \ldots\}$ be a permutation of $\{1, 2, 3, \ldots\}$. If $N \geqslant N_0$, if $M > N$ is so large that

$$\{1, 2, \ldots, N\} \subset \{n_1, n_2, \ldots, n_M\} \tag{3.1.8}$$

and

$$q_M(z) = \prod_{k=1}^{M}(1 + u_{n_k}) \tag{3.1.9}$$

denotes the Mth partial product of (3.1.5), then

$$q_M(z) - p_N(z) = p_N(z)\left(\prod_{n_k > N, 1 \leqslant k \leqslant M}(1 + u_{n_k}(z)) - 1\right). \tag{3.1.10}$$

The n_k which occur in (3.1.10) are all distinct and are large than N_0. Therefore, Lemma 3.1.3 and (3.1.7) show that

$$|q_M(z) - p_N(z)| \leqslant |p_N(z)|(e^\varepsilon - 1) \leqslant 2\varepsilon|p_N(z)| \leqslant 2\varepsilon C. \tag{3.1.11}$$

If $n_k = k$ $(k = 1, 2, \ldots)$, then $q_M(z) = p_M(z)$. Also, (3.1.11) shows that $\{p_M(z)\}$ converges uniformly to a limit function f on S. Also, (3.1.11) shows that

$$|p_M(z) - p_{N_0}(z)| \leqslant 2\varepsilon|p_{N_0}(z)| \quad (M > N_0), \tag{3.1.12}$$

so that $|p_M(z)| \geq (1-2\varepsilon)|p_{N_0}(z)|$. Hence,

$$|f(z)| \geq (1-2\varepsilon)|p_{N_0}(z)|,$$

which shows that $z_0 \in S$, $f(z_0) = 0$ if and only if there is N_0 such that $p_{N_0}(z_0) = 0$.

Finally, (3.1.11) also shows that $\{q_M(z)\}$ converges to the same limit $f(z)$ on S. □

Theorem 3.1.5. *Suppose $\{f_n(z)\}$ is a sequence of holomorphic functions on the open set Ω, no f_n is identically -1 in any component of Ω, and $\sum_{k=1}^{\infty} |f_k(z)|$ converges uniformly on compact subset of Ω. Then the product*

$$f(z) = \prod_{k=1}^{\infty}(1 + f_k(z)) \tag{3.1.13}$$

converges uniformly on compact subset of Ω, Hence, $f(z)$ is holomorphic on Ω. Furthermore, we have

$$m(f, z) = \sum_{n=1}^{\infty} m(1 + f_n, z), \quad z \in \Omega, \tag{3.1.14}$$

where $m(f, z)$ is defined to be the multiplicity of the zero of f at z. (if $f(z) \neq 0$, then $m(f, z) = 0$.)

Proof. The first part follows immediately from Theorem 3.1.5.4. For the second part, observe that each $z_0 \in \Omega$ has a $\delta > 0$ such that $\overline{D(z_0, \delta)} = \{z : |z - z_0| \leq \delta\} \subset \Omega$ and $\{f_n(z)\}$ converges uniformly to 0 on compact set $\overline{D(z_0, \delta)}$, so there is $N_0 > 0$ such that if $n \geq N_0$, $|f_n(z)| \leq \frac{1}{2}, z \in \overline{D(z_0, \delta)}$, thus $m(f, z) = \sum_{n=1}^{N_0} m(1 + f_n, z)$ ($z \in \overline{D(z_0, \delta)}$). □

3.2 The Weierstrass Factorization Theorem

Definition 3.2.1. Put $E_0(z) = 1 - z$ and for $p = 1, 2, \ldots$,

$$E_p(z) = (1-z)\exp\left\{z + \frac{z^2}{2} + \cdots + \frac{z^p}{p}\right\}.$$

These functions $E_p(z)$ are called elementary factors.

Lemma 3.2.2. *For $|z| \leq 1$ and $p = 0, 1, 2, \ldots,$*

$$|1 - E_p(z)| \leq |z|^{p+1}.$$

Proof. For $p = 0$, this is obvious. For $p \geq 1$, direct computation shows that

$$-E_p'(z) = z^p \exp\left\{z + \frac{z^2}{2} + \cdots \frac{z^p}{p}\right\}.$$

Then $|z| \leq 1$,

$$|1 - E_p(z)| = \left|\int_0^1 E_p'(tz) z \, dt\right| \leq |z|^{p+1} \left|\int_0^1 E_p'(t) dt\right| = |z|^{p+1}.$$

This gives the assertion of the lemma. □

Theorem 3.2.3. *Let $\{z_n\}$ be a sequence of complex numbers such that $z_n \neq 0$ and $|z_n| \to \infty$ as $n \to \infty$. If $\{p_n\}$ is a sequence of non-negative integers such that*

$$\sum_{n=1}^{\infty} \left(\frac{r}{|z_n|}\right)^{1+p_n} < \infty \tag{3.2.1}$$

for every positive $r > 0$, then the infinite product

$$P(z) = \prod_{n=1}^{\infty} E_{p_n}\left(\frac{z}{z_n}\right) \tag{3.2.2}$$

defines an entire function $P(z)$ which has a zero at each point z_n and has no other zeros in the plane.

More precisely, if α occurs m times in sequence $\{z_n\}$, then $P(z)$ has a zero of order m at α.

Condition (3.2.1) is always satisfied if $p_n = n - 1$, for instance.

Proof. For every $r > 0$, there is $N(r)$, such that if $n \geq N(r)$, $|z_n| \geq r$, so (3.2.1) holds with $p_n = n - 1$. Now, fix $r > 0$, if $|z| \leq r$, Lemma 3.2.2 shows that

$$\left|1 - E_{p_n}\left(\frac{z}{z_n}\right)\right| \leq \left|\frac{z}{z_n}\right|^{1+p_n} \leq \left(\frac{r}{|z_n|}\right)^{1+p_n}$$

for $n \geq N(r)$. It now follows from (3.2.1) that series

$$\sum_{n=1}^{\infty} \left|1 - E_{p_n}\left(\frac{z}{z_n}\right)\right|$$

converges uniformly on compact sets in the plane, and Theorem 3.1.5 gives the desired conclusion. □

Theorem 3.2.4 (Weierstrass Factorization Theorem). *Let $f(z)$ be an entire function, suppose that $f(z)$ has a zero of order k at 0, and let $z_1, z_2, z_3, \ldots$ be the zero of $f(z)$, listed according to their multiplicities, with $0 < |z_1| \leq |z_2| \leq |z_3| \leq \cdots$. Then there exists an entire function $h(z)$ and a sequence $\{p_n\}$ of non-negative integers such that (3.2.1) for every $r > 0$, and*

$$f(z) = z^k e^{h(z)} \prod_{n=1}^{\infty} E_{p_n}\left(\frac{z}{z_n}\right). \qquad (3.2.3)$$

Proof. Let $P(z)$ be the product in (3.2.2), formed with the zeros of $f(z)$ and $z_n \neq 0$. Then $\frac{f(z)}{z^k P(z)}$ has only removable singularities in the plane and hence is (or can be extended to be) an entire function. Also, $\frac{f(z)}{z^k P(z)}$ has no zero, and since the plane is simply connected, (3.2.3) holds for some entire function $h(z)$. □

Theorem 3.2.5. *Let Ω be an open set in plane, with its boundary $\partial\Omega$ not empty. Suppose $A = \{z_n\} \subset \Omega$ is a sequence of non-negative integers, in which each $\alpha \in A = \{z_n\}$ is listed precisely $m(\alpha)$, and $d(z_n, \partial\Omega) \to 0$ $(n \to \infty)$. There exists an holomorphic function P in Ω, all of whose zeros are in A, such that f has a zero of order $m(\alpha)$ at each $\alpha \in A$.*

Proof. Associate with z_n a point $\alpha_n \in \partial\Omega$ such that $|z_n - \alpha_n| = d(z_n, \partial\Omega)$, so $|z_n - \alpha_n| \to 0$ $(n \to \infty)$. Let K be a compact subset of Ω, there exists an N such that $|z_n - \alpha_n||z - \alpha_n|^{-1} \leq \frac{1}{2}$ for $n \geq N$ and $z \in K$, which implies, by Lemma 3.2.2, that

$$\left|1 - E_n\left(\frac{z_n - \alpha_n}{z - \alpha_n}\right)\right| \leq \left(\frac{1}{2}\right)^{n+1} \qquad (z \in K, n \geq N).$$

The function

$$P(z) = \prod_{n=1}^{\infty} E_n\left(\frac{z_n - a_n}{z - a_n}\right)$$

satisfies the assertion of Theorem 3.2.5. □

Theorem 3.2.6. *Suppose $A = \{a_n : n = 1, 2, \ldots\}$ is a infinite set such that $a_n \to \infty$ ($n \to \infty$), and to each $a \in A$, there are associated a non-negative integer $m(a)$ and a rational function*

$$Q_a(z) = \sum_{j=1}^{m(a)} c_{j,a}(z-a)^{-j}.$$

Then there exists a meromorphic function $Q(z)$, in which principal at $a \in A$ is $Q_a(z)$ and no other poles in $\mathbb{C}$.

Proof. We can assume, without loss of generality, that $0 < |a_1| \leq |a_2| \leq \cdots \leq |a_n| \leq |a_{n+1}| \leq \cdots$, then the function $Q_{a_n}(z)$ is analytic in the disk $D(0, |a_n|)$, and its Taylor series at the point 0 converges uniformly on $\overline{D(0, \frac{1}{2}|a_n|)}$. Therefore, there exists a polynomial $P_n(z)$ such that

$$|Q_{a_n}(z) - P_n(z)| \leq \frac{1}{n^2} \quad \left(|z| \leq \frac{1}{2}|a_n|\right).$$

Fixed $R > 0$, there is $N = N(R)$ such that $|a_n| > 2R$ for all $n \geq N$. Thus,

$$|Q_{a_n}(z) - P_n(z)| \leq \frac{1}{n^2} \quad (|z| \leq R)$$

for all $n \geq N$. The series

$$G_N(z) = \sum_{n=N+1}^{\infty} (Q_{a_n}(z) - P_n(z))$$

converges uniformly on the disk $\overline{D(0, R)}$. Since the poles of $Q_{a_n}(z)$ do not belong to $\overline{D(0, R)}$ for $n \geq N$, according to the Weierstrass theorem, $G_N(z)$ is analytic in $D(0, R)$. Thus, the poles of

$$Q(z) = \sum_{n=1}^{N} (Q_{a_n}(z) - P_n(z)) + G_N(z)$$

in $D(0, R)$ are the points $a_n : |a_n| < R$. Therefore, $Q(z)$ is a meromorphic in $\mathbb{C}$, the principal parts of the meromorphic function $Q(z)$ at each point $a_1, a_2, \ldots$ are
$$Q_{a_1}(z), Q_{a_2}(z), \ldots, Q_{a_n}(z), \ldots.$$
□

3.3 Order and Type of Entire Functions

A function that is analytic on the complex plane $\mathbb{C}$ is called an *entire function*. It is well known that entire functions can be expanded as power series
$$f(z) = \sum_{n=0}^{\infty} a_n z^n \tag{3.3.1}$$
on $\mathbb{C}$. Therefore, polynomials are entire functions. In fact, polynomials are the simplest form of entire functions. We refer to non-polynomial entire functions as *transcendental entire functions*.

Let $f(z)$ be a non-constant entire function, and let
$$M(f, r) = \max\{|f(z)| : |z| = r\}$$
be the maximum modulus of $f(z)$ on the circle $|z| = r$. It is one of the most important characteristics of $f(z)$. To study the growth behavior of $f(z)$, we start with the properties of $M(f, r)$. By the maximum modulus principle and Liouville's theorem, $M(f, r)$ is a monotonically increasing function tending to infinity on $[0, \infty)$. Therefore, we compare $M(f, r)$ with certain elementary functions of r to determine the rate of growth of $f(z)$. The definition of the order and type of an entire function, i.e., a transcendental entire function, is as follows.

Definition 3.3.1. (a) We say that an entire function $f(z)$ is of order ρ_f if
$$\rho_f = \limsup_{r \to \infty} \frac{\log(\log M(f, r))}{\log r}. \tag{3.3.2}$$
The order of an entire function $f(z)$ that is identically constant is defined as 0.

(b) Let $0 \leqslant \rho < \infty$, let f be an entire function if there are positive constants A and B such that
$$M(f, r) \leqslant A \exp\{Br^\rho\} \; (r > 0),$$

then we say that the entire function $f(z)$ is of finite type of the order ρ. The type $\tau(f,\rho)$ of the order ρ of a non-constant entire function $f(z)$ is defined by

$$\tau(f,\rho) = \limsup_{r\to\infty} \frac{\log M(f,r)}{r^\rho}. \qquad (3.3.3)$$

For a given non-constant entire function $f(z)$, its order ρ_f can be classified into the following two cases:

(i) If $\rho_f < \infty$, then $f(z)$ is called a finite order entire function with order ρ_f, (3.3.2) is equivalent for any arbitrary $\varepsilon > 0$, there is $r_0 > 0$,

$$M(f,r) \leqslant \exp\{r^{\rho+\varepsilon}\},$$

for all $r > r_0$, and there is a sequence of increasing positive numbers $\{r_n\}$ which $r_n \to \infty$ as $n \to \infty$ such that

$$M(f, r_n) > \exp\{r_n^{\rho-\varepsilon}\}$$

for $n \geqslant 1$.

(ii) if $\rho_f = \infty$, we refer to $f(z)$ as an entire function of infinite order.

Example 3.3.2.

(a) If P is a polynomial of degree n
$$P(z) = a_0 z^m + a_1 z^{m-1} + \cdots + a_m (a_0 \neq 0),$$
then its order $\rho_P = 0$.
(b) If $f(z) = e^{z^m}$, then $M(f,r) = e^{r^m}$, so the order of $f(z) = e^{z^m}$ is $\rho_f = m$. In particular, e^z is an entire function of order 1.
(c) The maximum modulus of $f(z) = e^{e^z}$ is $M(f,r) = e^{e^r}$, so the order of $f(z) = e^{e^z}$ is $\rho_f = \infty$.

Example 3.3.3. If $f(z) = \cos\sqrt{z}$ ($z = re^{i\theta}$), then

$$\frac{e^{\sqrt{r}}}{2} \leqslant M(f,r) \leqslant e^{\sqrt{r}}.$$

Therefore,

$$\rho_f = \limsup_{r\to\infty} \frac{\log(\log M(f,r))}{\log r} = \frac{1}{2},$$

the order of $\cos\sqrt{z}$ is $\frac{1}{2}$.

Theorem 3.3.4. *Suppose that the entire functions $f_1(z)$ and $f_2(z)$ have orders ρ_1, ρ_2, respectively,*

(a) *If ρ is the order of $f_1(z) + f_2(z)$, then $\rho \leq \max\{\rho_1, \rho_2\}$; if $\rho_1 < \rho_2$, then $\rho = \rho_2$.*
(b) *If ρ is the order of $f_1(z) \cdot f_2(z)$, then $\rho \leq \max\{\rho_1, \rho_2\}$.*

Proof. (a) We can assume that $\rho_0 = \max\{\rho_1, \rho_2\} < \infty$. For all $\varepsilon > 0$, there is $r_0 > 1$ such that

$$M(f_1, r) \leq \exp\{r^{\rho_1+\varepsilon}\}, \ M(f_2, r) \leq \exp\{r^{\rho_2+\varepsilon}\}$$

for $r > r_0$. Thus,

$$M(f_1 + f_2, r) \leq M(f_1, r) + M(f_2, r) \leq 2\exp\{r^{\rho_0+\varepsilon}\}.$$

This proves that $\rho \leq \rho_0 = \max\{\rho_1, \rho_2\}$.
Next, we prove that if $\rho_1 < \rho_2$, then $\rho = \rho_2$.
(i) When $\rho_2 < \infty$, we have already proved that $\rho \leq \rho_2$. It suffices to prove that $\rho \geq \rho_2$. By the definition of order, for any $\varepsilon > 0$ ($\varepsilon < (\rho_2 - \rho_1)/3$), there exists a sequence of positive numbers $\{r_n\}$ that tends to infinity such that

$$r_n > e, \ M(f_2, r_n) > \exp\{r_n^{\rho_2-\varepsilon}\}.$$

Taking $\{\theta_n\}$ such that $|f_2(r_n e^{i\theta_n})| = M(f_2, r_n)$, then

$$\begin{aligned} M(f_1 + f_2, r_n) &\geq |f_2(r_n e^{i\theta_n})| - |f_1(r_n e^{i\theta_n})| \\ &\geq M(r_n, f_2) - M(r_n, f_1) \\ &\geq e^{r_n^{\rho_2-\varepsilon}} - e^{r_n^{\rho_1+\varepsilon}} > e^{r_n^{\rho_2-2\varepsilon}}. \end{aligned}$$

Thus, $\rho \geq \rho_2 - 2\varepsilon$. Let ε tend to 0, $\rho \geq \rho_2$. This proves $\rho = \rho_2$.
(ii) When $\rho_1 < \rho_2 = \infty$, similarly, we can prove that $\rho = \infty$. This proves (a).
(b) In order to prove (b), we can assume that $\rho_1 \leq \rho_2 < \infty$ and $\rho > 0$. For all $\varepsilon > 0$, there is $r_0 > 1$

$$M(f_1 f_2, r) \leq M(f_1, r) M(r, f_2) \leq \exp\{r^{\rho_1+\varepsilon} + r^{\rho_2+\varepsilon}\} \leq e^{2r^{\rho_2+\varepsilon}}$$

for $r > r_0$. So, $\rho \leq \rho_2 + \varepsilon$. Let ε tend to 0, $\rho \leq \rho_2 = \max\{\rho_1, \rho_2\}$. □

Theorem 3.3.5. *The orders and types of an entire function $f(z)$ and of its derivative $f'(z)$ are the same.*

Proof. Since
$$f(z) = \int_0^z f'(\xi)d\xi + f(0),$$
where the integral contour is the segment from 0 to z, then
$$M(f,r) \leqslant M(f',r)r + |f(0)|,$$
i.e.,
$$M(f',r) \geqslant \frac{M(f,r) - |f(0)|}{r}. \tag{3.3.4}$$
On the other hand,
$$f'(z) = \frac{1}{2\pi i} \int_{|\zeta|=2|z|} \frac{f(\zeta)}{(\zeta-z)^2} d\zeta \quad (z \neq 0),$$
so
$$M(f',r) \leqslant \frac{2M(f,2r)}{r} \leqslant M(f,2r) \tag{3.3.5}$$
for $|z| = r > 2$. The conclusion of Theorem 3.3.5 follows from the inequalities (3.3.4) and (3.3.5). In fact, if $\rho_f = \infty$, $\rho_{f'} = \rho_f = \infty$ from the inequality (3.3.4). If $\rho_f < \infty$ for all $\varepsilon > 0$, there is $r_0 > 2$ such that
$$M(f,r) \leqslant \exp\{r^{\rho_f + \varepsilon}\}$$
for all $r > r_0$. By (3.3.5),
$$M(f',r) \leqslant \exp\{(2r)^{\rho_f + \varepsilon}\},$$
so $\rho_{f'} \leqslant \rho_f$. On the other hand, for all $\varepsilon > 0$, there is $r_0 > 2$ such that
$$M(f',r) \leqslant \exp\{r^{\rho_{f'} + \varepsilon}\}$$
for all $r > r_0$. By (3.3.4),
$$M(f,r) \leqslant |f(0)| + r\exp\{(2r)^{\rho_{f'} + \varepsilon}\},$$
so $\rho_f \leqslant \rho_{f'}$. This proves $\rho_{f'} = \rho_f$. □

Theorem 3.3.6. *The entire function $f(z) = \sum_{n=1}^{\infty} a_n z^n$ is of finite order ρ if and only if*

$$\mu = \limsup_{n \to \infty} \frac{n \log n}{-\log |a_n|} \quad \left(where \quad \frac{n \log n}{\log |a_n|^{-1}} = 0 \text{ for } a_n = 0 \right),$$

is finite. In this case, $\rho = \mu$.

Remark. This theorem establishes the relationship between the order of an entire function and the coefficients of its Taylor expansion.

Proof. Since f is an entire function, then $\sqrt[n]{|a_n|} \to 0$ $(n \to \infty)$. We first show that $\rho \leqslant \mu$. If $\mu =< \infty$, there is nothing to prove. Assume $\mu < \infty$. For any $\varepsilon > 0$, there is $N > 0$ such that

$$\frac{n \log n}{\log |a_n|^{-1}} \leqslant \mu + \varepsilon, \text{ i.e.,} \quad |a_n| < n^{-\frac{n}{\mu+\varepsilon}}$$

for all $n > N$. Note that this condition also implies that f is entire. Since the addition of a polynomial to f does not change the order of a function, we can assume that the last inequality holds for all values n and $a_0 = 0$. We have therefore, for $r \geq 1$,

$$M(f, r) \leqslant \sum_{n=1}^{\infty} |a_n| r^n \leqslant \sum_{n=1}^{\infty} r^n n^{-\frac{n}{\mu+\varepsilon}} = \Sigma_1 + \Sigma_2,$$

where

$$\Sigma_1 = \sum_{1 \leqslant n < (2r)^{\mu+\varepsilon}} r^n n^{\frac{-n}{\mu+\varepsilon}}, \quad \Sigma_2 = \sum_{n \geqslant (2r)^{\mu+\varepsilon}} r^n n^{\frac{-n}{\mu+\varepsilon}}.$$

In Σ_2, we have $rn^{-\frac{1}{\mu+\varepsilon}} < \frac{1}{2}$. Hence, $\Sigma_2 \leqslant \sum_{n=1}^{\infty} \frac{1}{2^n} = 1$. On the other hand,

$$\Sigma_1 \leqslant r^{(2r)^{\mu+\varepsilon}} \sum_{n=1}^{\infty} n^{-\frac{n}{\mu+\varepsilon}} \leqslant M \exp((2r)^{\mu+\varepsilon} \log r) \leq M_1 \exp(r^{\mu+2\varepsilon})$$

for some constants M and M_1. It follows that $\rho \leqslant \mu + 2\varepsilon$. Since $\varepsilon > 0$ is arbitrary, it follows that $\rho \leqslant \mu$.

Let us show now that $\rho \geqslant \mu$. We will use that $|a_n| \leqslant r^{-n} M(f, r)$, $r > 0$. If $\mu = 0$, we have clearly that $\rho \geqslant \mu$. Let us assume that $\mu > 0$. Let $\alpha \in (0, \mu)$, then there is a positive increasing unbounded sequence $\{n_k\}$ such that

$$\log |a_{n_k}|^{-1} < \frac{n_k}{\alpha} \log n_k,$$

i.e.,

$$|a_{n_k}| > n_k^{-\frac{n_k}{\mu - \varepsilon}}. \tag{3.3.6}$$

Set $r_k = (2n_k)^{\frac{1}{\alpha}}$. Then

$$|a_{n_k}| r_k^{n_k} > 2^{\frac{n_k}{\alpha}} = \exp\left\{\frac{n_k \log 2}{\alpha}\right\} = \exp\left\{\frac{r_k^\alpha \log 2}{2\alpha}\right\}.$$

The Cauchy inequality gives that

$$M(f, r_k) \geq |a_{n_k}| r_k^{n_k}.$$

Hence, there is a positive constant A such that

$$M(f, r_k) > \exp\{A r_k^\alpha\}. \tag{3.3.7}$$

So, $\rho_f \geq \alpha$. Let α be an arbitrary real number smaller than μ, then it follows that $\rho \geq \mu$. Theorem 3.3.6 has been proved. □

Example 3.3.7. Let α be a positive number, and $f(z) = \sum_{n=1}^{\infty} \frac{z^n}{n^{n\alpha}}$, where $|a_n|^{-1} = n^{n\alpha}$, then

$$\rho = \limsup_{n \to \infty} \frac{n \log n}{\log |a_n|^{-1}} = \limsup_{n \to \infty} \frac{n \log n}{n \alpha \log n} = \frac{1}{\alpha}.$$

$f(z)$ is an entire function of the order $\frac{1}{\alpha}$. □

Theorem 3.3.8. Let $f(z) = \sum_{n=1}^{\infty} a_n z^n$ an entire function of order ρ_f and $0 < \rho < \infty$. Let

$$\nu = \limsup_{n \to \infty} n |a_n|^{\rho/n}. \tag{3.3.8}$$

If $\nu < \infty$, then $\rho_f \leq \rho$, and if $\nu > 0$, then $\rho_f \geq \rho$.

Proof. Since f is an entire function, $\sqrt[n]{|a_n|} \to 0$ $(n \to \infty)$. If $\nu < \infty$, then f is at most of order ρ, and if $\mu > 0$, then f is at least of order ρ, i.e., let $\varepsilon > 0$, then if $\mu < \infty$, there is n_0 such that

$$-\log |a_n| \geqslant -\frac{n}{\rho} \log \left(\frac{\nu_f + \varepsilon}{n}\right).$$

For all $n \geqslant n_0$, it follows from Theorem 3.3.6 that the order of f,

$$\rho_f = \limsup_{n \to \infty} \frac{n \log n}{-\log |a_n|} \leqslant \limsup_{n \to \infty} \frac{n \log n}{-\frac{n}{\rho} \log \left(\frac{\nu_f + \varepsilon}{n}\right)} = \rho.$$

Analogously, if $\nu > 0$, by (3.3.8), for all $R \in (0, \nu_f)$, then there is a positive integer increasing unbounded sequence $\{n_k\}$ which tends to ∞ such that

$$-\log |a_{n_k}| \leqslant -\frac{n_k}{\rho} \log \left(\frac{R}{n_k}\right).$$

By Theorem 3.3.6, the order of f,

$$\rho_f = \limsup_{n \to \infty} \frac{n \log n}{-\log |a_n|} \geqslant \limsup_{k \to \infty} \frac{n_k \log n_k}{-\frac{n_k}{\rho} \log \left(\frac{R}{n_k}\right)} = \rho.$$

This completes the proof of Theorem 3.3.8. □

Theorem 3.3.9. Let $f(z) = \sum\limits_{n=0}^{\infty} a_n z^n$ be an entire function of order ρ_f, $0 < \rho < \infty$ and ν defined by (3.3.8). If $\nu < \infty$, then the type $\tau = \tau(f, \rho)$ of f satisfies

$$\tau = \limsup_{r \to \infty} \frac{\log M(f, r)}{r^\rho} \leqslant \frac{\nu}{e\rho}, \tag{3.3.9}$$

and if $\nu > 0$, then the type $\tau = \tau(f, \rho)$ of f satisfies

$$\tau = \limsup_{r \to \infty} \frac{\log M(f, r)}{r^\rho} \geqslant \frac{\nu}{e\rho}. \tag{3.3.10}$$

Proof. Since f is an entire function,

$$\sqrt[n]{|a_n|} \to 0 \ (n \to \infty).$$

Zeros of Holomorphic Functions

Assume now that $\nu < \infty$ and we will show that the type $\tau = \tau(f, \rho) \leq \frac{\nu}{e\rho}$. Let $\varepsilon > 0$, there is n_0 such that

$$-\log|a_n| \geq -\frac{n}{\rho} \log\left(\frac{\nu + \varepsilon}{n}\right)$$

for all $n \geq n_0$.

$$M(f, r) \leq \sum_{n=0}^{\infty} |a_n| r^n \leq r^{n_0} \sum_{n=0}^{n_0} |a_n| + \sum_{n=n_0+1}^{\infty} \left(\frac{(\nu + \varepsilon) r^\rho}{n}\right)^{\frac{n}{\rho}}$$

for all $r \geq 1$. Let

$$\Sigma_1 = \sum_{n \leq (\nu + 2\varepsilon) r^\rho} \left(\frac{(\nu + \varepsilon) r^\rho}{n}\right)^{\frac{n}{\rho}}.$$

Consider the function

$$\psi(t) = \frac{t}{\rho} \left(\log(r^\rho(\nu + \varepsilon)) - \log t\right)$$

for $t > 0$. Its maximum value is taken at $r^\rho(\nu + \varepsilon)e^{-1}$ and it is $r^\rho(\nu + \varepsilon)(\rho e)^{-1}$. We have thus, for some convenient $M_1 > 0$ and $M_2 > 0$,

$$\Sigma_1 \leq \sum_{n \leq (\nu + 2\varepsilon) r^\rho} \exp\left\{r^\rho(\nu + \varepsilon)(\rho e)^{-1}\right\}$$

$$\leq M_1 \exp\left\{r^\rho(\nu + 2\varepsilon)(\rho e)^{-1}\right\}$$

and

$$M(f, r) \leq M_1 + M_1 r^\rho \exp\left\{r^\rho(\nu + 2\varepsilon)(\rho e)^{-1}\right\}$$

$$M_1 + M_2 \exp\left\{r^\rho(\nu + 3\varepsilon)(\rho e)^{-1}\right\}.$$

It follows that $\tau = \tau(f, \rho) \leq \frac{\nu}{e\rho}$.

If $\nu > 0$, We will show the opposite inequality, note that if $R \in (0, \nu_f)$, there exists a sequence of positive integers $\{n_k\}$ tending to infinity such that

$$\log|a_{n_k}| \geq \frac{n_k}{\rho} \log\left(\frac{R}{n_k}\right).$$

In Cauchy's inequality $|a_n| \leq r^{-n} M(f,r)$, let us take $\log r_k = \rho^{-1}(\log(en_k) - \log R)$. One obtains

$$\log M(f, r_k) \geq \frac{n_k}{\rho} = \frac{r_k^\rho R}{e\rho},$$

so

$$\varlimsup_{r \to \infty} \frac{\log M(f,r)}{r^\rho} \geq \frac{R}{e\rho}.$$

Therefore, $\tau = \tau(f, \rho) \geq \frac{\nu}{e\rho}$. □

Theorem 3.3.10. Let

$$f(z) = \sum_{n=1}^{\infty} a_n z^n$$

be an entire function of order ρ_f, $0 < \rho < \infty$. Suppose that ν is defined by (3.3.8). If $0 < \nu < \infty$, then the order of f, ρ_f is equal to ρ and the type is $\tau(f, \rho) = \frac{\nu}{e\rho}$.

Proof. From Theorems 3.3.8 and 3.3.9, it follows that Theorem 3.3.10 holds. □

3.4 Exponent of Convergence, Genus and Canonical Product

Definition 3.4.1 (The Definition of Exponent of Convergence). Let $f(z)$ be an entire function with a zero of multiplicity m at $z = 0$; let $z_1, z_2, \ldots$ be the non-zeros of f, arranged so that a zero of multiplicity k is repeated in this sequence k times. Also, assume that $0 < |z_1| = r_1 \leq |z_2| = r_2 \leq \cdots$. Then $f(z)$ is of finite exponent of convergence if there is a negative number τ such that

$$\sum_{n=1}^{\infty} r_n^{-\tau} < \infty. \tag{3.4.1}$$

If ρ_1 is the infimum of τ such that (3.4.1) occurs, then f is said to be exponent of convergence ρ_1 (the number ρ_1 is also called the exponent of convergence of the sequence of non-zero numbers $\{z_n\}$); a function is of infinite exponent of convergence if it is not of finite convergence exponent, $\rho_1 = \infty$.

When the number of zeros of an entire function $f(z)$ is finite, it is clear that $\rho_1 = 0$. (In particular, if $f(z)$ has no non-zero zeros, then $\rho_1 = 0$.) Therefore, $\rho_1 > 0$ implies that the function has infinitely many (non-zero) zeros.

Example 3.4.2. The exponent of convergence of the entire function $f(z) = \cos\sqrt{z}$ is $\frac{1}{2}$.

In fact, the set of zeros of $f(z)$ is $\{(n\pi + \frac{\pi}{2})^2 : n \in \mathbb{Z}\}$,

$$\sum_{n=-\infty}^{\infty} \frac{1}{[(n\pi + \frac{\pi}{2})^2]^{\frac{1}{2}+\varepsilon}} = 2\sum_{n=0}^{\infty} \frac{1}{(n\pi + \frac{\pi}{2})^{1+2\varepsilon}} < 2\sum_{n=1}^{\infty} \frac{1}{n^{1+\varepsilon}} < \infty$$

for all $\varepsilon > 0$, and

$$\sum_{n=-\infty}^{\infty} \frac{1}{[(n\pi + \frac{\pi}{2})^2]^{\frac{1}{2}}} = 2\sum_{n=0}^{\infty} \frac{1}{n\pi + \frac{\pi}{2}} > \frac{4}{\pi} + \frac{2}{\pi}\sum_{n=1}^{\infty} \frac{1}{n} = \infty.$$

Combining Examples 3.3.3 and 3.4.2, the order ρ of $\cos\sqrt{z}$ is equal to the exponent of convergence ρ_1 of its zeros, both being $\frac{1}{2}$.

Theorem 3.4.3. *If $f(z)$ is an entire function of finite order ρ, then the exponent of convergence of its non-zeros satisfies $\rho_1 \leqslant \rho$.*

Proof. Let $f(z)$ be an entire function with a zero of multiplicity m at $z = 0$; let $z_1, z_2, \ldots$ be the zeros of f, $z_n \neq 0$, arranged so that a zero pf multiplicity k is repeated in this sequence k times. Also, assume that $0 < |z_1| = r_1 \leqslant |z_2| = r_2 \leqslant \cdots \lim_{n\to\infty} r_n = \infty$. Put

$$F_n(z) = \frac{f(z)}{z^m(z-z_1)(z-z_2)\cdots(z-z_n)},$$

then

$$|F_n(0)| = \frac{|f^{(m)}(0)|}{m!}\frac{1}{|z_1||z_2|\cdots|z_n|} \geqslant \frac{|f^{(m)}(0)|}{m!}r_n^{-n}.$$

On the other hand,

$$\max\{|F_n(z)| : |z| = 3r_n\} \leqslant \frac{M(f, 3r_n)}{(3r_n)^m(2r_n)^n}$$

for $|z| = 3r_n$ and $|z - z_k| \geqslant 2r_n$ ($k = 1, 2, \ldots, n$). Since $f(z)$ is analytic in disk $|z| \leqslant 3r_n$, we apply the maximum modulus principle to disk of radius $|z| = 3r_n$,

$$|F_n(0)| \leqslant \max\{|F_n(z)| : |z| = 3r_n\}.$$

Thus, there exists a constant $A_1 > 0$ such that

$$2^n (3r_n)^m \leqslant A_1 M(f, 3r_n). \tag{3.4.2}$$

Since $f(z)$ is an entire function of finite order ρ, for all $\varepsilon > 0$, there is $r_0 > 1$ such that $M(f, r) < \exp\{r^{\rho+\varepsilon}\}$ for all $r > r_0$. Combining (3.4.2), for sufficiently large n (with correspondingly large $r_n > r_0$), we have

$$2^n \leqslant A_1 (3r_n)^m \exp\{(3r_n)^{\rho+\varepsilon}\} \leqslant A_1 \exp\{(3r_n)^{\rho+\varepsilon}\} \quad (r_n > r_0),$$

hence there is constant $A_2 > 1$ such that

$$n \leqslant A_2 K r_n^{\rho+\varepsilon}.$$

Thus,

$$\sum_{n=1}^{\infty} r_n^{-\tau} \leqslant A_2^{\frac{\tau}{\rho+\varepsilon}} \sum_{n=1}^{\infty} n^{-\frac{\tau}{\rho+\varepsilon}} < \infty$$

for all $\tau > \rho + \varepsilon$, the exponent of convergence ρ_1 of the non-zeros of $f(z)$ satisfies that $\rho_1 \leqslant \rho + \varepsilon$, letting $\varepsilon \to 0$, we see that $\rho_1 \leqslant \rho$. □

Remark. The inequality $\rho_1 < \rho$ is possible. For example, consider the function $f(z) = e^z$. Here, we have $\rho = 1$, while since e^z has no zeros, we have $\rho_1 = 0$.

Definition 3.4.4. Let $f(z)$ be an entire function with a zero of multiplicity m at $z = 0$ and with zeros of the non-zeros $z_1, z_2, \ldots$, arranged so that a zero pf multiplicity k is repeated in this sequence k times. Also assume that $0 < |z_1| = r_1 \leqslant |z_2| = r_2 \leqslant \cdots$. Then $f(z)$ is of finite rank if there is an integer p such that

$$\sum_{n=1}^{\infty} r_n^{-(p+1)} < \infty. \tag{3.4.3}$$

If p is the smallest integer such that this (3.4.3) occurs, then f is said to be of rank p; a function with only number of zeros has rank 0. A function is of infinite rank if it is not of finite rank.

An entire function f has finite genus if $f(z)$ has finite rank, and if

$$f(z) = z^m P(z) e^{h(z)},$$

where m is the order of f at 0, $P(z)$ is in standard form $f(z)$ and $h(z)$ is a polynomial of degree q, then $\mu = \max\{p, q\}$ is called the genus of f.

If f is an entire function with the exponent convergence ρ_1 of zeros and with the finite order ρ, then $\rho_1 \leqslant \rho < \infty$. If ρ_1 is not an integer, then the range of $p = [\rho_1]$; If ρ_1 is an integer, then when $\sum_{n=1}^{\infty} r_n^{-\rho_1} = \infty$, $p = \rho_1$, otherwise when $\sum_{n=1}^{\infty} r_n^{-\rho_1} < \infty$, $p = \rho_1 - 1$. Thus, for a finite order entire function, the following relations always hold:

$$\rho_1 - 1 \leqslant p \leqslant \rho_1 \leqslant \rho.$$

Let $f(z)$ be an entire function and finite rank p, $z_1, z_2, \ldots$ a sequence of complex numbers $\neq 0$, the non-zeros of f, arranged so that a zero of multiplicity k is repeated in this sequence k times. In this case, we can take $p_n = p$ in the Weierstrass factorization of (3.2.2), and we refer to it as an infinite product:

$$P(z) = \prod_{n=1}^{\infty} E_p\left(\frac{z}{z_n}\right), \tag{3.4.4}$$

the canonical product determined by the sequence $\{z_n\}$ of the zeros of $f(z)$.

Theorem 3.4.5. *The order ρ of the typical product (3.4.4) is equal to the convergence exponent of its zeros ρ_1, i.e., $\rho = \rho_1$.*

Proof. It has been proven in Theorem 3.4.3 that $\rho_1 \leqslant \rho$. It suffices to prove that the order ρ of the entire function $P(z)$ defined by (3.4.4) satisfies $\rho \leqslant \rho_1$. Let $|z_n| = r_n$, $|z| = r > 1$,

$$\log |P(z)| = \sum_{n=1}^{\infty} \log \left| E_p\left(\frac{z}{z_n}\right) \right| = \Sigma_1 + \Sigma_2,$$

where

$$\Sigma_1 = \sum_{r_n \leqslant 2r} \log \left| E_p\left(\frac{z}{z_n}\right) \right|, \quad \Sigma_2 = \sum_{r_n > 2r} \log \left| E_p\left(\frac{z}{z_n}\right) \right|.$$

Next, we estimate Σ_1 and Σ_2 separately. For Σ_2, since $r_n > 2r$, we have

$$|\Sigma_2| = \left| \sum_{r_n > 2r} \sum_{k=p+1}^{\infty} \frac{1}{k} \operatorname{Re}\left(\frac{z^k}{z_n^k}\right) \right| \leq \sum_{r_n > 2r} \left(\frac{r}{r_n}\right)^{p+1} \sum_{k=p+1}^{\infty} \frac{1}{k 2^{k-p-1}}.$$

If $p > \rho_1 - 1$, then for all $\varepsilon \in (0, p+1-\rho_1)$, we have

$$|\Sigma_2| \leq 2 \sum_{r_n > 2r} \left(\frac{r}{r_n}\right)^{p+1} \leq 2 \sum_{r_n > 2r} \left(\frac{r}{r_n}\right)^{\rho_1 + \varepsilon} \leq 2 r^{\rho_1 + \varepsilon} \sum_{n=1}^{\infty} r_n^{-(\rho_1+\varepsilon)}.$$

If $p = \rho_1 - 1$, then $\sum_{n=1}^{\infty} r_n^{-\rho_1} < \infty$. Hence,

$$|\Sigma_2| \leq 2 r^{\rho_1} \sum_{n=1}^{\infty} r_n^{-\rho_1} \leq 2 r^{\rho_1 + \varepsilon} \sum_{n=1}^{\infty} r_n^{-\rho_1}.$$

Combining the above two cases, we know that for any $\varepsilon > 0$, there exists a constant $A_2 > 1$ such that

$$|\Sigma_2| \leq A_2 r^{\rho_1 + \varepsilon}. \tag{3.4.5}$$

For Σ_1, since $r_n \leq 2r$, we have

$$\log \left| E_p\left(\frac{z}{z_n}\right) \right| \leq \log\left(1 + \frac{r}{r_n}\right) + \frac{r}{r_n} + \left(\frac{r}{r_n}\right)^2 + \cdots + \left(\frac{r}{r_n}\right)^p$$

$$< 2\frac{r}{r_n} + \left(\frac{r}{r_n}\right)^2 + \cdots + \left(\frac{r}{r_n}\right)^p \leq 2^{p+1} \left(\frac{r}{r_n}\right)^p.$$

We obtain that $\rho_1 \geq p$, $\rho_1 + \varepsilon - p > 0$, so

$$\Sigma_1 < 2^{p+1} r^p \sum_{r_n \leq 2r} r_n^{-p} \leq 2^{\rho_1 + \varepsilon + 1} r^{\rho_1 + \varepsilon} \sum_{n=1}^{\infty} r_n^{-(\rho_1 + \varepsilon)}. \tag{3.4.6}$$

Combining (3.4.4), (3.4.5) and (3.4.6) gives that $\log |P(z)| \leq 2^{\rho_1 + \varepsilon + 2} A_2 r^{\rho_1 + \varepsilon}$. So, the order ρ of $P(z)$ satisfies that $\rho \leq \rho_1 + \varepsilon$, thus $\rho \leq \rho_1$. $\square$

Theorem 3.4.6 (Minimum Modulus Theorem). *Let $f(z)$ be an entire function of order $\leqslant \rho$. Let $z_1, z_2, \ldots$ be its sequence of zeros ($z_n \neq 0$), repeated according to their multiplicities. Let $(h > \rho_1)$. Let $P(z)$ be the canonical product determined by the sequence $\{z_n\}$ of the zeros of $f(z)$. Let Ω be the complement of the closed disk of radius $|z_n|^{-h}$ centered at z_n, for $|z_n| > 1$. Then for all $\varepsilon > 0$, there exists $vr_0 = r_0(\varepsilon, f) > 1$ such that for $z \in \Omega, |z| = r > r_0$, we have*

$$|P(z)| > e^{-r^{\rho_1+\varepsilon}}. \tag{3.4.7}$$

Proof. We first prove that, for all $\varepsilon > 0$, there exists a constant $A_2 > 1$ and $r_0 > 1$, such that (3.4.5) holds for all $r > r_0$. Let $z \in \Omega$ and $|z| = r > r_0$, similar to proof of Theorem 3.4.5, we have

$$\log|P(z)| = \left(\sum_{r_n \leqslant 2r} + \sum_{r_n > 2r}\right) \log\left|E_p\left(\frac{z}{z_n}\right)\right| = \Sigma_1 + \Sigma_2,$$

where $|\Sigma_2| \leqslant A_2 r^{\rho_1+\varepsilon}$, hence

$$\Sigma_2 \geqslant -A_2 r^{\rho_1+\varepsilon}. \tag{3.4.8}$$

Now, Σ_1 can be analogously estimated by the geometric series. Since $\frac{r_n}{r} \leqslant 2$, there exists constant $A > 1$ such that

$$\log\left|E_p\left(\frac{z}{z_n}\right)\right| \geqslant \log\left|1 - \frac{z}{z_n}\right| - \left(\frac{|z|}{|z_n|}\right)^p$$
$$\times \left[\left(\frac{|z_n|}{|z|}\right)^{p-1} + \left(\frac{|z_n|}{|z|}\right)^{p-2} + \cdots + 1\right]$$
$$\geqslant \log\left|1 - \frac{z}{z_n}\right| - A\left(\frac{|z|}{|z_n|}\right)^p.$$

We have the analogous estimate from Theorem 3.4.5,

$$2^{p+1} \sum_{r_n \leqslant 2r} \left(\frac{|z|}{|z_n|}\right)^p \leqslant 2^{\rho_1+\varepsilon+1} r^{\rho_1+\varepsilon} \sum_{n=1}^{\infty} r_n^{-(\rho_1+\varepsilon)},$$

so

$$\Sigma_1 \geqslant \sum_{r_n \leqslant 2r} \log\left|1 - \frac{z}{z_n}\right| - r^{\rho_1+\varepsilon} 2^{\rho_1+\varepsilon+1} \sum_{n=1}^{\infty} r_n^{-(\rho_1+\varepsilon)}. \tag{3.4.9}$$

Note that $|1 - \frac{z}{z_n}| = \frac{1}{|z_n|}|z_n - z| > |z_n|^{-(h+1)} \geqslant (2r)^{-(h+1)}$, one has

$$\sum_{r_n \leqslant 2r} \log\left|1 - \frac{z}{z_n}\right| > -(h+1)\log(2r) \sum_{r_n \leqslant 2r} 1$$

$$\geqslant -(h+1)\log(2r) \cdot (2r)^{\rho_1+\varepsilon} \sum_{r_n \leqslant 2r} \frac{1}{r_n^{\rho_1+\varepsilon}}$$

$$> -r^{\rho_1+2\varepsilon}. \tag{3.4.10}$$

Combining this with (3.4.8), (3.4.9) and (3.4.10), $\log|P(z)| > -r^{\rho_1+3\varepsilon}$ for $z \in \Omega$ and $r = |z| > r_0$. □

Remark. The inequality $h > \rho_1$ in the theorem is used to guarantee the convergence of the series $\sum_{n=1}^{\infty} |z_n|^{-h}$ (as can be seen from the above proof, the theorem holds for any $h > \rho$). Therefore, the sum of the radii of all exceptional disks is not infinite. Hence, there exists a sequence of positive real numbers $\{R_n\}$ tending to infinity such that the circles $|z| = R_n$ do not intersect with the exceptional disks. Thus, on the circle $|z| = R_n$ (for sufficiently large n), we have

$$|P(z)| > e^{-R_n^{\rho_1+\varepsilon}}.$$

Let $f(z)$ be an entire function, and let $z_1, z_2, \ldots$ the sequence of its zeros $\neq 0$, and m the order of f at 0. According to the Weierstrass factorization theorem, $f(z)$ can be decomposed as $f(z) = z^m P(z) e^{g(z)}$, where $P(z)$ is a typical product of the zeros of $f(z)$ and $g(z)$ is an entire function.

If $f(z)$ is an entire function of finite order, there exists a more detailed factorization.

Theorem 3.4.7 (Hadamard Factorization Theorem). *Let $f(z)$ be an entire function of finite order ρ, then $f(z)$ can be decomposed as*

$$f(z) = z^m P(z) e^{h(z)}, \tag{3.4.11}$$

where m is the order of f at 0, $P(z)$ is the canonical product determined by the the sequence $\{z_n\}$ of zeros of $f(z)$, and $h(z)$ is a polynomial of degree $\leqslant \rho_1$.

To prove this Hadamard factorization, we can utilize the following Carathéodory inequality (Lemma 3.4.8), which can be proven using the Schwarz lemma.

Lemma 3.4.8. *Suppose that $f(z)$ is analytic in the disk $\overline{D(0,R)}$. The following inequalities hold $0 \leq r < R$:*

$$M(r) \leq \frac{2r}{R-r} A(R) + \frac{R+r}{R-r} |f(0)| \qquad (3.4.12)$$

for $r \in [0, R]$, where $M(r)$ and $A(r)$ represent the maximum values of $|f(z)|$ and $\mathrm{Re} f(z)$ on the circle $|z| = r$, respectively. The inequality (3.4.12) is called the Carathéodory inequality.

Proof. It is easy to see that we can assume $f(0) = 0$, f is not constant. Hence, $\mathrm{Re} f(z)$ is not constant either and $A(R) > \mathrm{Re} f(z)$ if $|z| < R$. Consider the auxiliary function

$$F(z) = \frac{f(z)}{2A(R) - f(z)},$$

which is holomorphic in the disk $D(0, R)$ and continuous in the closed disk $\overline{D(0,R)}$. Furthermore, $|F(z)| \leq 1$. In fact,

$$|F(z)|^2 = \frac{u^2 + v^2}{(2A(R) - u)^2 + v^2} \leq 1,$$

where $f(z) = u + iv$. By Schwarz's lemma, $|F(z)| \leq \frac{|z|}{R}$. Writing f in terms of F, one obtains

$$|f(z)| = \left| \frac{2A(R)F(z)}{1 + F(z)} \right| \leq \frac{2A(R)r}{R-r}$$

if $|z| \leq r < R$. Lemma 3.4.8 follows immediately from this. $\square$

Remark. If we discuss $-f(z)$ or $\pm i f(z)$, we can obtain similar conclusions, i.e., in (3.4.12), $M(r)$ and $|f(0)|$ remain unchanged, while $A(R)$ is replaced by $-\min\{\mathrm{Re} f(z) : |z| = R\}$, $-\min\{\mathrm{Im} f(z) : |z| = R$, or $\max\{\mathrm{Im} f(z) : |z| = R\}$.

The following theorem uses $A(R)$ to estimate $|f^{(n)}(z)|$ ($|z| \leq R$).

Theorem 3.4.9. Let $f(z)$ be analytic in the disk $|z| \leqslant R$. Let $M(r)$ and $A(r)$ be as mentioned in Lemma 3.4.8, we have, when $0 < r = |z| < R, n \geq 1$,

$$\max\{|f^{(n)}(z)| : |z| = r\} < \frac{2e(n+1)!R}{(R-r)^{n+1}}(A(R) + |f(0)|).$$

Proof. When $|z| = r < R, n \geq 1$,

$$f^{(n)}(z) = \frac{n!}{2\pi i} \int_{C_\delta} \frac{f(\zeta) - f(0)}{(\zeta - z)^{n+1}} d\zeta,$$

where C_δ is the circle $|\zeta - z| = R - r - \delta, 0 < \delta < R - r$ (δ to be determined, $r = |z|$). When $\zeta \in C_\delta$, $|\zeta| \leqslant R - \delta$.

Applying inequality (3.4.12) to $f(\zeta) - f(0)$, we obtain

$$\max\{|f(\zeta) - f(0)| : \zeta \in C_\delta\} \leqslant \frac{2R}{\delta}(A(R) - \operatorname{Re} f(0)).$$

Therefore, using the nth order derivative formula mentioned above, we obtain

$$|f^{(n)}(z)| \leqslant \frac{2n!R}{\delta(R-r-\delta)^n}(A(R) + |f(0)|).$$

It is easy to see that when $\delta = \frac{R-r}{n+1}$, the denominator $\delta(R-r-\delta)^n$ on the right-hand side of the above expression attains its maximum, thus

$$|f^{(n)}(z)| \leqslant 2\left(1 + \frac{1}{n}\right)^n \frac{(n+1)!R}{(R-r)^{n+1}}(A(R) + |f(0)|)$$

$$\leqslant \frac{2e(n+1)!R}{(R-r)^{n+1}}(A(R) + |f(0)|). \qquad \square$$

Proof of Theorem 3.4.7. Without loss of generality, let's assume $f(0) \neq 0$. Otherwise, we can consider $f(z)/z^m$, where m is the multiplicity of f at the point $z = 0$. Then, $f(z)$ can be decomposed into

$$f(z) = f(0)P(z)e^{g(z)} = P(z)e^{G(z)},$$

where $P(z)$ is the canonical product determined by the the sequence $\{z_n\}$ of zeros of $f(z)$, $G(z) = g(z) + \log f(0)$ is an entire function.

Now, let's determine the order of $e^{G(z)}$. Let ρ_1 be the convergence exponent of the zeros of $f(z)$ and ρ be the order of $f(z)$. According to the remark in Theorem 3.4.6, there exists a positive real sequence $\{R_n\}$ tending to infinity such that on the circle $z = R_n$ with a sufficiently large radius R_n, we have

$$|P(z)| > e^{-R_n^{\rho_1+\varepsilon}} \geqslant e^{-R_n^{\rho+\varepsilon}}.$$

On the other hand,

$$|f(z)| \leqslant M(f, R_n) \leqslant e^{R_n^{\rho+\varepsilon}}$$

for all z on the circle $z = R_n$. Therefore, when n is sufficiently large, on the circle $|z| = R_n$, we have

$$|e^{G(z)}| = \frac{|f(z)|}{|P(z)|} < e^{2R_n^{\rho+\varepsilon}}.$$

This expression shows that the real part of $G(z)$, its maximum value on $|z| = R_n$, satisfies $A(R_n) < 2R_n^{\rho+\varepsilon} < R_n^{\rho+2\varepsilon}$.

Furthermore, by the Carathéodory inequality, when n is sufficiently large, on the circle $|z| = \frac{R_n}{2}$, we have

$$M\left(G, \frac{R_n}{2}\right) \leqslant \frac{R_n}{R_n - \frac{R_n}{2}} A(R_n) + \frac{R_n + \frac{R_n}{2}}{R_n - \frac{R_n}{2}} |G(0)|$$

$$= 2A(R_n) + 3|G(0)| \leqslant 2R_n^{\rho+3\varepsilon}.$$

Since $\frac{R_n}{2} \to \infty$ $(n \to \infty)$, $G(z)$ is a polynomial of degree $\leqslant \rho_1$. □

Remark. According to the Hadamard factorization theorem, if an entire function $f(z)$ is finite order ρ, then it can be decomposed as (3.4.11). Since the order of a canonical product is equal to the exponent of convergence ρ_1 of its zeros (Theorem 3.4.5), and when $h(z)$ is a polynomial of degree q ($q \leqslant \rho$), the order of $e^{h(z)}$ is q (Theorem 3.3.4), the order of $f(z)$ is $\rho \leqslant \max\{\rho_1, q\}$. If $\rho_1 \leqslant q$, then $\rho \leqslant \max\{\rho_1, q\} = q$, so $\rho = q$. If $\rho_1 > q$, then $\rho \leqslant \max\{\rho_1, q\} = \rho_1$, since $\rho \geqslant \rho_1$ (Theorem 3.4.3). Thus, $\rho = \rho_1$. In conclusion, $\rho = \max\{\rho_1, q\}$. Therefore, if the order of the entire function f is not an integer, then f must have infinitely many zeros. Combining this with Theorem 3.4.6, we can obtain the following corollary.

Corollary 3.4.10. If $f(z)$ is an entire function of finite order ρ, then for all $\varepsilon > 0$ and $h > \rho$, there exists $vr_0 = r_0(\varepsilon, f) > 1$ such that for $z \in \Omega, |z| = r > r_0$, we have

$$|f(z)| > e^{-r^{\rho_1+\varepsilon}},$$

where Ω is the complement of the closed disk of radius $|z_n|^{-h}$ centered at z_n for $|z_n| > 1$.

Proof. According to Theorem 3.4.7 (Hadamard factorization theorem), $f(z)$ can be expressed as $f(z) = z^m P(z) e^{h(z)}$, where m is the order of f at 0, and $P(z)$ is the canonical product determined by the the sequence $\{z_n\}$ of zeros of $f(z)$, $h(z)$ is a polynomial of degree at most ρ_1. As stated in Theorem 4.3.6, we only need to discuss $h(z)$. Let $h(z)$ be a polynomial of degree q ($q \leqslant \rho$), then there exists a positive constant $A > 0$ such that $|h(z)| \leqslant Ar^q$ ($|z| = r > 1$), so $\operatorname{Re} h(z) \geqslant -Ar^q$. Thus, $|e^{h(z)}| = e^{\operatorname{Re} h(z)} \geqslant e^{-Ar^q}$. By Theorem 3.4.6, for all $\varepsilon > 0$ and $h > \rho$, there exists $r_0 > 1$ such that

$$|f(z)| > e^{-r^{\rho+\varepsilon}} e^{-Ar^q} > e^{-r^{\rho+2\varepsilon}}$$

for all z with $|z| = r > r_0$. $\square$

Example 3.4.11. Let $f(z) = \frac{\sin \pi z}{\pi z}$, its zeros are $\pm 1, \pm 2, \ldots$ ($|z_{\pm n}| = r_{\pm n} = n$), $f(0) = 1$, then $f(z)$ is an entire function. Since

$$\sum_1^\infty \left(\frac{r}{r_n}\right)^{p+1} = 2 \sum_{n=1}^\infty \left(\frac{r}{n}\right)^{p+1},$$

when $p > 0$, this series is convergent. Taking $p = 1$, then

$$P(z) = \prod_{n=1}^\infty E_1\left(\frac{z}{n}\right) \prod_{n=1}^\infty E_1\left(\frac{z}{-n}\right) = \prod_{n=1}^\infty \left(1 - \frac{z^2}{n^2}\right).$$

Therefore,

$$\sin \pi z = \pi z e^{g(z)} \prod_{n=1}^\infty \left(1 - \frac{z^2}{n^2}\right),$$

where $g(z)$ is an entire function and $e^{g(0)} = 1$. The following explains why $\sin \pi z$ is an entire function of finite order. On one hand,

$$|\sin \pi z| = \left|\frac{1}{2i}(e^{i\pi z} - e^{-i\pi z})\right| \leqslant \frac{1}{2} \cdot 2e^{\pi|z|} = e^{\pi r}.$$

On the other hand, if $r > 1$ 时,
$$|\sin \pi i r| = \left|\frac{1}{2}(e^{\pi r} - e^{-\pi r})\right| > \frac{1}{4}e^{\pi r}.$$

Thus,
$$\frac{1}{4}e^{\pi r} < M(r) \leqslant e^{\pi r}$$

for all $r > 1$. Thus, $\rho = 1$. By applying the Hadamard factorization theorem and the properties of $\sin \pi z$, we have $g(z) = Az$. Since $\sin \pi z$ is an odd function, it follows that $A = 0$. Therefore,
$$\sin \pi z = \pi z \prod_{n=1}^{\infty}\left(1 - \frac{z^2}{n^2}\right).$$

As an application of the Hadamard factorization theorem, we prove the following two propositions.

Proposition 3.4.12. *If $f(z)$ is a non-constant entire function of finite order, then $f(z)$ takes on every finite complex number, with at most one exceptional value.*

Proof. Otherwise, suppose $f(z)$ does not take on α or β, $(\alpha \neq \beta)$. $f(z) - \alpha$ is also an entire function of finite order that is not identically zero. By the Hadamard factorization theorem, we have
$$f(z) - \alpha = e^{h(z)},$$
where $h(z)$ is a polynomial. $e^{h(z)}$ does not take on $\beta - \alpha$. Therefore, $h(z)$ does not take on every value of $\log(\beta - \alpha)$, i.e.,
$$h(z) \neq \log|\beta - \alpha| + i\arg(\beta - \alpha) + 2k\pi i \quad (k \in \mathbb{Z}).$$

However, for any integer k, the function $H_k(z) = h(z) - \log|\beta - \alpha| + i\arg(\beta - \alpha) + 2k\pi i$ is also a polynomial. According to the fundamental theorem of algebra, $H_k(z)$ must have zeros. This leads to a contradiction. □

Proposition 3.4.13. *Suppose $f(z)$ is an entire function of finite order ρ and ρ is not an integer. Then, $f(z)$ takes on every finite complex number infinitely many times.*

Proof. If there exists a finite complex number α such that α, $f(z) - \alpha$ has only finitely many zeros, $a_1, a_2, \ldots, a_m$ (with no zeros when $m = 0$), then by the Hadamard factorization theorem,

$$f(z) - \alpha = (z - a_1)(z - a_2) \cdots (z - a_m)e^{h(z)},$$

where $h(z)$ is a polynomial of degree $\deg(h) < \rho$ (it is easy to see that the order of $e^{h(z)}$ is precisely $\deg(h)$, which is an integer, and ρ is not an integer. On the other hand, by the remark in the Hadamard factorization theorem, $f(z)$ and $e^{h(z)}$ have the same order, i.e., $\rho = \deg h$, This leads to a contradiction. $\square$

3.5 The Gamma Function, the Beta Function and the Riemann Zeta Function

Definition 3.5.1 (Γ-function). The function

$$\Gamma(z) = \int_0^\infty e^{-t} t^{z-1} dt \tag{3.5.1}$$

is called the Gamma function.

When $z = x > 0$, it coincides with the function $\Gamma(x)$ in mathematical analysis. Therefore, $\Gamma(z)$ is an extension of $\Gamma(x)$ from the positive real axis to the right half-plane $z : \operatorname{Re} z = x > 0$. Since the integrand $\left|e^{-t}t^{z-1}\right| = e^{-t}t^{x-1}$ $(z = x + iy)$, the integral (3.5.1) is absolutely converging for $\operatorname{Re} z > 0$, the integral $\Gamma(z)$ is well defined and continuous in the right half-plane $\operatorname{Re} z > 0$. Morera's theorem implies that the function $\Gamma(z)$ defined by (3.5.1) is holomorphic in the right half-plane $\operatorname{Re} z > 0$, and

$$\Gamma'(z) = \int_0^\infty e^{-t} t^{z-1} \log t \, dt, \quad \operatorname{Re} z > 0.$$

To derive a functional equation for $\Gamma(z)$, we integrate by parts

$$\Gamma(z+1) = \int_0^\infty e^{-t} t^z dt = z\Gamma(z). \tag{3.5.2}$$

We apply the functional equation n times to obtain

$$\Gamma(n+z) = z(z+1)\cdots(z+n-1)\Gamma(z) \quad (\operatorname{Re} z > 0).$$

In particular,
$$\Gamma(n+1) = n!\Gamma(1)$$
for any positive n, and we declare that $0! = \Gamma(1) = \int_0^\infty e^{-t}dt = 1$, so
$$\Gamma(n+1) = n!. \tag{3.5.3}$$

The functional equation (3.5.2) allows us to rewrite as $\Gamma(z)$ to extend the left plane as follows. (3.5.2) can be rewritten as
$$\Gamma(z) = \frac{\Gamma(z+n)}{z(z+1)\cdots(z+n-1)}.$$

The right-hand side is defined and meromorphic for Re $z > -n$, with simple poles at $z = 0, -1, \ldots, -(n-1)$. By the uniqueness principle, the meromorphic extension is unique and it satisfies the functional equation (3.5.2). Passing to the limit as $n \to \infty$, we see that the function $\Gamma(z)$ extends to be meromorphic on the entire complex plane $\mathbb{C}$, where it satisfies the functional equation $\Gamma(z+1) = z\Gamma(z)$. Its poles are simple poles $z = -n\,(n = 0, 1, 2, \ldots)$, the residue of $\Gamma(z)$ at $z = -n$ is
$$\text{Res}(\Gamma(z), -n) = \frac{(-1)^n}{n!}. \tag{3.5.4}$$

Now, we will express $\Gamma(z)$ in terms of an infinite product. Define
$$f_n(z) = \int_0^n \left(1 - \frac{t}{n}\right)^n t^{z-1}dt \quad (\text{Re}\,z > 0), \tag{3.5.5}$$

the substitution $s = \dfrac{t}{n}$, and we integrate by parts n times, and this yields
$$f_n(z) = n^z \int_0^1 (1-s)^n s^{z-1}ds = \frac{n!n^z}{z(z+1)\cdots(z+n)}. \tag{3.5.6}$$

On the other hand, we shall prove that
$$\lim_{n\to\infty} f_n(z) = \Gamma(z) \quad (\text{Re}\,z > 0). \tag{3.5.7}$$

So, we shall prove that

$$\lim_{n\to\infty}\left[\int_0^n e^{-t}t^{z-1}dt - \int_0^n \left(1-\frac{t}{n}\right)^n t^{z-1}dt\right]$$
$$= \lim_{n\to\infty}\int_0^n e^{-t}t^{z-1}\left[1-e^t\left(1-\frac{t}{n}\right)^n\right]dt = 0$$

for Re$z > 0$. First, we shall prove that

$$0 \leqslant 1 - e^t\left(1-\frac{t}{n}\right)^n \leqslant \frac{et^2}{4n} \qquad (3.5.8)$$

for $0 \leqslant t \leqslant n, n > 1$. In fact, since

$$\left\{e^t\left(1-\frac{t}{n}\right)^n\right\}' = -\frac{1}{n}e^t t\left(1-\frac{t}{n}\right)^{n-1},$$

so

$$1 - e^t\left(1-\frac{t}{n}\right)^n = \frac{1}{n}\int_0^t e^x x\left(1-\frac{x}{n}\right)^{n-1}dx.$$

The function $e^x\left(1-\frac{x}{n}\right)^{n-1}$ attains its maximum at the point $x = 1$,

$$\sup\left\{e^x\left(1-\frac{x}{n}\right)^{n-1} : x \geqslant 0\right\} = e\left(1-\frac{1}{n}\right)^{n-1}$$
$$= \frac{e}{\left(1+\frac{1}{n-1}\right)^{n-1}} \leqslant \frac{e}{2} \quad (n > 1).$$

Thus,

$$1 - e^t\left(1-\frac{t}{n}\right)^n \leqslant \frac{e}{2n}\int_0^t x\,dx = \frac{et^2}{4n} \quad (n > 1).$$

Further, $e^{-x} \geqslant 1 - x$ for $x > 0$, then $e^{-t/n} \geqslant 1 - \frac{t}{n}$, thus $e^{-t} \geqslant \left(1-\frac{t}{n}\right)^n$, therefore

$$1 - e^t\left(1-\frac{t}{n}\right)^n \geqslant 0.$$

Zeros of Holomorphic Functions

The inequality (3.5.8) yields

$$\left| \int_0^n e^{-t} t^{z-1} dt - \int_0^n \left(1 - \frac{t}{n}\right)^n t^{z-1} dt \right|$$

$$= \left| \int_0^n e^{-t} t^{z-1} \left[1 - e^t \left(1 - \frac{t}{n}\right)^n\right] dt \right|$$

$$\leq \frac{e}{4n} \int_0^n e^{-t} t^{x+1} dt \leq \frac{e}{4n} \Gamma(x+2),$$

so

$$\lim_{n \to \infty} \int_0^n e^{-t} t^{z-1} \left[1 - e^t \left(1 - \frac{t}{n}\right)^n\right] dt = 0.$$

Combining (3.5.6) with (3.5.7), we obtain

$$\Gamma(z) = \lim_{n \to \infty} \frac{n! n^z}{z(z+1) \cdots (z+n)}. \tag{3.5.9}$$

Since the function $\Gamma(z)$ extends to be meromorphic on the entire complex plane $\mathbb{C}$, by the uniqueness principle, the meromorphic extension is unique, (3.5.9) holds for $z \neq 0, -1, \ldots$. (3.5.9) is called the Gauss formula.

Since

$$\frac{n! n^z}{z(z+1) \cdots (z+n)} = \frac{\exp\{z \log n\}}{z \left(1 + \frac{z}{1}\right) \left(1 + \frac{z}{2}\right) \cdots \left(1 + \frac{z}{n}\right)}$$

$$= \frac{\exp\{z \log n\}}{z \prod_{k=1}^n \left(1 + \frac{z}{k}\right)} = \frac{e^{z \gamma_n}}{z \prod_{k=1}^n \left(1 + \frac{z}{k}\right) e^{-z/k}},$$

where $\gamma_n = \sum_{k=1}^n \frac{1}{k} - \log n$. The sequence $\{\gamma_n\}$ decreases to Euler's constant ($\gamma = 0.5772\ldots$) as $n \to \infty$. Thus, when passed to the limit, we obtain the following:

$$\frac{1}{\Gamma(z)} = \lim_{n \to \infty} \frac{z(z+1) \cdots (z+n)}{n! n^z} = e^{\gamma z} z \prod_{n=1}^{\infty} \left(1 + \frac{z}{n}\right) e^{-z/n}. \tag{3.5.10}$$

Initially, the identity is established only in the right half-plane. However, both sides are meromorphic, so the identity persists for all

$z \in \mathbb{C}$ by the uniqueness. (3.5.10) is called the Weierstrass formula. The function

$$ze^{\gamma z} \prod_{n=1}^{\infty} \left(1 + \frac{z}{n}\right) e^{-z/n}$$

is an entire function, denoted by $\varphi(z)$, and its zeros are $z = 0, -1, -2, \ldots$, so $\Gamma(z) \neq 0$. Since the entire function $\varphi(z)$ has convergence index $\rho_1 = 1$, and $q = 1$, we have that the order of $\varphi(z) = 1/\Gamma(z)$ $\rho = \max\{\rho_1, q\} = 1$. The Weierstrass formula (3.5.10) implies that

$$\frac{1}{\Gamma(-z)} = -e^{-\gamma z} z \prod_{n=1}^{\infty} \left(1 - \frac{z}{n}\right) e^{z/n},$$

$$\frac{1}{\Gamma(z)\Gamma(-z)} = -z^2 \prod_{n=1}^{\infty} \left(1 - \frac{z^2}{n^2}\right) = \frac{-z}{\pi} \sin \pi z.$$

The (3.5.2) implies that $\Gamma(1-z) = -z\Gamma(-z)$, so

$$\Gamma(z)\Gamma(1-z) = \frac{\pi}{\sin \pi z}, \qquad (3.5.11)$$

In particular, by putting $z = 1/2$, we find that

$$\Gamma\left(\frac{1}{2}\right) = \sqrt{\pi}. \qquad (3.5.12)$$

On the other hand, we have

$$\Gamma\left(\frac{1}{2}\right) = \int_0^{\infty} e^{-t} t^{-1/2} dt = 2 \int_0^{\infty} e^{-x^2} dx,$$

which gives us the familiar probability integral

$$\int_0^{\infty} e^{-x^2} dx = \frac{\sqrt{\pi}}{2}. \qquad (3.5.13)$$

Now, let's discuss the asymptotic properties of $\Gamma(z)$ as $z \to \infty$. Let

$$p_1(t) = t - [t] - 1/2$$

be the sawtooth function, where $[t]$ denotes the largest integer $\leqslant t$. Let $f(t)$ be any $C^1[0, \infty)$, then we need a Euler summation formula:

$$\sum_{k=1}^{n} f(k) = \int_0^n f(t) dt + \frac{1}{2}(f(n) - f(0)) + \int_0^n f'(t) p_1(t) dt. \quad (3.5.14)$$

In fact, note that $p'_1(t) = 1$ for t which is not an integer. Integrating by parts,

$$\int_{k-1}^{k} f'(t)p_1(t)dx = \int_{k-1}^{k} f'(t)\left(t - k + \frac{1}{2}\right)dt$$

$$= \frac{1}{2}[f(k) + f(k-1)] - \int_{k-1}^{k} f(t)dt,$$

we take the sum from $k = 1$ to $k = n$,

$$\int_0^n f'(t)p_1(t)dt = \sum_{k=0}^{n} f(k) - \frac{1}{2}[f(n) + f(0)] - \int_0^n f(t)dt,$$

then yielding the sum

$$\sum_{k=0}^{n} f(k) = \int_0^n f(t)dt + \frac{1}{2}[f(n) + f(0)] + \int_0^n f'(t)p_1(t)dt$$

on the right side and proving the formula.

We now apply the Euler formula to the function $f(t) = \log(t + z)$ and $f(t) = \log(1 + t)$, and assume until further notice that z satisfies $-\pi + \delta \leq \arg z \leq \pi - \delta$ ($\delta > 0$). Then we have no difficulty dealing with the log and its properties from freshman calculus:

$$\log z(z+1)\cdots(z+n)$$

$$= \int_0^n \log(t+z)dt + \frac{1}{2}[\log(z+n) + \log z] + \int_0^n \frac{p_1(t)}{t+z}dt$$

$$= z\log(n+z) - n - z\log z + \left(n + \frac{1}{2}\right)\log(z+n)$$

$$+ \frac{1}{2}\log z + \int_0^n \frac{p_1(t)}{t+z}dt, \qquad (3.5.15)$$

and taking $z = 1$,

$$\log n! = n\log(n+1) - n + \frac{1}{2}\log(1+n) + \int_0^n \frac{p_1(t)}{1+t}dt. \qquad (3.5.16)$$

Subtracting the expressions for these, (3.5.15) and (3.5.16),

$$\log \frac{z(z+1)\cdots(z+n)}{n!n^z} = z\log(n+z) + \left(n+\frac{1}{2}\right)\log(z+n) - z\log z$$
$$- \left(n+\frac{1}{2}\right)\log(1+n) + \frac{1}{2}\log z + \int_0^n \frac{p_1(t)}{1+t}dt$$
$$- \int_0^n \frac{p_1(t)}{z+t}dt - z\log n.$$

None of this is so bad. We write

$$z\log(z+n) = z\log n + z\log\left(1+\frac{z}{n}\right)$$
$$= z\log n + O\left(\frac{1}{n}\right), \quad \left(n+\frac{1}{2}\right)\log\left(\frac{n+z}{n+1}\right)$$
$$= \left(n+\frac{1}{2}\right)\log\left(1+\frac{z-1}{n+1}\right) = z - 1 + O\left(\frac{1}{n}\right).$$

By Gauss formula (3.5.9),

$$\log \Gamma(z) = \left(z - \frac{1}{2}\right)\log z - (z-1) + \int_0^\infty \frac{p_1(t)}{1+t}dt - \int_0^\infty \frac{p_1(t)}{t+z}dt.$$
(3.5.17)

Let $p_2(s) = \int_0^s p_1(t)dt = \frac{1}{2}(s^2-s)$ for $0 \leq s < 1$ and extend $p_2(s)$ by periodicity to all of $\mathbb{R}$ (period 1). Then function $p_2(s)$ is a continuous function with the period 1, $-1/8 \leq p_2(s) \leq 0$. Furthermore, $p_2'(s) = p_1(s)$. Integrating by parts,

$$\left|\int_0^\infty \frac{p_1(t)}{t+z}dt\right| = \left|\int_0^\infty \frac{p_2(t)}{(t+z)^2}dt\right| \leq \frac{1}{8}\int_0^\infty \frac{dt}{(t+x)^2+y^2}.$$

where $z = x + iy$, so

$$\left|\int_0^\infty \frac{p_2(t)}{(t+z)^2}dt\right| \leq \frac{1}{8|y|}\left(\frac{\pi}{2} - \arctan\left(\frac{x}{|y|}\right)\right) \leq \frac{\pi}{8|y|} \quad (y \neq 0).$$
(3.5.18)

Since $0 \leqslant t\left(\frac{\pi}{2} - \arctan t\right) \leqslant 1$ $(t \geqslant 0)$,

$$\left|\int_0^\infty \frac{p_2(t)}{(t+z)^2} dt\right| \leqslant \frac{1}{8x} \quad (x > 0).$$

The constant in the first integral in (3.5.17) remains to be evaluated; by taking $z = iy$ in (3.5.17),

$$1 + \int_0^\infty \frac{p_1(t)}{1+t} dt = \log|\Gamma(iy)| + \frac{1}{2}\log y + \frac{\pi}{2}y + \operatorname{Re}\int_0^\infty \frac{p_1(t)dx}{t+iy}, \tag{3.5.19}$$

$$1 + \int_0^\infty \frac{p_1(t)}{1+t} dt = \lim_{y\to+\infty}\left(\log|\Gamma(iy)| + \frac{1}{2}(\log y + \pi y)\right) \tag{3.5.20}$$

by combining (3.5.18) and letting $y \to \infty$. On the other hand, by (3.5.11),

$$\Gamma(iy)\Gamma(-iy) = \frac{\pi}{y\sin\pi yi},$$

and note that $\overline{\Gamma(z)} = \Gamma(\bar{z})$, we get

$$|\Gamma(iy)|^2 = \frac{2\pi}{y(e^{\pi y} - e^{-\pi y})},$$

so

$$\log|\Gamma(iy)| + \frac{1}{2}\log y + \frac{\pi}{2}y = \frac{1}{2}\log(2\pi) - \frac{1}{2}\log(1 - e^{-2\pi y}).$$

By (3.5.20),

$$1 + \int_0^\infty \frac{p_1(t)}{1+t} dt = \frac{1}{2}\log(2\pi). \tag{3.5.21}$$

Finally, from (3.5.17), (3.5.18) and (3.5.21) we get

$$\log\Gamma(z) = \left(z - \tfrac{1}{2}\right)\log z - z + \tfrac{1}{2}\log(2\pi) - \int_0^\infty \frac{p_2(t)}{(t+z)^2} dt \tag{3.5.22}$$

and

$$\log\Gamma(z) = \left(z - \frac{1}{2}\right)\log z - z + \frac{1}{2}\log(2\pi) + O\left(\frac{1}{|z|}\right) \quad (|\arg z| \leqslant \pi - \delta), \tag{3.5.23}$$

where O is dependent only δ. These two formulas can be called the Stirling formula. By (3.5.23), $|\Gamma(z)| = O(e^{|z|\log|z|})$. Furthermore, this order is exact, emphasizing the precision of the result. This once again highlights the fact that the entire function $1/\Gamma(z)$ is one of order 1.

Definition 3.5.2 (Definition of Beta Function). The function

$$B(a,b) = \int_0^1 x^{a-1}(1-x)^{b-1}dx \quad (a,b \in \mathbb{R}) \tag{3.5.24}$$

is called the beta function, the improper integral converges and absolutely for $\operatorname{Re} a > 0, \operatorname{Re} b > 0$. The Γ function and and the beta function are called Euler's integral.

The main result of Γ and beta function is Euler's identity.

Theorem 3.5.3.

$$B(a,b) = \frac{\Gamma(a)\Gamma(b)}{\Gamma(a+b)} \tag{3.5.25}$$

for $\operatorname{Re} a > 0$, $\operatorname{Re} b > 0$.

Proof. First, we have

$$\Gamma(a) = \int_0^\infty x^{a-1}e^{-x}dx$$

for $\operatorname{Re} a > 1$, $\operatorname{Re} b > 1$. Substituting $ty(t > 0)$ for x, we find that

$$\frac{\Gamma(a)}{t^a} = \int_0^\infty y^{a-1}e^{-ty}dy \quad (t > 0),$$

substituting a for $a+b$, t for $t+1$, we obtain that

$$\frac{\Gamma(a+b)}{(1+t)^{a+b}} = \int_0^\infty y^{a+b-1}e^{-(1+t)y}dy,$$

Multiplying by t^{a-1} in the equality, integral t from 0 to ∞,

$$\Gamma(a+b)\int_0^\infty \frac{t^{a-1}}{(1+t)^{a+b}}dt = \int_0^\infty \left(\int_0^\infty t^{a-1}y^{a+b-1}e^{-(1+t)y}dy\right)dt.$$

The integration in the left-hand equality

$$\int_0^\infty \frac{t^{a-1}}{(1+t)^{a+b}} dt = B(a,b),$$

Substituting $\frac{t}{1+t}$ for x, $B(a,b) = \int_0^1 x^{a-1}(1-x)^{b-1}dx$, reversing the order of integration is legitimate here by Fubini's theorem:

$$\int_0^\infty \int_0^\infty t^{a-1}y^{a+b-1}e^{-(1+t)y}dydt = \int_0^\infty \int_0^\infty t^{a-1}y^{a+b-1}e^{-(1+t)y}dtdy$$

$$= \int_0^\infty y^{a+b-1}e^{-y}\frac{\Gamma(a)}{y^a}dy = \Gamma(a)\int_0^\infty y^{b-1}e^{-y}dy = \Gamma(a)\Gamma(b).$$

For the general case where $\operatorname{Re} a > 0$ and $\operatorname{Re} b > 0$, by the previous conclusion, we have

$$B(a+1, b+1) = \frac{\Gamma(a+1)\Gamma(b+1)}{\Gamma(a+b+2)}.$$

From the recurrence relations of Γ and B, we have

$$\frac{\Gamma(a+1)\Gamma(b+1)}{\Gamma(a+b+2)} = \frac{a\Gamma(a)b\Gamma(b)}{(a+b+1)(a+b)\Gamma(a+b)}$$

and

$$(a+b+1)B(a+1, b+1) = aB(a, b+1) = \frac{ab}{a+b}B(a,b).$$

From this, we immediately deduce the validity of (3.5.25). □

We know that the series $\sum_{n=1}^\infty n^{-z}$ uniformly converges on $z : x = \operatorname{Re} z \geqslant x_0 > 1$. Note that

$$\left|n^{-z}\right| = \left|e^{-z\log n}\right| = n^{-x},$$

The series $\sum_{n=1}^\infty n^{-z}$ converges absolutely for $\operatorname{Re} z > 1$.

Definition 3.5.4 (Riemann Zeta Function). The Riemann zeta function is defined by

$$\zeta(z) = \sum_{n=1}^\infty n^{-z} \tag{3.5.26}$$

for $\operatorname{Re} z > 1$. The series converges absolutely for $x = \operatorname{Re} z > 1$ and it converges uniformly for $x = \operatorname{Re} z \geqslant 1+\varepsilon$ for any $\varepsilon > 0$. Next, we wish to extend $\zeta(z)$ to be a meromorphic function on the entire complex plane. To do this, we represent $\zeta(z)$ as a contour integral. Performing a change of variable in (3.5.1) by letting $t = nu$ gives

$$n^{-z}\Gamma(z) = \int_0^\infty t^{z-1} e^{-nt}\,dt,$$

If $x = \operatorname{Re} z > 1$ and we sum this equation over all positive n, then

$$\zeta(z)\Gamma(z) = \int_0^\infty \frac{t^{z-1}dt}{e^t - 1} = \int_{-\infty}^\infty g_z(s)ds. \qquad (3.5.27)$$

It is straightforward to justify interchanging the summation and integration for $x > 1$, where

$$g_z(\zeta) = \frac{e^{z\zeta}}{\exp\{e^\zeta\} - 1}.$$

Let $R > 0$, $D_R = \{\zeta : 0 < \operatorname{Im}\zeta < 2\pi, \operatorname{Re}\zeta > -R\}$ a semi-band of width 2π, whose contour of boundary of semi-band ∂D_R indicated below, following the top edge $L_{1,R}^-$ of the semi-band from $+\infty$ to $-R$, detouring along the interval I_R, and returning from $-R$ to $+\infty$ along the bottom edge $L_{2,R}$ of the semi-band in the half line, where $L_{1,R} = \{s+2\pi i : s \geqslant -R\}$, $L_{2,R} = \{s : s \geqslant -R\}$ and $I_R = \{-R+it : 0 \leqslant t \leqslant 2\pi\}$. We consider the function

$$F_R(z) = \int_{\partial D_R} g_z(\zeta)d\zeta.$$

The integral converges, and it represents an entire function of z independent on $R > 0$, and

$$F_R(z) = -i\int_0^{2\pi} g_z(-R+it)dt + (1 - e^{2\pi i z})\int_{-R}^\infty g_z(s)ds.$$

Passing to the limit, we obtain

$$F(z) = \lim_{R \to \infty} F_R(z)$$

is an entire function of z, which satisfies $F(n) = 0$, $n = 2, 3, \ldots$, $F(1) = -2\pi i$ and

$$F(z) = (1 - e^{2\pi i z})\int_{-\infty}^\infty g_z(s)ds$$

for Re $z > 1$. (3.5.27) implies that $F(z) = (1 - e^{2\pi i z})\zeta(z)\Gamma(z)$. Using $\Gamma(z)\Gamma(1-z) = \dfrac{\pi}{\sin \pi z}$,

$$F(z) = \frac{-2\pi i e^{\pi i z}\zeta(z)}{\Gamma(1-z)} \tag{3.5.28}$$

for $z:$ Re $z > 1$, so we can use this identity to extend $\zeta(z)$ to be a meromorphic function in the entire complex plane. The extended function is holomorphic except for possibly pole at $z = 1$ with residue 1. Now, we can calculate the values of $\zeta(z)$ at zero and negative integer points $\zeta(-n)$. In fact, from (3.5.28),

$$\zeta(-n) = \frac{(-1)^{n+1}n!F(-n)}{2\pi i} = \frac{(-1)^n n!}{2\pi i} \lim_{R\to\infty} i \int_0^{2\pi} g_{-n}(-R+it)dt.$$

From this, we can see that

$$\zeta(-n) = \frac{(-1)^n n!}{2\pi i}\int_{|z|=\varepsilon} \frac{dz}{(e^z-1)z^{n+1}} \quad (0 < \varepsilon < 1)$$

is equal to the coefficient of z^n in the Laurent expansion of the function $\dfrac{1}{e^z - 1}$ at $z = 0$, multiplied by $(-1)^n n!$. To do this, consider the Laurent expansion of

$$\frac{1}{e^z - 1} = \frac{1}{z} - \frac{1}{2} + \sum_{k=1}^{\infty}(-1)^{k-1}\frac{B_k}{(2k)!}z^{2k-1} \tag{3.5.29}$$

for some constants $B_1, B_2, B_3, \ldots$. The constants B_k is called the Bernoulli constants. We can calculate several terms of this expression, e.g.,

$$\zeta(0) = -\frac{1}{2}, \quad \zeta(-2n) = 0, \quad \zeta(-2n+1) = (-1)^n \frac{B_n}{2n},$$

where n is positive integer. These points $z = -2, -4, -6, \ldots$ are called the trivial zeros of ζ function.

Next, we derive a functional equation for the Riemann ζ function.

$$\zeta(z) = 2^z \pi^{z-1} \sin \tfrac{\pi z}{2} \Gamma(1-z)\zeta(1-z). \tag{3.5.30}$$

For the proof, we modify the contour of boundary of semi-band ∂D_R to define $F(z)$. Fix z be real and negative. Let

$$\Omega_{R,n} = \{\zeta : 0 < \text{Im } \zeta < 2\pi, -R < \text{Re } \zeta < \ln(2n\pi + \pi)\}$$

be a rectangle with the width 2π, the length $\ln(2n\pi + \pi) + R$ and vertices at $-R + i2\pi, -R, \ln(2n\pi + \pi)$ and $\ln(2n\pi + \pi) + i2\pi$, $(R > 0$, $n = 1, 2, \ldots)$. Define

$$F_{R,n}(z) = \int_{\partial \Omega_{R,n}} g_z(\zeta) d\zeta.$$

The contour $\partial \Omega_{R,n}$ encloses only the poles of the integrand at the point $\ln(2k\pi) + i(\pi \pm \frac{\pi}{2})$ $(1 \leqslant k \leqslant n)$. Each pole is simple pole of $g_z(\zeta)$, with residue

$$\left(2k\pi e^{(\pi \pm \frac{\pi}{2})i}\right)^{z-1}.$$

We combine the residue for $k(1 \leqslant k \leqslant n)$ and sum and obtain from the residue theorem

$$F_{R,n}(z) = 2\pi i \sum_{k=1}^{n} \left(\left(2k\pi e^{\frac{\pi i}{2}}\right)^{z-1} + \left(2k\pi e^{\frac{3\pi i}{2}}\right)^{z-1}\right) \quad (3.5.31)$$

$$= -4\pi i e^{\pi i z} \sin \frac{\pi z}{2} \sum_{k=1}^{n} (2k\pi)^{z-1}.$$

We have $|\exp(e^{\zeta}) - 1| \geqslant \frac{1}{2}$ on $C_n = \{ln(2n\pi + it : 0 \leqslant t \leqslant 2\pi\}$ on the edge $C_n = \{\ln(2n\pi + \pi) + it : 0 \leqslant t \leqslant 2\pi\}$ of the rectangle. In fact, the function

$$\varphi(t) = |\exp\{(2n+1)\pi e^{it}\} - 1|, t \in \mathbb{R},$$

satisfies that

$$\varphi(t + 2\pi) = \varphi(t), \ \varphi(-t) = \varphi(t) \text{ and } \varphi(\pi - t)$$
$$= \exp\{-(2n+1)\pi \cos t\}\varphi(t).$$

So,

$$\min\{\varphi(t) : t \in \mathbb{R}\} = \min \left\{\varphi\left(t + \frac{\pi}{2}\right) : t \in \left[0, \frac{\pi}{2}\right]\right\}$$

and $-(2n\pi + \pi) \sin t \leqslant -\log 2$ for $2t(2n+1) \geqslant \log 2, t \leqslant \frac{\pi}{2}$, thus

$$\varphi\left(t + \frac{\pi}{2}\right) = |1 - \exp\{(2n+1)\pi(-\sin t + i \cos t)\}| \geqslant 1 - \frac{1}{2} = \frac{1}{2}$$

and
$$(2n+1)\pi(1-\cos t) = 2(2n+1)\pi \sin^2 \frac{t}{2} \leqslant \frac{\pi}{2}$$
for $0 \leqslant 2t(2n+1) \leqslant \log 2$, thus
$$\varphi\left(t+\frac{\pi}{2}\right) = |1+\exp\{(2n+1)\pi(-\sin t + i(\cos t - 1))\}|$$
$$\geqslant 1 + \exp\{-(2n+1)\pi \sin t\}\cos\{(2n+1)\pi(\cos t - 1)\} \geqslant 1,$$
$$|e^{z\zeta}| \leqslant \left(n+\frac{1}{2}\right)^x \exp\{2\pi|y| + x\log(2\pi)\}$$
for $\zeta \in C_n = \{\ln(2n\pi + \pi) + it : 0 \leqslant t \leqslant 2\pi\}$. Therefore,
$$\left|\int_{C_n} \frac{e^{z\zeta}d\zeta}{\exp(e^\zeta)-1}\right| \leqslant 4\pi\left(n+\frac{1}{2}\right)^x \exp\{2\pi|y| + x\log(2\pi)\},$$
and this tends to 0 as $n \to \infty$ and the left-hand side of (3.5.31) tends to
$$F_R(z) = F(z) = \frac{-2\pi i \zeta(z)e^{\pi i z}}{\Gamma(1-z)},$$
and the series
$$\sum_{k=1}^{\infty} k^{z-1}$$
converges to $\zeta(1-z)$ for $x = \operatorname{Re} z < 0$. We obtain the functional equation (3.5.30) at least for $x = \operatorname{Re} z < 0$ by letting $n \to \infty$. Since both sides of the identity (3.5.31) are meromorphic, (3.5.30) holds for all z by the uniqueness principle. From the functional equation (3.5.30), we can determine $\zeta(2n)$. In fact, due to
$$2\sin\frac{\pi z}{2}\cos\frac{\pi z}{2}\Gamma(z)\Gamma(1-z) = \pi,$$
$$\zeta(2n) = 2^{2n}\pi^{2n-1}\zeta(1-2n)\lim_{z\to 2n}\left(\sin\frac{\pi z}{2}\Gamma(1-z)\right)$$
$$= 2^{2n}\pi^{2n-1}\frac{(-1)^n B_n}{2n}\frac{(-1)^n \pi}{2(2n-1)!} = \frac{2^{2n-1}\pi^{2n}B_n}{(2n)!}.$$

Following is the proof of Legendre's duplication formula:

$$\sqrt{\pi}\Gamma(2z) = 2^{2z-1}\Gamma(z)\Gamma\left(z + \frac{1}{2}\right). \tag{3.5.32}$$

In fact, putting

$$\varphi(z) = \frac{\Gamma(z)\Gamma\left(z + \frac{1}{2}\right)}{\Gamma(2z)},$$

then

$$\frac{\Gamma'(z)}{\Gamma(z)} = -\frac{1}{z} - \gamma + \sum_{n=1}^{\infty}\left(\frac{1}{n} - \frac{1}{z+n}\right),$$

so

$$\left(\frac{\Gamma'(z)}{\Gamma(z)}\right)' = \frac{1}{z^2} + \sum_{n=1}^{\infty}\frac{1}{(z+n)^2}.$$

We have

$$\left(\frac{\varphi'(z)}{\varphi(z)}\right)' = 0,$$

hence there are two constants α and β such that $\varphi(z) = \beta e^{\alpha z}$. Since

$$\varphi(1) = \frac{\Gamma(1+1/2)}{\Gamma(2)} = \frac{\sqrt{\pi}}{2}, \quad \varphi(2) = \frac{\Gamma(2+1/2)}{\Gamma(4)} = \frac{\sqrt{\pi}}{8},$$

then $e^{-\alpha} = 4, \beta = 2\sqrt{\pi}$. Therefore, $\Gamma(z)\Gamma\left(z + \frac{1}{2}\right) = 2\sqrt{\pi}2^{-2z}\Gamma(2z)$. This proves (3.5.32).

Theorem 3.5.5 (Euler's Theorem). *If* $\operatorname{Re} z > 1$, *then Riemann zeta has the representation of the follow Euler product*

$$\zeta(z) = \prod_{n=1}^{\infty}\frac{1}{\left(1 - \frac{1}{p_n^z}\right)}, \tag{3.5.33}$$

where $\{p_n\}$ is the sequence of prime numbers $2 = p_1 < p_2 = 3 < p_4 = 5 < \cdots < p_n < p_{n+1} < \cdots$.

Proof. The series $\sum_{n=1}^{\infty}\frac{1}{(p_n)^z}$ converges absolutely for $\operatorname{Re} z > 1$, and it converges uniformly any half-plane $\{z : \operatorname{Re} z \geqq 1 + \varepsilon\}$ for any $\varepsilon > 0$. Hence, the product (3.5.33)

$$\zeta(z)(1-2^{-z}) = \sum_{n=1}^{\infty} \frac{1}{n^z} - \sum_{n=1}^{\infty} \frac{1}{(2n)^z} = \sum_{m \in \Theta_1} \frac{1}{m^z},$$

where Θ_1 is the set of all positive odd integers more than 2. Similarly,

$$\zeta(z)(1-2^{-z})(1-3^{-z}) = \sum_{m \in \Theta_2} \frac{1}{m^z},$$

where Θ_2 is the set of all positive integers which cannot be divided by either 2 or 3. Furthermore,

$$\zeta(z)(1-2^{-z})(1-3^{-z})\cdots(1-p_N^{-z}) = \sum_{m \in \Theta_N} \frac{1}{m^z},$$

where Θ_N is the set of all positive integers which cannot be divided by any of the primes $2, 3, \ldots, p_N$. We know that there are infinitely many prime numbers. In fact, if there were only finitely many primes, let's say $2 = p_1, 3 = p_2, \ldots, p_N$ (with 2 being the first prime and p_N being the Nth prime), then by definition of prime numbers, $p_N! + 1$ would be divisible by one of the numbers in the set $\{p_1, p_2, \ldots, p_N\}$. Consequently, 1 would also be divisible by one of the numbers in the set $\{p_1, p_2, \ldots, p_N\}$, which is impossible. Therefore, we conclude that there must be infinitely many prime numbers. Since every integer $n \geqslant 1$ has a unique representation as a product of powers of distinct primes, a summand $\frac{1}{n^z}$ appears at most once in this sum, the sum is a subsum of the series $\sum_{n \geqslant 1} \frac{1}{n^z}$. As we incorporate more primes into the product, we eventually capture all terms n^{-z}, and in limit, we have (3.5.33). □

The functional equation (3.5.30) show that the zeros of $\zeta(z)$ in the left half-plane $\operatorname{Re} z < 0$ are the zeros of $\sin(z\pi/2)$, which are simple zeros at $z = -2, -4, -6, \ldots$. As mentioned above, these zeros are called the **trivial zeros** of ζ function. The nontrivial zeros of the zeta function $\zeta(z)$ lie in the strip $0 \leqslant \operatorname{Re} z \leqslant 1$, which is called the **critical strip**. Many outstanding mathematicians have worked on the **Riemann hypothesis** [29] that the non-trival zeros of the zeta function $\zeta(z)$ lies on the critical line $\operatorname{Re} z = \frac{1}{2}$. It is known that ζ has infinitely many zeros in the critical strip. Meanwhile, whether the Riemann hypothesis is true remains a famous unresolved problem.

Exercises III

1. **(Carathéodory Inequality).** Using the lemma, maximum modulus principle, and linear transformation, the following conclusion can be translated as follows: If the function $f(z)$ is analytic in the disk $D(0, R) = \{z : |z| < R\}$ and continuous in the closed disk $\overline{D(0, R)} = \{z : |z| \leqslant R\}$, where $M(r)$ and $A(r)$ are the maximum values of $|f(z)|$ and Re $f(z)$, respectively, on the circle $|z| = r$ $(0 \leqslant r \leqslant R)$, prove the following:

   (a) $M(r)$ and $A(r)$ are monotonically increasing on $[0, R]$.
   (b) When $0 < r < R$,
   $$M(r) \leqslant \frac{2r}{R-r} A(R) + \frac{R+r}{R-r} |f(0)|.$$
   (c) When $0 < r < R$,
   $$A(r) \leqslant \frac{2r}{R+r} A(R) + \frac{R-r}{R+r} A(0).$$
   (d) If $f(z)$ is nonzero everywhere, then for $0 < r < R$,
   $$\log M(r) \leqslant \frac{R-r}{R+r} \log M(0) + \frac{2r}{R+r} \log M(R).$$

2. Let $\{a_n\}$ and $\{b_n\}$ be two sequences of complex numbers such that $\sum\limits_{n=1}^{\infty} |a_n - b_n| < \infty$. Explain where the infinite product
$$\prod_{n=1}^{\infty} \frac{z - a_n}{z - b_n}$$
uniformly converges and where it represents an analytic function.

3. Expand the following function into an infinite product:
$$\cos \pi z, \quad (e^z - e^{iz}) \sin z, \quad \frac{e^z - e^{-z}}{2}, \quad \frac{e^z + e^{-z}}{2} - \cos z.$$

4. If $f(z)$ is an entire function of finite order ρ and $g(z)$ is a polynomial, show that the order of $g(z)f(z)$ is also ρ.

5. Construct an entire function $f(z)$ such that the set of zeros of $f(z)$ is $\{n^2 : n = 0, 1, 2, \ldots\}$ and each zero is a simple zero.

6. Let $f(z)$ be an entire function such that for any $a \in \mathbb{C}$, there exists a positive integer $n = n(a)$ such that $f^{(n)}(a) = 0$. Prove that $f(z)$ is a polynomial.
7. Let $f(z)$ be an entire function with $f(0) \neq 0$. Let $\{z_n\}$ be the set of zeros (counted with multiplicity) of $f(z)$, and let $n(r)$ denote the number of $\{z_n\}$ such that $|z_n| \leqslant r$. Show that
$$M(f,r) \geqslant \frac{|f(0)|r^{n(r)}}{|z_1||z_2|\cdots|z_{n(r)}|}.$$
8. Let $f(z)$ be an entire function with $f(0) \neq 0$. Let $\{z_n\}$ be the set of zeros (counted with multiplicity) of $f(z)$, and let $n(r)$ denote the number of $\{z_n\}$ such that $|z_n| \leqslant r$. Show that
$$-(\log \delta)n(\delta r) \leqslant \log M(f,r) - \log|f(0)|.$$
9. Let $f(z) = \sum_{n=0}^{\infty} a_n z^n$ be a non-constant entire function with $f(0) = a_0 \neq 0$ and $\lambda > 0$ such that $|f(z)| \leqslant e^{|z|^\lambda}$ ($|z| \geqslant 1$). Let $\{z_n\}$ be the set of zeros (counted with multiplicity) of $f(z)$, and let $n(r)$ denote the number of $\{z_n\}$ such that $|z_n| \leqslant r$. Show that there exists $n_0 \geqslant 1$ such that
$$|a_n| \leqslant \left(\frac{e\lambda}{n}\right)^{n/\lambda} \quad (n \geqslant n_0) \quad \text{and} \quad \limsup_{r\to\infty} \frac{\log n(r)}{\log r} \leqslant \lambda.$$
10. (a) Let $f(z)$ be an entire function such that for any $\mathbb{C}$,
$$|f(z)| \leqslant A + B|z|^k,$$
where A, B and k are positive constants. Show that $f(z)$ is a polynomial of degree not exceeding k.
    (b) Let $f(z)$ and $g(z)$ be entire functions such that for any $z \in \mathbb{C}$, $|f(z)| \leqslant |g(z)|$. Based on this inequality regarding the zeros of $f(z)$ and $g(z)$, what conclusions can you draw?
11. Find the order of the following entire functions:

(a) $\sum_{n=1}^{\infty} \frac{z^n}{n^{\alpha n}} \ (\alpha > 0)$,   (c) $\prod_{n=1}^{\infty}\left(1 + \frac{z}{n^p}\right) \ (p > 1)$,

(b) $\sum_{n=0}^{\infty} \frac{z^n}{(n!)^\beta} \ (\beta > 0)$,   (d) $\prod_{n=1}^{\infty}\left(1 + \frac{z}{p^n}\right) \ (p > 1)$.

12. Find the order, type, and convergence exponent of the entire function $\cos\sqrt{z}$.

13. Suppose $f(z)$ and $g(z)$ are entire functions of exponential type less than 2 and satisfy the following conditions: $(f(z))^2 + (g(z))^2 \equiv 1$. Show that there are two complex numbers α and β such that $f(z) = \cos(\alpha z + \beta), g(z) = \sin(\alpha z + \beta)$.

14. Let $f(z)$ be a non-constant entire function that does not take on complex values α. Show that there is a sequence of complex number $\{z_n\}$ such that
$$\lim_{n\to\infty} f(z_n) = \alpha.$$

15. Show that for all given real number x, when $|y| \to +\infty$,
$$|\Gamma(x+iy)| \sim \sqrt{2\pi}\exp\left\{-\frac{\pi}{2}|y| + \left(x-\frac{1}{2}\right)\log|y|\right\}.$$

16. Show that
$$\frac{\Gamma'(z)}{\Gamma(z)} = -\frac{1}{z} - \gamma + \sum_{n=1}^{\infty}\left(\frac{1}{n} - \frac{1}{z+n}\right),$$
$$\left(\frac{\Gamma'(z)}{\Gamma(z)}\right)' = \frac{1}{z^2} + \sum_{n=1}^{\infty}\frac{1}{(z+n)^2}.$$

17. Show that, for all z with $\operatorname{Re} z > 0$,
$$\frac{\Gamma'(z)}{\Gamma(z)} = \int_0^{\infty}\left(\frac{e^{-t}}{t} - \frac{e^{-zt}}{1-e^{-t}}\right)dt.$$

18. Prove Legendre's duplication formula for $(n = 2, 3, \ldots)$
$$(2\pi)^{(n-1)/2}n^{-nz+1/2}\Gamma(nz) = \Gamma(z)\Gamma\left(z+\frac{1}{n}\right)\cdots\Gamma\left(z+\frac{n-1}{n}\right).$$

19. Show that when $\operatorname{Re}\alpha < 1$ and α are not integers, then
$$\sum_{n=0}^{\infty}(-1)^n\frac{\Gamma(\alpha+n)}{\Gamma(1+n)} = 2^{-\alpha}\Gamma(\alpha).$$

20. Show that when $0 < \operatorname{Re} z < 1$,
$$\zeta(z)\Gamma(z) = \int_0^{\infty}\left(\frac{1}{e^t-1} - \frac{1}{t}\right)t^{z-1}dt.$$

21. Show that the function $\xi(z)$ defined by

$$\xi(z) = \frac{1}{2}z(z-1)(\pi)^{-z/2}\zeta(z)\Gamma(z/2)$$

is an entire function and satisfies the equation $\xi(1-z) = \xi(z)$.

22. Show that the function $(1 - 2^{1-z})\zeta(z)$ is an entire function and when $\operatorname{Re} z > 1$,

$$(1 - 2^{1-z})\zeta(z) = 1 - \sum_{n=2}^{\infty} \frac{(-1)^n}{n^z}.$$

Chapter 4

Harmonic and Subharmonic Functions

4.1 Property of Harmonic Functions

In this chapter, Ω denotes a region in $\mathbb{C} = \mathbb{R}^2$, i.e., a connected open subset of $\mathbb{C}$.

Definition 4.1.1. Suppose that u is a complex function in a plane open set Ω such that $u \in C^2(\Omega)$, and

$$\Delta u = \frac{\partial^2 u}{\partial x^2} + \frac{\partial^2 u}{\partial y^2} = 0 \qquad (4.1.1)$$

at every point $z = x + iy \in \Omega$, then u is said to be harmonic in Ω, where $\Delta = \frac{\partial^2}{\partial x^2} + \frac{\partial^2}{\partial y^2}$ is called the Laplace operator.

Since

$$\frac{\partial}{\partial z} = \frac{1}{2}\left(\frac{\partial}{\partial x} - i\frac{\partial}{\partial y}\right), \quad \frac{\partial}{\partial \bar{z}} = \frac{1}{2}\left(\frac{\partial}{\partial x} + i\frac{\partial}{\partial y}\right),$$

$$\Delta = \frac{\partial^2}{\partial x^2} + \frac{\partial^2}{\partial y^2} = 4\frac{\partial^2}{\partial z \partial \bar{z}}. \qquad (4.1.2)$$

We know that a function $f(z) = u(x, y) + iv(x, y)$ (where $z = x + iy$) is holomorphic in an open set Ω if and only if both its real part $u(x, y)$ and its imaginary part $v(x, y)$ in $C^2(\Omega)$ and the Cauchy–Riemann equations $\frac{\partial f}{\partial \bar{z}} = 0$, or

$$\frac{\partial u}{\partial x} = \frac{\partial v}{\partial y}, \quad \frac{\partial u}{\partial y} = -\frac{\partial v}{\partial x} \qquad (4.1.3)$$

holds for every $z \in \Omega$. Therefore, if the function $f(z) = u(x,y) + iv(x,y)$ is holomorphic in an open set Ω, then both $u(x,y)$ and $v(x,y)$ are harmonic in Ω. Its real part $u(x,y)$ and its imaginary part $v(x,y)$ in $C^2(\Omega)$ satisfy the Cauchy–Riemann equations (4.1.3), then the function v is called the harmonic conjugate of u.

Suppose $u(x,y)$ is harmonic in simply connected regions Ω, is there another harmonic function $v(x,y)$ in Ω such that $f(z) = u(x,y) + iv(x,y)$ is holomorphic in Ω? The answer is affirmative.

Theorem 4.1.2. *If $u(x,y)$ is harmonic in simply connected regions Ω, there is another harmonic function $v(x,y)$ in Ω such that $f(z) = u(x,y) + iv(x,y)$ is holomorphic in Ω.*

Proof. Suppose $u(x,y)$ is harmonic in Ω and $\frac{\partial u}{\partial z}$ is holomorphic in Ω. Since Ω is simply connected, there is a holomorphic function $g(z)$ in Ω such that $g'(z) = \frac{\partial u}{\partial z}$, thus $u - 2\text{Re}\, g$ is a constant a in Ω. Therefore, there is a harmonic function $v(x,y) = 2\text{Im}\, g(z)$ in simply connected regions Ω such that $f(z) = 2g(z) + a = u(x,y) + iv(x,y)$ is holomorphic in Ω. $\square$

For analytic functions, we have the important Cauchy formula, which allows us to determine the value of an analytic function at any point in a region based on its values on certain points in that region. Now, we extend this idea to the case of a disk-shaped region and derive a similar formula for harmonic functions.

Theorem 4.1.3. *Let $0 < R < \infty$, suppose $u(z) = u(x,y)$ is a continuous real function on the closed disk $|z| \leqslant R$, and suppose u is harmonic in the open disk $|z| < R$, then*

$$u(z) = \frac{1}{2\pi} \int_0^{2\pi} u(Re^{i\theta}) P(Re^{i\theta}, z) d\theta \qquad (4.1.4)$$

for each z with $|z| < R$, where ($\zeta = Re^{i\theta}$, $z = Re^{i\varphi}$):

$$P(\zeta, z) = \text{Re}\left(\frac{\zeta + z}{\zeta - z}\right) = \frac{|\zeta|^2 - |z|^2}{|\zeta - z|^2} = \frac{R^2 - r^2}{R^2 - 2Rr\cos(\theta - \varphi) + r^2}.$$

$P(\zeta, z)$ is called Poisson kernel in $|z| < R$, (4.1.4) is called the Poisson formula.

Proof. By Theorem 4.1.2, We can construct a function $v(x, y)$ such that $f(z) = u + iv$ is holomorphic in $|z| < R$, $0 < \rho < R$. By Cauchy formula,

$$f(z) = \frac{1}{2\pi i} \int_{|\zeta|=\rho} \frac{f(\zeta)}{\zeta - z} d\zeta, \quad |z| < \rho. \tag{4.1.5}$$

Let $z^* = \rho^2/\bar{z}$ be the point symmetric to z with respect to the circle $|z| = \rho$, then $|z^*| > \rho$, thus

$$\frac{1}{2\pi i} \int_{|\zeta|=\rho} \frac{f(\zeta)}{\zeta - \rho^2/\bar{z}} d\zeta = -\frac{1}{2\pi i} \int_{|\zeta|=\rho} \frac{\bar{z} f(\zeta)}{\rho^2 - \zeta \bar{z}} d\zeta = 0. \tag{4.1.6}$$

Since

$$\frac{1}{\zeta - z} + \frac{\bar{z}}{\zeta \bar{\zeta} - \zeta \bar{z}} = \frac{|\zeta|^2 - |z|^2}{\zeta |\zeta - z|^2} = \frac{1}{\zeta} \operatorname{Re}\left(\frac{\zeta + z}{\zeta - z}\right),$$

adding equations (4.1.5) and (4.1.6), we obtain the following result:

$$f(z) = \frac{1}{2\pi} \int_0^{2\pi} f(\rho e^{i\theta}) \operatorname{Re}\left(\frac{\rho e^{i\theta} + z}{\rho e^{i\theta} - z}\right) d\theta, \quad |z| < \rho.$$

Taking the real part of both sides of the above equation, we obtain

$$u(z) = \frac{1}{2\pi} \int_0^{2\pi} u(\rho e^{i\theta}) \operatorname{Re}\left(\frac{\rho e^{i\theta} + z}{\rho e^{i\theta} - z}\right) d\theta, \quad |z| < \rho.$$

Put $z = Re^{i\varphi}, r < \rho$, then

$$u(Re^{i\varphi}) = \frac{1}{2\pi} \int_0^{2\pi} \frac{(\rho^2 - r^2) u(\rho e^{i\theta})}{\rho^2 + r^2 - 2\rho r \cos(\theta - \varphi)} d\theta, \quad r < \rho.$$

Letting $\rho \to R$, we see that (4.1.4) holds. □

Substituting $z = 0$ into (4.1.4), we get

$$u(0) = \frac{1}{2\pi} \int_0^{2\pi} u(Re^{i\theta}) d\theta.$$

Corollary 4.1.4. *If $0 < R < \infty$, and $u(z)$ is harmonic in the disk $|z - z_0| < R$, then*

$$u(z_0) = \frac{1}{2\pi} \int_0^{2\pi} u(z_0 + Re^{i\theta}) d\theta \quad (0 \leqslant r < R). \quad (4.1.7)$$

(4.1.7) is known as the mean value formula for harmonic functions.

Corollary 4.1.5. *If $0 < R < \infty$, $u(z)$ is a real continuous function on the closed disk $|z| \leqslant R$, $f(z) = u(x,y) + iv(x,y)$, which is an analytic function on the open disk $|z| < R$, then*

$$f(z) = \frac{1}{\pi i} \int_{|\zeta|=R} \frac{u(\zeta) d\zeta}{\zeta - z} - \overline{f(0)}, \quad |z| < R. \quad (4.1.8)$$

(4.1.8) is called the Schwarz formula.

Proof. Taking the conjugate of (4.1.6) and noting that $\overline{d\zeta} = d\bar{\zeta} = -\frac{\rho^2}{\zeta^2} d\zeta$, we have

$$\frac{1}{2\pi i} \int_{|\zeta|=\rho} \frac{z \overline{f(\zeta)}}{(\zeta - z)\zeta} d\zeta = 0.$$

Adding it to (4.1.5), we obtain

$$f(z) = \frac{1}{2\pi i} \int_{|\zeta|=\rho} \left(\frac{f(\zeta)}{\zeta - z} + \frac{z \overline{f(\zeta)}}{\zeta - z} \frac{1}{\zeta} \right) d\zeta$$

$$= \frac{1}{2\pi i} \int_{|\zeta|=\rho} u(\zeta) \frac{\zeta + z}{\zeta - z} \frac{d\zeta}{\zeta} + \frac{1}{2\pi} \int_{|\zeta|=\rho} \frac{v(\zeta)}{\zeta} d\zeta.$$

By the mean value formula (4.1.7),

$$f(z) = \frac{1}{2\pi} \int_0^{2\pi} u(\rho e^{i\theta}) \frac{\rho e^{i\theta} + z}{\rho e^{i\theta} - z} d\theta + iv(0)$$

$$= \frac{1}{\pi} \int_0^{2\pi} u(\rho e^{i\theta}) \frac{\rho e^{i\theta} d\theta}{\rho e^{i\theta} - z} - \overline{f(0)}.$$

This proves (4.1.8). □

Analytic functions have derivatives of any order in their region, while real harmonic functions are the real or imaginary parts of some analytic function. Therefore, we can conclude that the partial derivatives of a harmonic function of any order are still harmonic functions.

Corresponding to the maximum modulus principle for analytic functions, we have the maximum principle for harmonic functions.

Theorem 4.1.6 (Extreme Value Principle). *If $u(z) = u(x,y)$ is a non-constant real harmonic function in the region Ω, then $u(z)$ cannot attain a maximum or minimum value in Ω.*

Proof. Let $u(z)$ be a harmonic function in the region Ω. We will first prove that it cannot attain a maximum $M = \sup\{u(z) : z \in \Omega\}$.

Assume, to the contrary, that $u(z)$ achieves a maximum value M at some point z_0 in Ω and let $D = \{z \in \Omega : u(z) = M\}$ and $D_1 = \{z \in \Omega : u(z) < M\}$. Consider a small disk $|z - z_0| < R$ centered at z_0 and contained entirely in Ω. By the mean value property for harmonic functions, we have

$$M = u(z_0) = \frac{1}{2\pi} \int_0^{2\pi} u(z_0 + Re^{i\theta})d\theta \leqslant M$$

for all $0 < r < R$. Since $u(z)$ achieves its maximum at z_0, we have

$$M = \frac{1}{R^2} \int_0^R \frac{1}{2\pi} \int_0^{2\pi} u(z_0 + Re^{i\theta})d\theta r dr.$$

Therefore, we can conclude that

$$u(z_0 + Re^{i\theta}) \equiv M \quad (0 < r < R, \quad 0 \leqslant \theta \leqslant 2\pi). \tag{4.1.9}$$

This proves that the set D is a non-empty open set. Let $z_0 \in \Omega$ be in $\overline{D}$, there is a sequence of points $z_n \in D$ such that z_n approaches z_0 as $n \to \infty$. Since $u(z)$ is continuous, we have $u(z_0) = \lim_{n\to\infty} u(z_n) = u(z_0)$. This proves that $D \cup D_1 = \Omega, D \cap D_1 = \emptyset$. So, $\overline{D} \cap \Omega = \overline{D} \cap (D \cup D_1) = D \cup (\overline{D} \cap D_1) = D \subset \Omega$. Since Ω is a region, Theorem 1.1.12 implies that $D = \Omega$, $u(z)$ is a constant M in Ω, which contradicts our assumption that u is a non-constant real harmonic function in the region Ω. Similarly, for $-u(z)$, we can show that it cannot attain a minimum value. Hence, the proof is complete. $\square$

The classical Dirichlet problem is as follows: Let $\Omega \subset \mathbb{C}$ be a bounded region in $\mathbb{C}$, and let f be a continuous function defined on its boundary $\partial\Omega$. The question is whether there exists a continuous function u defined on the closure of Ω that satisfies

$$\begin{cases} \Delta u(z) = 0, & z \in \Omega, \\ u(z) = f(z), & z \in \partial\Omega. \end{cases}$$

In the case where $\Omega = U$ is a disk, we will demonstrate that the Dirichlet problem in the disk can be solved using the Poisson kernel on the unit disk.

Theorem 4.1.7 (Poisson Kernel). *The Poisson kernel satisfies the following:*

(a) $\frac{1}{2\pi}\int_0^{2\pi} P(\mathrm{Re}^{i\theta}, z)d\theta = 1$ \hfill $(|z| < R)$;

(b) $P(\mathrm{Re}^{i\theta}, z) > 0$ \hfill $(|z| < R)$;

(c) $\frac{R-|z|}{R+|z|} \leq P(\mathrm{Re}^{i\theta}, z) \leq \frac{R+|z|}{R-|z|}$;

(d) *for each $\theta_0 \in [0, 2\pi]$, $\varepsilon > 0$ and $\delta \in (0, \frac{\pi}{2}]$, there is $r_0 > 0$ such that when $R > r > r_0$, $|\varphi - \theta_0| < \delta$, for all θ with $\pi \geq |\theta - \theta_0| > 2\delta$, we have $P(\mathrm{Re}^{i\theta}, \mathrm{Re}^{i\varphi}) < \varepsilon$.*

Proof. Using the Poisson formula (4.1.4) by setting $u(z) \equiv 1$, we obtain (a).

From the Poisson kernel definition and the triangle inequality, we immediately obtain statements (b) and (c). Now, let's prove statement (d): When $|\varphi - \theta_0| < \delta$ and $|\theta - \theta_0| > 2\delta$, we have $|\theta - \varphi| > \delta$. So, when $r_0 > \frac{R}{2}$, $R - r_0 < R\varepsilon \sin^2 \frac{\delta}{2}$, we get

$$0 \leq P(\mathrm{Re}^{i\theta}, \mathrm{Re}^{i\varphi}) \leq \frac{R^2 - r^2}{R^2 - 2Rr\cos\delta + r^2}$$

$$= \frac{R^2 - r^2}{(R-r)^2 + 4Rr\sin^2\frac{\delta}{2}} \leq \frac{R^2 - r^2}{4Rr\sin^2\frac{\delta}{2}} < \varepsilon \quad (r > r_0).$$

This proves (d). $\square$

By utilizing the properties of the Poisson kernel $P(\zeta, z)$ on the unit disk, one can solve the Dirichlet problem within the disk.

Theorem 4.1.8. *Let $0 < R < +\infty$, let $h(\zeta)$ be a continuous function defined on $|\zeta| = R$, then the function u is defined by*

$$u(z) = \begin{cases} \dfrac{1}{2\pi} \displaystyle\int_0^{2\pi} h(\mathrm{Re}^{i\theta}) P(\mathrm{Re}^{i\theta}, z) d\theta, & |z| < R, \\ h(z), & |z| = R, \end{cases} \quad (4.1.10)$$

which is harmonic in the disk $|z| < R$ and continuous in closed disk $|z| \leqslant R$.

Proof. Note that the function $u(z)$ is the real part of the analytic function f defined by

$$f(z) = \frac{1}{2\pi} \int_0^{2\pi} h(\mathrm{Re}^{i\theta}) \left(\frac{\mathrm{Re}^{i\theta} + z}{\mathrm{Re}^{i\theta} - z} \right) d\theta,$$

thus $u(z)$ is harmonic in the disk: $|z| < R$. By Theorem 4.1.7(a), we have

$$u(z) - h(\mathrm{Re}^{i\theta_0}) = \frac{1}{2\pi} \int_0^{2\pi} \left(h(\mathrm{Re}^{i\theta}) - h(\mathrm{Re}^{i\theta_0}) \right) P(\mathrm{Re}^{i\theta}, z) d\theta.$$

For all $\varepsilon > 0$, there is $\delta \in (0, \frac{\pi}{2}]$ such that, when $|\theta - \theta_0| < 2\delta$, we have $|h(\mathrm{Re}^{i\theta}) - h(\mathrm{Re}^{i\theta_0})| < \varepsilon$. Thus,

$$|u(z) - h(\mathrm{Re}^{i\theta_0})|$$

$$\leqslant \frac{\varepsilon}{2\pi} \int_{|\theta - \theta_0| < 2\delta} P(\mathrm{Re}^{i\theta}, z) d\theta$$

$$+ \frac{1}{2\pi} \int_{2\delta \leqslant |\theta - \theta_0| \leqslant \pi} \left| h(\mathrm{Re}^{i\theta}) - h(\mathrm{Re}^{i\theta_0}) \right| P(\mathrm{Re}^{i\theta}, z) d\theta.$$

By utilizing the properties (a) and (d) of the Poisson kernel, there exists $r_0 > 0$ such that for $R > r > r_0$ and $|\varphi - \theta_0| < \delta$, for all θ satisfying $|\theta - \theta_0| > 2\delta$, we have $P(\mathrm{Re}^{i\theta}, \mathrm{Re}^{i\varphi}) < \varepsilon$. Thus,

$$|u(z) - h(\mathrm{Re}^{i\theta_0})| \leqslant \varepsilon + 2M\varepsilon,$$

where $M = \max\{|h(\zeta)| : |\zeta| = R\}$. Note that $\zeta_0 = \mathrm{Re}^{i\theta_0}$ is any arbitrary point on $|\zeta| = R$, $u(z)$ is continuous up to the entire boundary of the disk $|z| < R$ and takes $h(\zeta)$ at $|\zeta| = R$. $\square$

By applying Theorem 4.1.8, we can obtain the solution to the Dirichlet problem in any disk in $\mathbb{C}$.

Corollary 4.1.9. *Let $0 < R < \infty$, $z_0 \in \mathbb{C}$, h be continuous on the circle $\partial D(z_0, R)$, then the function u is defined by*

$$u(z) = \begin{cases} \dfrac{1}{2\pi} \displaystyle\int_0^{2\pi} \dfrac{R^2 - |z - z_0|^2}{|\mathrm{Re}^{i\theta} + z_0 - z|^2} h(z_0 + \mathrm{Re}^{i\theta}) d\theta, & |z - z_0| < R, \\ h(z), & |z - z_0| = R, \end{cases}$$

which is harmonic in the disk $D(z_0, R)$ and continuous in the closed disk $\overline{D(z_0, R)}$.

Just as there is the famous Cauchy theorem for analytic functions, there is a corresponding theorem for harmonic functions.

Theorem 4.1.10. *Let Ω be a bounded region with boundary C consisting of a finite number of piecewise smooth, simple closed curves. If $u(x, y)$ is harmonic inside Ω and $u \in C^1(\overline{\Omega})$, then*

$$\int_C \frac{\partial u}{\partial n} ds = 0, \tag{4.1.11}$$

where n denotes the outward unit normal to C and ds denotes a differential of arc length.

To prove Theorem 4.1.10, we first establish the following lemma.

Lemma 4.1.11. *Let Ω be a bounded region with boundary C consisting of a finite number of piecewise smooth, simple closed curves. If $u(x, y)$ and $v(x, y)$ have continuous second-order partial derivatives on $\overline{\Omega}$, then*

$$\int_\Omega (u\Delta v - v\Delta u) dx dy = \int_C \left(u \frac{\partial v}{\partial n} - v \frac{\partial u}{\partial n} \right) ds, \tag{4.1.12}$$

where n denotes the outward unit normal to C and ds denotes a differential of arc length.

Proof. According to Green's theorem, if $P(x, y)$ and $Q(x, y)$ have continuous first-order partial derivatives on $\overline{\Omega}$, then

$$\int_\Omega \left(\frac{\partial Q}{\partial x} - \frac{\partial P}{\partial y} \right) dx dy = \int_C P dx + Q dy. \tag{4.1.13}$$

Taking $P(x,y) = -u\frac{\partial v}{\partial y}$, $Q(x,y) = u\frac{\partial v}{\partial x}$. Substituting the previous expression into (4.1.13), we have

$$\int_\Omega (u_x v_x + u_y v_y)dxdy + \int_\Omega u\Delta v\, dxdy + \int_C u(v_y dx - v_x dy) = 0. \tag{4.1.14}$$

Let $\widehat{x,n}$ and $\widehat{y,n}$ denote the angles between the outward normal vector n to the curve C and the x-axis and y-axis, respectively. Then we have

$$-(v_y dx - v_x dy) = [v_y \cos(\widehat{y,n}) + v_x \cos(\widehat{x,n})]ds.$$

Substituting the previous expression into (4.1.14), we obtain

$$\int_\Omega (u_x v_x + u_y v_y)dxdy + \int_\Omega u\Delta v\, dxdy - \int_C u\frac{\partial v}{\partial n}ds = 0. \tag{4.1.15}$$

(4.1.15) is called Green's preparatory formula. By interchanging u and v in (4.1.15), and subtracting the resulting equation from (4.1.15), we obtain (4.1.12). □

Proof of Theorem 4.1.10. If $u(x,y)$ and $v(x,y)$ are harmonic in Ω and $u, v \in C^1(\Omega)$, then (4.1.12) changes into

$$\int_C \left(u\frac{\partial v}{\partial n} - v\frac{\partial u}{\partial n}\right)ds = 0. \tag{4.1.16}$$

If $v \equiv 1$, then

$$\int_C \frac{\partial u}{\partial n}ds = 0,$$

which completes the proof of Theorem 4.1.10. □

For analytic functions, we have the inverse theorem of Cauchy's theorem, known as Morera's theorem. Now, we will prove a similar theorem for harmonic functions

Theorem 4.1.12. *Suppose the function $u(z) = u(x,y)$ is continuous in a region Ω. If for any point a within Ω, there exists a decreasing sequence of positive numbers $\{r_n\}$ tending to zero such*

that $\overline{D(a, r_1)} \subset \Omega$ and

$$u(a) = \frac{1}{2\pi} \int_0^{2\pi} u(a + r_n e^{i\theta}) d\theta, \quad n = 1, 2, \ldots,$$

then $u(z)$ is harmonic in Ω.

Proof. For all $a \in \Omega$, there is a positive number $r_a > 0$ such that $\overline{D(a, r_a)} \subset \Omega$. According Corollary 4.1.9, the function h_a is defined by

$$h_a(z) = \frac{1}{2\pi} \int_0^{2\pi} u(a + r_a e^{i\theta}) P(\operatorname{Re}^{i\theta}, z - a) d\theta,$$

which is continuous in $D(a, r_a)$. For $z \in \partial D(a, r_a)$, put $h_a(z) = u(z)$, then $h_a(z)$ is continuous in $\overline{D(a, r_a)}$. We will now prove that $u(z)$ is equal to $h_a(z)$ in $\overline{D(a, r_a)}$. Consider the function $\varphi(z) = u(z) - h_a(z)$. Letting $M = \sup\{\varphi(z) : |z - a| \leqslant r_a\}$, if $M > 0$, we put $K = \{z \in \Omega : u(z) = M\}$, then $K \subset D(a, r_a)$ is compact and there is $z_0 \in K$ such that $\operatorname{Re} z_0 = \max\{\operatorname{Re} z : z \in K\}$. The condition of Theorem 4.1.12 implies that there is $\rho > 0$ such that $D(z_0, \rho) \subset D(a, r_a)$, and

$$M = \varphi(z_0) = \frac{1}{2\pi} \int_0^{2\pi} \varphi(z_0 + \rho e^{i\theta}) d\theta.$$

Thus, $M = \varphi(z_0) = \varphi(z_0 + \rho e^{i\theta})$, $\theta \in [0, 2\pi]$. Hence, $z_0 + \rho \in K$, which contradicts our assumption that $\operatorname{Re} z_0 = \max\{\operatorname{Re} z : z \in K\}$. Therefore, $M \leqslant 0$. Similarly, we can prove that $\sup\{-\varphi(z) : |z - a| \leqslant r_a\} \leqslant 0$. This shows that $u(z) = h_a(z)$ for all $z \in \overline{D(a, r_a)}$, so $u(z) = h_a(z)$ is harmonic in $D(a, r_a)$. The point a in Ω is arbitrary, so $u(z)$ is harmonic in Ω. □

Corollary 4.1.13. *Suppose $\{u_k\}$ is a sequence of harmonic functions defined in a region Ω. If $\{u_k\}$ uniformly converges to u on every compact subset of Ω (referred to as uniform convergence on compact subsets in Ω), then u is harmonic in Ω.*

Proof. Since Ω is a domain and u_k is harmonic on Ω, each u_k is continuous on Ω. Moreover, because $\{u_k\}$ uniformly converges to u on compact subsets of Ω, it follows that u is continuous on Ω. For any

$a \in \Omega$, there exists $r_0 > 0$ such that $\overline{D(z, r_0)} \subset \Omega$. By the mean value theorem for harmonic functions (Corollary 4.1.4), we know that

$$u_k(a) = \frac{1}{2\pi} \int_0^{2\pi} u_k(a + Re^{i\theta}) d\theta, \quad 0 < r < r_0.$$

Since $\{u_k\}$ uniformly converges to u on $\partial D(a, r)$, then

$$u(a) = \frac{1}{2\pi} \int_0^{2\pi} u(a + Re^{i\theta}) d\theta, \quad 0 < r < r_0.$$

According to Theorem 4.1.12, u is harmonic in Ω. □

Theorem 4.1.14. *If $u(\zeta)$ is a harmonic function in a region D, $f(z)$ is an analytic function in a region Ω, and $f(\Omega) \subset D$, then the composite function $u(f(z))$ is a harmonic function in Ω.*

Proof. By the chain rule for differentiation, we have

$$\frac{\partial u}{\partial z} = \frac{\partial u}{\partial \zeta} \frac{\partial f}{\partial z} + \frac{\partial u}{\partial \overline{\zeta}} \frac{\partial \overline{f}}{\partial z} = \frac{\partial u}{\partial \zeta} \overline{f'(z)},$$

$$\frac{\partial^2 u}{\partial z \partial \overline{z}} = \left[\frac{\partial^2 u}{\partial \zeta \partial \overline{\zeta}} \frac{\partial f}{\partial \overline{z}} + \frac{\partial^2 u}{\partial \overline{\zeta} \partial \overline{\zeta}} \frac{\partial \overline{f}}{\partial \overline{z}}\right] \overline{f'(z)} + \frac{\partial u}{\partial \zeta} \frac{\partial \overline{f'}}{\partial \overline{z}} = \frac{\partial^2 u}{\partial \zeta \partial \overline{\zeta}} |f'(z)|^2.$$

By (4.1.2), $\dfrac{\partial^2 u}{\partial \zeta \partial \overline{\zeta}} = 0$. Thus, $u(f(z))$ is harmonic in D. □

Theorem 4.1.15. *If $z = 0$ does not belong to a region Ω, then the necessary and sufficient condition for $u(z)$ to be harmonic in Ω is that for any $z = Re^{i\theta} \in \Omega$, we have*

$$\frac{\partial^2 u}{\partial (\log r)^2} + \frac{\partial^2 u}{\partial \theta^2} = 0. \tag{4.1.17}$$

(4.1.17) is referred to as the polar coordinate form of the Laplace equation.

Proof. It is sufficient to prove that at every point $z_0 \in \Omega$, $z_0 \neq 0$, there exists a neighborhood $D(z_0, \rho)$ $(0 < \rho < |z_0|)$ such that in $D(z_0, \rho)$, (4.1.17) is equivalent to $\Delta u(z) = 0$.

Since there exists a simply connected domain W such that the function $z = e^\zeta$ maps W conformally onto $D(z_0, \rho)$, according to

Theorem 4.1.14, the function $u(z)$ is harmonic in $D(z_0, \rho)$ if and only if the function $u(e^\zeta)$ is harmonic in W, i.e., that

$$\frac{\partial^2 u}{\partial x^2} + \frac{\partial^2 u}{\partial y^2} = 0 \quad (z = x + iy \in D(z_0, \rho))$$

holds in $D(z_0, \rho)$ is equivalent to

$$\frac{\partial^2 u(e^\zeta)}{\partial \xi^2} + \frac{\partial^2 u(e^\zeta)}{\partial \eta^2} = 0 \quad (\zeta = \xi + i\eta \in W)$$

holds in $D(z_0, \rho)$. By substituting $\zeta = \log z = \log r + i\theta$ into the previous equation, we obtain that

$$\frac{\partial^2 u}{\partial (\log r)^2} + \frac{\partial^2 u}{\partial \theta^2} = 0$$

holds in $D(z_0, \rho)$. □

Theorem 4.1.16. *Let $\{u_k\}$ be a sequence of harmonic functions in a region Ω. If for any compact subset $K \subset \Omega$, the sequence $\{u_k\}$ is uniformly bounded on K, then there exists a subsequence $\{u_{k_j}\}$ of $\{u_k\}$ and a harmonic function u in Ω such that for any compact subset $K \subset \Omega$, the subsequence $\{u_{k_j}\}$ converges uniformly to u on K.*

Proof. For all $z_0 \in \Omega$, there is $R > 0$ such that the closed disk $\overline{D(z_0, 2R)} \subset \Omega$. So, $\{u_k\}$ is uniformly bounded in $K = \overline{D(z_0, 2R)}$. Since $D(z_0, 2R)$ is simply connected, there are $f_k \in H(D(z_0, 2R))$ such that $u_k = \operatorname{Re} f_k$, both $\{e^{f_k}\}$ and $\{e^{-f_k}\}$ are uniformly bounded in $\overline{D(z_0, 2R)}$. By Theorem 2.1.4, there exists a subsequence of $\{f_k\}$ that converges uniformly on $\overline{D(z_0, R)}$. Therefore, there exists a subsequence of $\{f_k\}$ that converges uniformly on $\overline{D(z_0, R)}$. By Borel's lemma on finite coverings, for any compact subset $K \subset \Omega$, there exists a subsequence of $\{u_k\}$ that converges uniformly on K. Construct a sequence of compact sets $K_n = \{z : d(z, \Omega^c) \geq 1/n, |z| \leq n\}$, where $d(z, \Omega)$ represents the distance between z and the complement of Ω. The sequence satisfies the following properties:

$$K_1 \subset \operatorname{int} K_2 \subset \cdots \subset \Omega, \text{ and } \bigcup_{n=1}^{\infty} K_n = \Omega.$$

For K_1, there exists a subsequence $\{u_k^{(1)}\}$ of $\{u_k\}$ that converges uniformly to v_1 on K_1. Furthermore, there exists a subsequence $\{u_k^{(2)}\}$ of $\{u_k^{(1)}\}$ that converges uniformly to a function v_2 on K_2, with the restriction of v_2 on K_1 being equal to $_1$. By repeating this process, we obtain a nested subsequence chain $\{u_k\} \supset \{u_k^{(1)}\} \supset \cdots \supset \{u_k^{(i)}\} \supset \cdots$, such that for each $i = 1, 2, \ldots$, the subsequence $\{u_k^{(i)}\}$ converges uniformly to v_i on K_i, and v_i restricted to $K_j (1 \leqslant j \leqslant i-1)$ is equal to v_j. Now, for each $i = 1, 2, \ldots$, let $b_i(z) = u_i^i(z)$, where $u_i^i(z)$ is the ith function in the sequence $\{u_k^{(i)}(z)\}$. For any j, we have constructed a subsequence $\{b_i\}$ of $\{u_k^j\}$ (possibly after removing the first $j-1$ terms) such that $\{b_i\}$ converges uniformly to v_j on K_j. Therefore, we obtain a function $u(z)$ defined on Ω such that its restriction to K_j is equal to v_j for all $j = 1, 2, \ldots$. By construction, we also have that $\{b_i(z)\}$ converges uniformly to u on any compact subset of Ω. By Corollary 4.1.13, we conclude that u is harmonic in Ω. $\square$

Theorem 4.1.17 (Harnack's Theorem). *Let $\{u_n(z)\}$ be a sequence of harmonic functions in a disk $D(z_0, R)$ such that $u_n(z) \leqslant u_{n+1}(z)$, then either $u_n(z) \to \infty$ converges uniformly on compact subsets of $D(z_0, R)$ or $\{u_n(z)\}$ converges to a harmonic function $u(z)$ uniformly on compact subsets of $D(z_0, R)$.*

Proof. Consider an increasing sequence $\{u_n(z_0)\}$. We will prove the following statements:

(i) If $\lim_{n \to \infty} u_n(z_0) = \infty$, then $u_n(z)$ converges to infinity uniformly on compact subsets of $D(z_0, R)$;
(ii) If $\lim_{n \to \infty} u_n(z_0) < \infty$, then $u_n(z)$ converges to a harmonic function u uniformly on compact subsets of $D(z_0, R)$.

For all $r < R, r > 0$, there is ρ such that $r < \rho < R$. When $|z - z_0| \leqslant r$ and $n > m$, the Poisson formula (4.1.4) yields

$$u_n(z) - u_m(z) = \frac{1}{2\pi} \int_0^{2\pi} \frac{\rho^2 - |z - z_0|^2}{|\rho e^{i\varphi} + z_0 - z|^2} (u_n(z_0 + \rho e^{i\varphi}) - u_m(z_0 + \rho e^{i\varphi})) d\varphi.$$

By utilizing Theorem 4.1.7(b) and (c), as well as the monotonicity of the sequence of functions $\{u_n(z)\}$, we can obtain

$$\frac{\rho - |z - z_0|}{\rho + |z - z_0|}(u_n(z_0) - u_m(z_0))$$

$$\leqslant u_n(z) - u_m(z) \leqslant \frac{\rho + |z - z_0|}{\rho - |z - z_0|}(u_n(z_0) - u_m(z_0)).$$

So, when $|z - z_0| \leqslant r$,

$$\frac{\rho - r}{\rho + r}(u_n(z_0) - u_m(z_0))$$

$$\leqslant u_n(z) - u_m(z) \leqslant \frac{\rho + r}{\rho - r}(u_n(z_0) - u_m(z_0)). \quad (4.1.18)$$

(i) If $\lim_{n\to\infty} u_n(z_0) = \infty$, for a fixed m, since $u_n(z)$ is bounded in $\overline{D(z_0, r)}$, by considering the inequality on the left-hand side of (4.1.18), we can observe that the sequence $\{u_n(z)\}$ converges to ∞ uniformly on $\overline{D(z_0, r)}$. Therefore, the sequence $\{u_n(z)\}$ converges to ∞ uniformly on compact subsets of $D(z_0, R)$.

(ii) If $\lim_{n\to\infty} u_n(z_0) < \infty$, by considering the inequality on the right-hand side of (4.1.18) and utilizing the convergence principle, we can observe that the sequence $\{u_n(z)\}$ converges uniformly on $\overline{D(z_0, r)}$. Therefore, the sequence $\{u_n(z)\}$ converges uniformly on compact subsets of $D(z_0, R)$ to a function $u(z)$. Furthermore, by Theorem 4.1.13, $u(z)$ is harmonic in $D(z_0, R)$. □

Theorem 4.1.18 (The Schwar Reflection Principle). *Let Ω be a region in $\mathbb{C}$ that is symmetric with respect to the real axis, i.e., if $z = x + iy \in \Omega$, then $\overline{z} = x - iy \in \Omega$. If a continuous function u defined on Ω satisfies $u(z) = -u(\overline{z})$ and is harmonic in $\Omega_+ =: \{z = x + iy \in \Omega : y > 0\}$, then u is harmonic in Ω.*

Proof. Since u is harmonic in Ω_+, $(\Delta u)(z) = 0$ for all $z \in \Omega_+$, also

$$-(\Delta u)(\overline{z}) = (\Delta u)(z) = 0$$

for all $z \in \Omega_- =: \{z = x + iy \in \Omega : y < 0\}$, so u is harmonic in Ω_-. Given $a \in \Omega \cap \partial\Omega_+$, there is $r_0 > 0$ such that $\overline{D(a, r_0)} \subset \Omega$, since u is continuous on Ω and $u(a) = -u(a)$, so $u(a) = 0$, and since $u(z) + u(\bar{z}) = 0$ for all $z \in D(a, r_0)$. Since

$$\int_{-\pi}^{\pi} u(a + \mathrm{Re}^{i\theta})d\theta = \int_{0}^{\pi} (u(a + \mathrm{Re}^{i\theta}) + u(a + \mathrm{Re}^{-i\theta}))d\theta$$
$$= 0 \quad (0 < r < r_0),$$

then

$$u(a) = 0 = \frac{1}{2\pi} \int_{0}^{2\pi} u(a + \mathrm{Re}^{i\theta})d\theta \quad (0 < r < r_0).$$

By Theorem 4.1.12, u is harmonic in Ω. □

Corollary 4.1.19. *Let u be harmonic $\mathbb{C}_+$, continuous $\overline{\mathbb{C}}_+ = \{z = x + iy \in \mathbb{C} : y \geqslant 0\}$. Furthermore, if u is zero on $\partial \mathbb{C}_+$ and bounded in $\mathbb{C}_+$, then u is identically zero.*

Proof. By reflecting u to extend it onto $\mathbb{C}$, let us denote the extension as u_0. According to the reflection principle for harmonic functions (Theorem 4.1.18), we know that u_0 is harmonic on $\mathbb{C}$. Since $\mathbb{C}$ is simply connected, there exists an entire function f such that $u_0 = \operatorname{Re} f$. As u_0 is bounded in $\mathbb{C}$, the entire function $\{e^f\}$ is bounded in $\mathbb{C}$. By Liouville's theorem, it follows that f is a constant function on $\mathbb{C}$, hence u_0 is identically zero on $\mathbb{C}$. Consequently, u is also identically zero on $\mathbb{C}_+$. □

4.2 Upper Semicontinuous Functions

Definition 4.2.1. Let $u(z)$ be a real-valued function defined in an open set Ω that can take the value $-\infty$. Let $a \in \Omega$. If

$$\limsup_{z \to a} u(z) \leqslant u(a), \qquad (4.2.1)$$

then we say that $u(z)$ is upper semicontinuous at the point a. If (4.2.1) holds for every point $a \in \Omega$, then we say that $u(z)$ is upper semicontinuous in Ω.

The upper semicontinuous functions have the following proposition.

Theorem 4.2.2. *The function $u(z)$ is upper semicontinuous in an open set Ω if and only if, for every real number α, the set $U(\alpha) = \{z \in \Omega : u(z) < \alpha\}$ is open.*

Proof. Suppose that for all real numbers α, $U(\alpha)$ is an open set. For all $a \in \Omega$ and $\alpha > u(a)$, we have $a \in U(\alpha)$, and $U(\alpha)$ is open set. So, there is $\delta > 0$ such that the disk $D(a, \delta) \subset U(\alpha)$, hence

$$\limsup_{z \to a} u(z) \leqslant \alpha.$$

This proves that the function $u(z)$ is upper semicontinuous in Ω.

Conversely, suppose $u(z)$ is upper semicontinuous in Ω. For any real number α and any $a \in U(\alpha)$, since $\alpha > u(a)$, by (4.2.1), there exists $\delta > 0$ such that $D(a, \delta) \subset U(\alpha)$, and for any $z \in D(a, \delta)$, we have $u(z) < \alpha$. Therefore, $U(\alpha)$ is an open set. □

Theorem 4.2.3. *Suppose $u(z)$ is upper semicontinuous in an open set Ω. If K is a compact subset of Ω, then $u(z)$ has an upper bound on K.*

Proof. Since K is compact, for any $a \in K$, there is a real number $\alpha(a) > u(z)$. Let $U(a) = \{z \in \Omega : u(z) < \alpha(a)\}$, Theorem 4.2.2 implies that each $U(a)$ is open, the family of sets $\{U(a) : a \in K\}$ is an open cover of the compact K. Therefore, there exists a finite subcover of K:

$$\{U(a_1), U(a_2), \ldots, U(a_N)\}.$$

The number $\max\{\alpha(a_1), \alpha(a_2), \ldots, \alpha(a_N)\}$ is an upper bound of $u(z)$ on K. □

Theorem 4.2.4. *Let Ω be an open set. Then a function $u(z)$ is upper semicontinuous in Ω if and only if, for any compact subset $K \subset \Omega$, there exists a sequence of continuous functions $\{u_k(z)\}$ defined on K such that $\{u_k(z)\}$ monotonically decreases to $u(z)$ on K.*

Proof. For any compact subset $K \subset \Omega$, there exists a sequence of continuous functions $\{u_k(z)\}$ defined on K such that $\{u_k(z)\}$ monotonically decreases to $u(z)$ on K. For any $a \in \Omega$, there exists $\delta > 0$ such that the compact set $K = \overline{D(a, \delta)} \subset \Omega$. For all α, the set $\{z \in D(a, \delta) : u_j(z) < \alpha\}$ is an open set, so

$$\{z : |z - a| < \delta, u(z) < \alpha\} = \bigcup_{j=1}^{\infty} \{z : |z - a| < \delta, u_j(z) < \alpha\}$$

is also open. The function $u(z)$ is upper semicontinuous at point a. Hence, function $u(z)$ is upper semicontinuous in an open set Ω.

Conversely, suppose $u(z)$ is upper semicontinuous in Ω. By Theorem 4.2.3, for any compact subset $K \subset \Omega$, $u(z)$ has an upper bound M on K. Consider the sequence of functions

$$u_j(z) = \sup\{u(\zeta) - j|\zeta - z| : \zeta \in K\}, \quad j = 1, 2, \ldots, \quad z \in K.$$

It is easy to see that $u_j(z) \geqslant u_{j+1}(z)$, $j = 1, 2, \ldots$ and u_j are real-valued. It is here one uses that u does not take the value $+\infty$, and hence,

$$z \in K, -\infty < u_j(z) < +\infty;$$
$$\leqslant M = \sup\{u(\zeta) : \zeta \in K\} \geqslant u_j(z) \geqslant u(z)$$

by the upper semicontinuity. To show that each u_j is continuous in K, it is enough to show that for all $z, z' \in K$ and $\zeta \in K$, we have $|u_j(z) - u_j(z')| \leqslant j|z' - z|$. Since

$$u_j(z) \geqslant u(\zeta) - j|\zeta - z| \geqslant u(\zeta) - j|\zeta - z'| - j|z' - z|,$$

then $u_j(z) \geqslant u_j(z') - j|z' - z|$. Interchanging the role of z and z', we obtain $|u_j(z') - u_j(z)| \leqslant j|z' - z|$.

Finally, we show that $\{u_k(a)\}$ monotonically decreases to $u(a)$ at each point $a \in K$. For any $a \in K$ and $\alpha > u(a)$, the set

$$U(\alpha) = \{z : u(z) < \alpha, z \in U\}$$

is open. Since $a \in U(\alpha)$, there is $\delta > 0$ such that $D(a, 2\delta) \subset U(\alpha)$. Taking $j_0 > 0$ such that $M - j_0\delta < \alpha$,

$$u_j(a) \leqslant \max\{\sup\{u(\zeta) : \zeta \in K, |\zeta - a| \leqslant \delta\}, M - j\delta\} \leqslant \alpha$$

for all $j > j_0$. Therefore, $\{u_k(a)\}$ monotonically decreases to $u(a)$ at $a \in K$. □

4.3 Subharmonic Functions

Definition 4.3.1. Let Ω be an open set. Let $u(z)$ be upper semicontinuous in Ω.

(a) $u(z)$ is said to be subharmonic in Ω if it has the following property: For each compact set $K \subset \Omega$, a continuous real function

$u(z)$ on K is harmonic in the interior of K, and $u(z) \leqslant v(z)$ for all $z \in \partial K$, then $u(z) \leqslant v(z)$ for all $z \in K$.

(b) Suppose that the function $u(z)$ is subharmonic in Ω, if $v(z)$ is harmonic in Ω and $u(z) \leqslant v(z)$ for $z \in \Omega$, then v is called a harmonic majorant function for u in Ω.

(c) Let $u(z)$ be a subharmonic in Ω, if v_0 is a harmonic majorant function for u in Ω, which satisfies the condition, if v is a majorant function for u in Ω, then $v_0(z) \leqslant v(z)$ for all $z \in \Omega$, then v_0 is called the least harmonic majorant function for u in Ω.

Theorem 4.3.2. *Let $u(z)$ be defined on an open set Ω and assume that u is upper semicontinuous in Ω, $z \in \Omega$, $\Omega^c = \mathbb{C}\backslash\Omega, d(z,\Omega^c) = \inf\{|z - z'| : z' \in \Omega^c\}$, $\Omega_\delta = \{z : d(z,\Omega^c) > \delta\}$ $(\delta > 0)$. Then each of the following conditions is necessary and sufficient for u to be subharmonic on Ω:*

(a) $u(z)$ is subharmonic in Ω.

(b) If $\overline{D(a,R)}$ is a closed disk $\subset \Omega$ and
$$f(z) = \sum_{k=0}^{n} c_k z^k$$
is an analytic polynomial such that
$$u(a + Re^{i\theta}) \leqslant \operatorname{Re} f(a + Re^{i\theta}),$$
it follows that $u(z) \leqslant \operatorname{Re} f(z)$ for all $z \in \overline{D(a,R)} \subset \Omega$.

(c) For all $\delta > 0, z \in \Omega_\delta$ and every positive measure $d\mu$ on the interval $[0,\delta]$, we have
$$2\pi u(z) \int_0^\delta d\mu(r) \leqslant \int_0^\delta \int_0^{2\pi} u(z + Re^{i\theta}) d\theta d\mu(r). \quad (4.3.1)$$

(d) For all $\delta > 0$ and every $z \in \Omega_\delta$, there exists some positive measure $d\mu$ with support in $[0,\delta]$ such that $d\mu_1$ has some mass outside the origin $\mu_1\{(0,\delta]\} > 0$ and (4.3.1) is valid.

(e) For all $a \in \Omega$, there exists a decreasing sequence $\{r_n\}$ which tends to 0 and $\overline{D(a,r_1)} \subset \Omega$ such that
$$u(a) \leqslant \frac{1}{2\pi} \int_0^{2\pi} u(a + r_n e^{i\theta}) d\theta, \quad n = 1, 2, \ldots \quad (4.3.2)$$
for all n.

(f) For all $a \in \Omega$, there exists a positive number $\delta > 0$ such that $\overline{D(a,\delta)} \subset \Omega$ and

$$u(a) \leqslant \frac{1}{2\pi} \int_0^{2\pi} u(a + te^{i\theta})d\theta, \quad 0 < t < \delta. \qquad (4.3.3)$$

(g) If $z = g(w)$ is an analytic function from the open set $\widetilde{\Omega}$ into Ω, then $\widetilde{u}(w) = u(g(w))$ is subharmonic in $\widetilde{\Omega}$.

Note: By Theorems 4.2.2 and 4.2.3, the integrals in (4.3.1)–(4.3.3) are well defined.

Proof. (a) implies (b) and (g), and (g) implies (a). This is evident from the definition of subharmonic functions.

(b) implies (f). Let $a \in \Omega$, $\delta > 0$ such that $\overline{D(a,\delta)} \subset \Omega$. If

$$0 < t < \delta, \quad \varphi(\theta) = \sum_{k=-n}^{n} c_k e^{ik\theta}, \quad c_{-k} = \overline{c_k}\,(k = 0, 1, \ldots, n)$$

is a real trigonometrical polynomial such that

$$u(a + te^{i\theta}) \leqslant \varphi(\theta) \qquad (4.3.4)$$

for all $\theta \in [0, 2\pi]$, the polynomial

$$f(z) = c_0 + 2\sum_{k=1}^{n} \frac{c_k}{t^k}(z-a)^k$$

has a real part which is an upper bound for u on $\partial D(a,t)$,

$$\mathrm{Re} f(a + te^{i\theta}) = \varphi(\theta) \geqslant u(a + te^{i\theta}).$$

Hence, $u(z) \leqslant \mathrm{Re} f(z)$ for all $z \in \overline{D(a,t)}$, and in particular,

$$u(a) \leqslant c_0 = \frac{1}{2\pi} \int_0^{2\pi} \varphi(\theta)d\theta. \qquad (4.3.5)$$

By Weierstrass's theorem, the space consisting of all real trigonometric polynomials is dense in the space of real continuous functions $C_0([0, 2\pi]) = \{h : h(\theta)$ is continuous on $[0, 2\pi], h(0) = h(2\pi)\}$. Now, $\varphi(\theta) \in C_0([0, 2\pi])$ is the arbitrary continuous function such that the inequality (4.3.4) holds, therefore (4.3.5) holds. By Theorem 4.2.4 and the monotone convergence theorem, we know that (4.3.3) holds.

(f) implies (c). Suppose that $\delta > 0, 0 < t \leqslant \delta, a \in \Omega_\delta$ and $d\mu$ is a positive measure on $[0, \delta]$, integration in (4.3.3) with respect to $d\mu$ over $[0, \delta]$ now gives (4.3.1).

(c) implies (d), which is obvious.

(d) implies (e). Given $a \in \Omega$ and for all $\rho_1 > 0$ such that $\overline{D(a, \rho_1)} \subset \Omega$, there exists some positive measure $d\mu_1$ with support in $(0, \rho_1]$ such that $\mu_1\{(0, \rho_1]\} > 0$ and (4.3.1) holds for μ_1. So there exists $r_1 \in (0, \rho_1]$, and (4.3.2) holds for $n = 1$. By the principle of induction, it is known that there exists a sequence of positive numbers $\{r_n\}$ approaching zero such that (4.3.2) holds.

(e) implies (a). Let K be a compact subset of Ω and $f(z)$ a continuous function on K which is harmonic in the interior of K, and assume that $u(z) \leqslant f(z)$ on ∂K. Letting

$$\varphi(z) = u(z) - f(z), \quad M = \sup\{\varphi(z) : z \in K\},$$
$$E = \{z \in K : \varphi(z) = M\}.$$

If $M > 0$, the semicontinuity of u shows that $E \subset \text{int} K$ is a non-empty compact subset of the interior of K. In fact, by the definition of M, there is a sequence $\{z_n\}$ ($z_n \in K$) and $z_0 \in K$ such that

$$z_n \to z_0 \quad (n \to \infty), \quad \varphi(z_n) \to M \quad (n \to \infty).$$

So,

$$M \geqslant \varphi(z_0) \geqslant \lim_{n \to \infty} \varphi(z_n) = M.$$

If $b_n \in E, b \in K$ such that

$$b_n \to b \quad (n \to \infty), \quad \varphi(b_n) = M,$$

then

$$M \geqslant \varphi(b) \geqslant \lim_{n \to \infty} \varphi(b_n) = M.$$

Therefore, E is a non-empty compact subset of the interior of K.m. Let $z_0 \in E$ such that $\text{Re } z_0 = \max\{\text{Re } z : z \in E\}$. So, there is a positive number $r_1 > 0$ such that $\overline{D(a, r_1)} \subset \text{int} K$ and

$$M = \varphi(z_0) \leqslant \frac{1}{2\pi} \int_0^{2\pi} \varphi(z_0 + r_1 e^{i\theta}) d\theta \leqslant M.$$

Thus, $M = \varphi(z_0) = \varphi(z_0 + r_1 e^{i\theta})$ for almost everywhere $\theta \in [0, 2\pi]$. Since $\varphi(z)$ is upper semicontinuous in K, so $M = \varphi(z_0) = \varphi(z_0 + r_1 e^{i\theta})$ for all $\theta \in [0, 2\pi]$, thus $z_0 + r_1 \in E \subset K$. But this contradicts $\operatorname{Re} z_0 = \max\{\operatorname{Re} z : z \in K\}$. Therefore, $M \leqslant 0$, so $u(z) \leqslant f(z)$ for $z \in K$. $\square$

Theorem 4.3.3. *Suppose that $u(z)$ is subharmonic in a region Ω and $M = \sup\{u(z) : z \in \Omega\}$, if there is $z_0 \in \Omega$ such that $u(z_0) = M$, then $u(z) \equiv M$.*

Proof. Let $D = \{z \in \Omega : u(z) = M\}$. If $b \in \overline{D} \cap \Omega$, there are $z_n \in D$ such that $z_n \to b$ $(n \to \infty), u(z_n) = M$. So,

$$M \geqslant u(b) \geqslant \lim_{n \to \infty} u(z_n) = M,$$

therefore $\overline{D} \cap \Omega \subset D$. If $a \in D$, then there exists $d > 0$ such that $D(a, d) \subset \Omega$. Let $\delta \in (0, d)$ and $d\mu(r) = rdr$ a positive measure on $[0, \delta]$. By Theorem 4.3.2(c),

$$M = u(a) \leqslant \frac{1}{\pi \delta^2} \int_0^{2\pi} \int_0^{\delta} u(a + \operatorname{Re}^{i\theta}) d\theta r dr \leqslant M. \qquad (4.3.6)$$

So, $M - u(a + \operatorname{Re}^{i\theta}) = 0$ for almost everywhere $\theta \in [0, 2\pi]$. Since $u(z)$ is upper semicontinuous in Ω, thus $M - u(z) = 0$ for all z with $|z - b| \leqslant \delta$. This proves that D is non-empty, Theorem 1.1.12 implies that $D = \Omega$, so $u(z) \equiv M$ in Ω. $\square$

Theorem 4.3.4. (a) *Let $u_k(z)$ $(k = 1, 2, \ldots, n)$ be a subharmonic function in an open set Ω, c_k $(k = 1, 2, \ldots, n)$ positive constant, then*

$$\sum_{k=1}^{n} c_k u_k(z), \quad \max\{u_1, u_2, \ldots, u_n\}$$

is subharmonic in Ω.

(b) *u_α $(\alpha \in \Lambda)$ is a family of subharmonic functions in an open set Ω, then $u(z) = \sup\{u_\alpha(z) : \alpha \in \Lambda\} < \infty$ is subharmonic in Ω if $u(z) < \infty$ for all $z \in \Omega$ and $u(z)$ is semicontinuous in Ω.*

(c) Let $\{u_n(z)\}$ be a sequence of continuous subharmonic functions in Ω and $\{u_n(z)\}$ converges uniformly in each compact of Ω to $u(z)$, then $u(z)$ is subharmonic in Ω.

(d) $\{u_n(z)\}$ is a decreasing sequence of subharmonic functions in an open set Ω, then $u(z) = \lim_{n\to\infty} u_n(z)$ is also subharmonic in Ω.

(e) A function $u(z)$ defined in an open set Ω is harmonic if and only if every point $a \in \Omega$ has a neighborhood $D(a, \delta) \subset \Omega$, where $u(z)$ is subharmonic in $D(a, \delta)$.

(f) If Ω is an open set, $f(z) \in H(\Omega)$ and $f(z) \not\equiv 0$, it follows that $\log|f(z)|$ is harmonic in Ω.

Proof. (a), (b), (c), (e) and (f) are obviously true. To prove (d), note that

$$\{z \in \Omega : u(z) < \alpha\} = \bigcup_{n=1}^{\infty} \{z \in \Omega : u_n(z) < \alpha\}$$

is an open set, $u(z)$ is upper semicontinuous in Ω. For each compact set $K \subset \Omega$, a continuous real function $f(z)$ on K is harmonic in the interior of K, and $u(z) \leqslant f(z)$ at all $z \in \partial K$, then for all $\varepsilon > 0$, the set

$$K_n = \{z \in \partial K : u_n(z) \geqslant f(z) + \varepsilon\}$$

is a compact and decreasing $K_n \supset K_{n+1}$. Since $\bigcap_{n=1}^{\infty} K_n = \varnothing$, there is N such that the set K_n must be empty for large $n \geqslant N$, which implies $u_n(z) \leqslant f(z) + \varepsilon$ if $n \geqslant N$ and $z \in K$, $u_n(z) \leqslant f(z) + \varepsilon$. Hence, $u(z) \leqslant f(z)$ and $u(z)$ are subharmonic in Ω. □

Theorem 4.3.5. Let $\varphi(t)$ be a convex increasing function on $\mathbb{R}$ and set $\varphi(-\infty) = \lim_{t\to-\infty} \varphi(t)$. Then $\varphi(u(z))$ is subharmonic in Ω if $u(z)$ is subharmonic in Ω.

Proof. If $z_0 \in \Omega$, then

$$\limsup_{z \to z_0} \varphi(u(z)) = \varphi(\limsup_{z \to z_0} u(z)) \leqslant \varphi(u(z_0)),$$

i.e., $\varphi(u(z))$ is supper semicontinuous in Ω. Let $D(z_0, r_0) \subset \Omega$, $0 < r < r_0$, and

$$t_0 = \frac{1}{2\pi} \int_0^{2\pi} u(z + Re^{i\theta}) d\theta,$$

there is a real number $\alpha \in \mathbb{R}$ such that
$$\varphi(u(z_0 + Re^{i\theta})) \geq \varphi(t_0) + \alpha(u(z_0 + Re^{i\theta}) - t_0).$$
This gives
$$\frac{1}{2\pi}\int_0^{2\pi} \varphi(u(z_0 + Re^{i\theta}))d\theta \geq \varphi(t_0).$$
We use the subharmonicity of u in Ω and the fact that $\varphi(t)$ is increasing in $\mathbb{R}$, it follows that
$$\frac{1}{2\pi}\int_0^{2\pi} \varphi(u(z_0 + Re^{i\theta}))d\theta \geq \varphi(t_0) \geq \varphi(u(z_0)).$$
This proves the theorem. $\square$

Theorem 4.3.6. Let $u_k(z) \geq 0$ ($k = 1, 2, \ldots, n$). If $\log u_k(z)$ ($k = 1, 2, \ldots, n$) is subharmonic in the open set Ω, where ($\log 0 = -\infty$), then
$$\log\left(\sum_{k=1}^n u_k(z)\right)$$
is subharmonic in the open set Ω.

Proof. Let a disk $\overline{D(a, R)} \subset \Omega$ and
$$f(z) = \sum_{k=0}^m c_k z^k$$
be a polynomial in z such that
$$\log\left(\sum_{k=1}^n u_k(a + Re^{i\theta})\right) \leq \operatorname{Re} f(a + Re^{i\theta}),$$
i.e.,
$$\sum_{k=1}^n u_k(a + Re^{i\theta}) \leq |e^{f(a+Re^{i\theta})}|.$$
Since $\log u_k(z) - \operatorname{Re} f(z)$ is subharmonic in Ω, Theorem 4.3.5 shows that $u_k(z)|e^{-f(z)}|$ is subharmonic in Ω. By Theorem 4.3.4,
$$\sum_{k=1}^n u_k(z)|e^{-f(z)}|$$

is subharmonic in Ω and
$$\sum_{k=1}^{n} u_k(z)|e^{-f(z)}| \leqslant 1$$
for all $z \in \overline{D(a,R)}$, i.e.,
$$\log\left(\sum_{k=1}^{n} u_k(z)\right) \leqslant \operatorname{Re} f(z)$$
for all $z \in \overline{D(a,R)}$, Theorem 4.3.2 shows that
$$\log\left(\sum_{k=1}^{n} u_k(z)\right)$$
is subharmonic in Ω. This proves the theorem. $\square$

Theorem 4.3.7. *Let $u(z)$ be subharmonic in the open set Ω and not $-\infty$ identically in any component of Ω. Then*

(a) $E = \{z \in \Omega : u(z) = -\infty\}$ *does not contain any non-empty open set;*

(b) *if $a \in \Omega, \overline{D(a,R)} \subset \Omega$, then*
$$\int_0^{2\pi} |u(a + \operatorname{Re}^{i\theta})| d\theta < \infty;$$

(c) *u is Lebesgue-integrable on all compact subset $K \subset \Omega$ (we write $u \in L^1_{loc}(\Omega)$).*

Proof. If the conclusion of (a) is false, i.e., if E contains a non-empty open set. Let $G \neq \emptyset$ be a component of $\operatorname{int} E$ and $\Omega(G)$, the component of Ω which contains G, since $G \neq \Omega(G), G \subset \Omega(G)$, there is $a \in (\partial G) \cap \Omega(G)$ such that $K = \overline{D(a,R)} \subset \Omega(G)$, so there exists a curve $\gamma \subset \partial K = \{z : |z-a| = R\}$ such that $u(z) = -\infty$ for all $z \in \gamma$, and also, there is $z_0 \in D(a,R)$ such that $u(z_0) > -\infty$.

Suppose that $\{u_k(z)\}$ is a decreasing continuous sequence which converges to $u(z)$ in K. For $z = a + \operatorname{Re}^{i\theta} \in D(a,R)$, let
$$h_k(z) = \frac{1}{2\pi}\int_0^{2\pi} u_k(a + \operatorname{Re}^{i\varphi})\operatorname{Re}\left(\frac{\operatorname{Re}^{i\varphi} + z - a}{\operatorname{Re}^{i\varphi} - (z-a)}\right) d\varphi,$$

then $h_k(z)$ is harmonic in $D(a, R)$ and can be extended to the boundary ∂K, so $h_k(z) = u_k(z) \geqslant u(z)$ for all $z \in \partial K$. Hence,
$$h_k(z) \geqslant u(z), z \in K, k = 1, 2, \ldots.$$
Since $u_k(z)$ is decreasing and converge to $u(z)$ and
$$P_a(R, \varphi; z) = \frac{1}{2\pi} \operatorname{Re} \left(\frac{Re^{i\varphi} + z - a}{Re^{i\varphi} - (z - a)} \right) \geqslant 0,$$
we have
$$-\infty < u(z_0) \leqslant \lim_{k \to \infty} h_k(z_0) = \int_0^{2\pi} u(a + Re^{i\varphi}) P_a(R, \varphi; z_0) d\varphi.$$
$u(z)$ have an upper bound on ∂K and is equal to $-\infty$ on the curve γ, therefore the integral value on the right-hand side of the equation is $-\infty$, leading to a contradiction. Hence, (a) holds.

(b) Let $a \in \Omega$, $K = \overline{D(a, R)} \subset \Omega$ and $M = \sup\{u(z) : z \in K\}$, then M is a real number and there is $z_0 \in D(a, R)$ such that $u(z_0) > -\infty$. Similarly,
$$\infty > M - u(z_0) \geqslant \int_0^{2\pi} (M - u(a + Re^{i\varphi})) P_a(R, \varphi; z_0) d\varphi$$
$$\geqslant \frac{R - |z_0 - a|}{R + |z_0 - a|} \left(M - \frac{1}{2\pi} \int_0^{2\pi} u(a + Re^{i\varphi}) d\varphi \right) \geqslant 0,$$
hence
$$-\infty < \int_0^{2\pi} u(a + Re^{i\varphi}) d\varphi \leqslant M.$$
This proves (b).

(c) For any compact $K \subset \Omega$, there is $R > 0$ such that $D(z, 3R) \subset \Omega$ for $z \in K$, and also, there is $\zeta(z) \in \Omega \cap E^c$ such that $|\zeta(z) - z| < R$, so
$$\bigcup_{z \in K} D(\zeta(z), R) \supset K$$
and
$$-\infty < u(\zeta(z)) \leqslant \frac{1}{\pi R^2} \int_0^R \int_0^{2\pi} u(\zeta(z) + Re^{i\varphi}) r d\varphi dr < \infty.$$

By the finite cover theorem, there are $\zeta_1, \zeta_2, \ldots, \zeta_N \in \Omega \cap E^c$ such that
$$\bigcup_{n=1}^{N} D(\zeta_k, R) \supset K,$$
so
$$-\infty < R^2 \pi \sum_{k=1}^{N} u(\zeta_k) \leqslant \sum_{k=1}^{N} \int_{D(\zeta_k, R)} u(z) d\lambda(z) < \infty.$$
Therefore, u is Lebesgue-integrable on all compact subsets $K \subset \Omega$. □

Theorem 4.3.8. Let $u(z)$ be subharmonic in an open set Ω and
$$M(r, z; u) = \frac{1}{2\pi} \int_0^{2\pi} u(z + Re^{i\varphi}) d\varphi, \quad 0 < r < d(z, \Omega^c),$$
where
$$d(z, \Omega^c) = \inf\{|z - z'| : z' \in \Omega^c\}, \Omega_\delta = \{z : d(z, \Omega^c) > \delta\} \ (\delta > 0).$$
Then

(a) $M(r_1, z; u) \leqslant M(r_2, z; u)$ for $0 < r_1 < r < r_2 < d(z, \Omega^c)$ and
$$M(r, z; u) \leqslant \frac{\log r - \log r_1}{\log r_2 - \log r_1} M(r_2, z; u)$$
$$+ \frac{\log r_2 - \log r}{\log r_2 - \log r_1} M(r_1, z; u); \quad (4.3.7)$$

(b) $z \in \Omega$, then
$$\lim_{r \to 0} M(r, z; u) = u(z).$$

Proof. Fixed $z \in \Omega$, for $0 < r_1 < r_2 < d(z, \Omega^c)$, the function
$$M(\zeta, z; u) = \frac{1}{2\pi} \int_0^{2\pi} u(z + \zeta e^{i\varphi}) d\varphi$$
satisfies $M(\zeta, z; u) = M(|\zeta|, z; u)$, $|\zeta| \leqslant r_2$. Theorem 4.2.2 shows that the function $u(\zeta)$ have an upper bound. By Fatou's lemma, when $|\zeta_0| < r_2$, we have that
$$\limsup_{\zeta \to \zeta_0} M(\zeta, z; u) \leqslant \frac{1}{2\pi} \int_0^{2\pi} \limsup_{\zeta \to \zeta_0} u(z + \zeta e^{i\varphi}) d\varphi \leqslant M(\zeta_0, z; u).$$

Therefore, $\varphi_z(\zeta) = M(\zeta, z; u)$ is upper semicontinuous in $D(0, r_2)$. If
$$|\zeta_0| < r_2, 0 < \rho < r_2 - |\zeta_0|,$$
then
$$\frac{1}{2\pi} \int_0^{2\pi} \varphi_z(\zeta_0 + \rho e^{i\theta}) d\theta = \frac{1}{(2\pi)^2} \int_0^{2\pi} \int_0^{2\pi} u(z + (\zeta_0 + \rho e^{i\theta})e^{i\varphi}) d\varphi d\theta$$
$$\geq \frac{1}{2\pi} \int_0^{2\pi} u(z + \zeta_0 e^{i\varphi}) d\varphi = \varphi_z(\zeta_0).$$

This proves that $\varphi_z(\zeta) = M(\zeta, z; u)$ is subharmonic in $D(0, r_2)$. If $0 < r_1 < r_2 < d(z, \Omega^c)$, and let $K = \partial D(z, r_2)$, suppose that $\{u_k(\zeta)\}$ is continuous decreasing and converge to $u(\zeta)$ on K. For $z + \zeta \in D(z, r_2)$, put
$$h_k(z + \zeta) = \frac{1}{2\pi} \int_0^{2\pi} u_k(z + r_2 e^{i\varphi}) \operatorname{Re}\left(\frac{r_2 e^{i\varphi} + \zeta}{r_2 e^{i\varphi} - \zeta}\right) d\varphi,$$
then $h_k(\zeta)$ is harmonic on $D(z, r_2)$ and can be extended to the boundary ∂K, $h_k(\zeta) = u_k(\zeta) \geq u(\zeta)$ for ζ on ∂K. Thus, $h_k(\zeta) \geq u(\zeta)$ for all ζ in $D(z, r_2)$. Therefore,
$$M(r_1, z; u) \leq \frac{1}{2\pi} \int_0^{2\pi} h_k(z + r_1 e^{i\varphi}) d\varphi$$
$$= h_k(z) = \frac{1}{2\pi} \int_0^{2\pi} u_k(z + r_2 e^{i\varphi}) d\varphi.$$

Since $u_k(\zeta)$ is decreasing to converge to $u(\zeta)$ for $\zeta \in K$, by the monotone convergence theorem, it is known that $M(r_1, z; u) \leq M(r_2, z; u)$. The function
$$H(\zeta) = \frac{\log|\zeta| - \log r_1}{\log r_2 - \log r_1} M(r_2, z; u) + \frac{\log r_2 - \log|\zeta|}{\log r_2 - \log r_1} M(r_1, z; u)$$
is continuous in the compact $K : r_1 \leq |\zeta| \leq r_2$ and harmonic in the interior $r_1 < |\zeta| < r_2$ of K. On the boundary ∂K of a compact set K, it is equal to $H(\zeta) = M(\zeta, z; u)$, thus (4.3.7) holds. Moreover, since
$$u(z) \leq M(r, z; u) \leq \max\{u(\zeta) : |\zeta - z| \leq r\},$$
by upper semicontinuity, it can be known that (b) holds. □

Theorem 4.3.9. *Suppose that $u(z)$ is subharmonic in an open set Ω, $\alpha(t)$ is a non-negative infinitely differentiable function with support $\operatorname{supp}\alpha \subset [0,1]$ such that*

$$2\pi \int_0^1 t\alpha(t^2)dt = 1.$$

Let $\alpha_\delta(z) = \frac{1}{\delta^2}\alpha\left(\frac{|z|^2}{\delta^2}\right)$, then

$$u_\delta(z) = \int_{D(z,\delta)} u(\zeta)\alpha_\delta(z-\zeta)d\lambda(\zeta) \tag{4.3.8}$$

is infinitely differentiable subharmonic in $\Omega_\delta = \{z : d(z,\Omega^c) > \delta\}$ ($\delta > 0$), and as δ monotonically approaches zero, $u_\delta(z)$ converges pointwise monotonically decreasing to $u(z)$ for $z \in K \subset \Omega_\delta = \{z : d(z,\Omega^c) > \delta\}(\delta > 0)$, where $d\lambda(\zeta)$ is a Lebesgue measure.

Proof. Theorem 4.3.8 and

$$u_\delta(z) = \int_0^\delta t\alpha_\delta(t)\int_0^{2\pi} u(z+te^{i\theta})d\theta dt = 2\pi\int_0^1 t\alpha(t^2)M(\delta t, z; u)dt$$

shows that as δ monotonically approaches zero, $u_\delta(z)$ decreases pointwise toward $u(z)$. By Proposition 1.1.4 (differentiation under the integral sign), it is known that $u_\delta(z)$ is infinitely differentiable and subharmonic in Ω_δ ($\delta > 0$). Since

$$\frac{1}{2\pi}\int_0^{2\pi} u_\delta(z+\rho e^{i\varphi})d\varphi = \int_0^1 t\alpha(t^2)\int_0^{2\pi} M(\delta t, z+\rho e^{i\varphi}; u)d\varphi dt$$

$$\geq u_\delta(z),$$

then $u_\delta(z)$ is infinitely differentiable subharmonic in $\Omega_\delta = \{z : d(z,\Omega^c) > \delta\}(\delta > 0)$. □

Theorem 4.3.10. (a) *Let $u(z)$ be subharmonic in the open set Ω and not $-\infty$ identically in any component of Ω. Then*

$$\int_\Omega u(z)\Delta v(z)d\lambda(z) \geqslant 0 \tag{4.3.9}$$

for all $v(z) \in C_0^2(\Omega) = \{v : v$ is a two times continuously differentiable function in Ω vanishing outside a compact subset $\operatorname{supp} v$ of Ω, here $d\lambda(z)$ denotes the Lebesgue measure.

(b) If $u(z) \in C^2(\Omega)$, then $u(z)$ is subharmonic in Ω if and only if $\Delta u \geq 0$.

(c) If $z = g(w)$ is an analytic function from the open set $\widetilde{\Omega}$ into Ω and $g(\widetilde{\Omega}) \subset \Omega$, then $\widetilde{u}(w) = u(g(w))$ is subharmonic in $\widetilde{\Omega}$.

Proof. (a) If $0 < r < d(\operatorname{supp} v, \Omega^c)$, then

$$2\pi u(z) \leq \int_0^{2\pi} u(z + Re^{i\theta}) d\theta$$

for all $z \in \operatorname{supp} v$. Multiplication by v and integration with respect to $d\lambda$ gives

$$\frac{1}{r^2} \int_\Omega u(z) \left(\int_0^{2\pi} v(z - Re^{i\theta}) d\theta - 2\pi v(z) \right) d\lambda(z) \geq 0. \qquad (4.3.10)$$

Substituting the Taylor expansion of v,

$$v(z+\zeta) = v(z) + \frac{\partial v}{\partial z}\zeta + \frac{\partial v}{\partial \bar{z}}\bar\zeta + \frac{1}{2}\frac{\partial^2 v}{\partial z^2}\zeta^2 + \frac{\partial^2 v}{\partial z \partial \bar{z}}|\zeta|^2 + \frac{1}{2}\frac{\partial^2 v}{\partial \bar{z}^2}\bar\zeta^2$$
$$+ o(|\zeta|^2), \quad \zeta \to 0,$$

into (4.3.10) and letting $r \to 0$, it yields (4.3.9). This proves (a).

From (a), the necessity of (b) can be obtained. To prove the sufficiency, by Theorem 4.3.2, it suffices to show that if $\overline{D(a,R)} \subset \Omega$ is a closed disk, $f(z) = \sum_{k=0}^n c_k z^k$ is an analytic polynomial such that

$$u(a + Re^{i\theta}) \leq \operatorname{Re} f(a + Re^{i\theta}),$$

we want to prove that $u(z) \leq \operatorname{Re} f(z)$ for all $z \in \overline{D(a,R)} \subset \Omega$. Otherwise, there exist a closed disk $\overline{D(a,R)} \subset \Omega$ and an analytic polynomial $f(z) = \sum_{k=0}^n c_k z^k$ such that $u(a + Re^{i\theta}) \leq \operatorname{Re} f(a + Re^{i\theta})$, and there is $z_0 \in \overline{D(a,R)} \subset \Omega$, such that $\varepsilon_0 = u(z_0) - \operatorname{Re} f(z_0) > 0$. Taking $\varepsilon > 0$ such that $R^2 \varepsilon < \varepsilon_0$ and putting $u_\varepsilon(z) = u(z) - \operatorname{Re} f(z) + \varepsilon |z-a|^2$,

$$\max\{u_\varepsilon(z) : |z - a| = R\} \leq \varepsilon R^2 < \varepsilon_0 \leq u_\varepsilon(z_0).$$

Since there is $z_\varepsilon \in \overline{D(a,R)}$ such that

$$u_\varepsilon(z_\varepsilon) = \max\{u_\varepsilon(z) : |z - a| \leq R\},$$

then $z_\varepsilon \in D(a, R)$, thus
$$\frac{\partial u_\varepsilon(z_\varepsilon)}{\partial x} = 0, \quad \frac{\partial u_\varepsilon(z_\varepsilon)}{\partial y} = 0, \quad \frac{\partial^2 u_\varepsilon(z_\varepsilon)}{\partial x^2} \leq 0, \quad \frac{\partial^2 u_\varepsilon(z_\varepsilon)}{\partial y^2} \leq 0,$$
hence $\Delta u_\varepsilon(z_\varepsilon) \leq 0$. But $\Delta u_\varepsilon(z_\varepsilon) = \Delta u(z_\varepsilon) + 4\varepsilon \geq 4\varepsilon > 0$. This is a contradiction, thus (b) holds.

According to Definition (4.3.8) of $u_\delta(z)$, $u_\delta(z)$ is an infinitely differentiable subharmonic function on $\Omega_\delta = \{z : d(z, \Omega^c) > \delta\}(\delta > 0)$. This implies that $\Delta u_\delta(z) \geq 0$ ($z \in \Omega_\delta$, so
$$4\frac{\partial^2}{\partial w \partial \overline{w}} u_\delta(g(w)) = \Delta u_\delta(g(w))|g'(w)|^2 \geq 0,$$
thus $u_\delta(g(w))$ is subharmonic in $g^{-1}(\Omega_\delta) = \{w : g(w) \in \Omega_\delta\}$. Therefore,
$$2\pi u_\delta(g(w)) \leq \int_0^{2\pi} u_\delta(g(w + \rho e^{i\varphi}))d\varphi.$$
Letting $\delta \to 0$, by Theorem 4.3.9,
$$2\pi u(g(w)) \leq \int_0^{2\pi} u(g(w + \rho e^{i\varphi}))d\varphi.$$
Therefore, $\tilde{u}(w) = u(g(w))$ is subharmonic in $\tilde{\Omega}$. □

Theorem 4.3.11. *Suppose that u is a subharmonic function with an upper bound in an open set Ω, $\zeta_1, \zeta_2, \ldots, \zeta_n \in \partial\Omega$ and $\partial\Omega \setminus \{\zeta_1, \zeta_2, \ldots, \zeta_n\} \neq \varnothing$,*
$$\limsup_{z \in \Omega, z \to \zeta} u(z) \leq M,$$
for all $\zeta \in \partial\Omega \setminus \{\zeta_1, \zeta_2, \ldots, \zeta_n\}$, then $u(z) \leq M$ for all $z \in \Omega$. If there exists a point in Ω such that the equality holds, then $u(z) \equiv M$.

Proof. According to Theorem 4.3.3, it suffices to prove that there exists M such that $u(z) \leq M$ for all $z \in \Omega$. We will consider three cases. The first case is when Ω is a bounded region. In this case, let us assume that $d = \sup\{|z - z'| : z, z' \in \Omega\}$ is the diameter of Ω. For any $\varepsilon > 0$, we consider a subharmonic function in Ω
$$u_\varepsilon(z) = u(z) + \varepsilon \sum_{k=1}^n \log\left(\frac{|z - \zeta_k|}{d}\right),$$

then
$$\limsup_{z \to \zeta} u_\varepsilon(z) \leqslant M$$
for all $\zeta \in \partial\Omega$, by Theorem 4.3.3, $u_\varepsilon(z) \leqslant M$ for all $z \in \Omega$. This proves that $u(z) \leqslant M$ by letting $\varepsilon \to 0$.

The second case is that Ω is an unbounded region and $\infty \in \partial\Omega \setminus \{\zeta_1, \zeta_2, \ldots, \zeta_n\}$. In this case, for all $\varepsilon > 0$, there is $R_\varepsilon > \max\{|\zeta_1|, |\zeta_2|, \ldots, |\zeta_n|\}$ such that $R_\varepsilon \to \infty$ ($\varepsilon \to 0$) and $u(z) \leqslant M + \varepsilon$ for all $\zeta \in \Omega, |\zeta| \geqslant R_\varepsilon$. Let Ω_ε be a connected component of $\Omega \cap D(0, R_\varepsilon)$, except for finitely many points, for any point ζ on $\partial\Omega_\varepsilon$, we have
$$\limsup_{z \in \Omega_\varepsilon, z \to \zeta} u(z) \leqslant M + \varepsilon.$$
Therefore, as the proof in the first case, we have $u(z) \leqslant M + \varepsilon$ for all $z \in \Omega_\varepsilon$. Thus, $u(z) \leqslant M + \varepsilon$ for all $z \in \Omega, |z| < R_\varepsilon$. This proves that $u(z) \leqslant M$ by $\varepsilon \to 0$.

The third case, when Ω is an unbounded region and $\infty \in \{\zeta_1, \zeta_2, \ldots, \zeta_n\}$. We take $\zeta_n = \infty$, since there is $a \in \partial\Omega \setminus \{\zeta_1, \zeta_2, \ldots, \zeta_n\}$, for all $\varepsilon > 0$, there is
$$\delta_\varepsilon > 0, \quad 2\delta_\varepsilon < \min\{|\zeta_1 - a|, |\zeta_2 - a|, \ldots, |\zeta_{n-1} - a|\},$$
such that $\delta_\varepsilon \to 0$ ($\varepsilon \to 0$) and $u(z) \leqslant M + \varepsilon$ for all $\zeta \in \Omega, |\zeta - a| \leqslant \delta_\varepsilon$. Let Ω_ε be a connected component of $\Omega \setminus D(a, \delta_\varepsilon)$ and $\eta > 0$
$$\limsup_{z \in \Omega_\varepsilon, z \to \zeta} \left(u(z) + \eta \log \frac{\delta_\varepsilon}{|z - a|} \right) \leqslant M + \varepsilon$$
for all $\zeta \in \partial\Omega_\varepsilon \setminus \{\zeta_1, \zeta_2, \ldots, \zeta_{n-1}\}$. Therefore, as the proof in the second case,
$$u(z) \leqslant M + \varepsilon - \eta \log \frac{\delta_\varepsilon}{|z - a|}$$
for all $z \in \Omega_\varepsilon$. We obtain that $u(z) \leqslant M + \varepsilon$ for all $z \in \Omega_\varepsilon$ by letting $\eta \to 0$, so $u(z) \leqslant M + \varepsilon$ for all $z \in \Omega$ with $|z - a| > \delta_\varepsilon$. Finally, this proves that $u(z) \leqslant M$ by letting $\varepsilon \to 0$. □

Theorem 4.3.12. *Suppose that u is subharmonic in a region Ω, and the disk $\overline{D(a, R)} \subset \Omega$. If*
$$P_{R,u}(z) = \frac{1}{2\pi} \int_0^{2\pi} u(a + Re^{i\varphi}) \operatorname{Re} \left(\frac{Re^{i\varphi} + z - a}{Re^{i\varphi} - (z - a)} \right) d\varphi$$

is a harmonic function in the disk $D(a, R)$ defined by the Poisson integral of u, then the function

$$v(z) = \begin{cases} P_{R,u}(z), & |z-a| < R, \\ u(z), & |z-a| \geqslant R, z \in \Omega \end{cases}$$

is subharmonic in Ω and $u(z) \leqslant v(z)$ for all $z \in \Omega$. If $f \in C(\overline{D(a,R)})$ is a subharmonic majorant function for u in the disk $D(a,R)$, then within the disk $D(a,R)$, we have $P_{R,u}(z) \leqslant f(z)$. In particular,

$$P_{R_1,u}(z) \leqslant P_{R_2,u}(z) \quad (|z-a| < R_1 < R_2 \leqslant R).$$

Proof. From the proof of Theorem 4.3.7(a), we have $u(z) \leqslant v(z)$ for all $z \in D(a, R)$, and therefore, $u(z) \leqslant v(z)$ for all $z \in \Omega$. For $z_0 \in \Omega \setminus \partial D(a, R)$, the inequality (4.3.3) holds for appropriately small $\delta > 0$. If z_0 lies on the boundary $\partial D(a, R)$,

$$v(z_0) = u(z_0) \leqslant \frac{1}{2\pi} \int_0^{2\pi} u(z_0 + Re^{i\theta})d\theta \leqslant \frac{1}{2\pi} \int_0^{2\pi} v(z_0 + Re^{i\theta})d\theta.$$

Therefore, (4.3.3) holds for appropriately small $\delta > 0$ respect to v. Following the proof of Theorem 4.1.8, when $z_0 = a + Re^{i\theta_0} \in \partial D(a, R), z \in D(a, R)$, we have

$$v(z) - v(z_0) = \frac{1}{2\pi} \int_0^{2\pi} \left(u(a + Re^{i\theta}) - u(a + Re^{i\theta_0}) \right) P(Re^{i\theta}, z-a) d\theta.$$

For all $\varepsilon > 0$, there is $\delta \in (0, \frac{\pi}{2}]$ such that $u(a+Re^{i\theta}) - u(a+Re^{i\theta_0}) < \varepsilon$ for $\theta : |\theta - \theta_0| < 2\delta$, so

$$v(z) - v(a + Re^{i\theta_0})$$

$$\leqslant \frac{\varepsilon}{2\pi} \int_{|\theta - \theta_0| < 2\delta} P(Re^{i\theta}, z-a) d\theta$$

$$+ \frac{1}{2\pi} \int_{2\delta \leqslant |\theta - \theta_0| \leqslant \pi} \left(u(a+Re^{i\theta}) - u(a+Re^{i\theta_0}) \right) P(Re^{i\theta}, z) d\theta.$$

By the properties (a) and (d) of the Poisson kernel, there exists $r_0 > 0$ such that when $R > r > r_0, |\varphi - \theta_0| < \delta$ $P(Re^{i\theta}, Re^{i\varphi}) < \varepsilon$,

$$v(a + Re^{i\varphi}) - v(a + Re^{i\theta_0}) \leqslant \varepsilon + 2M\varepsilon$$

for all $\theta \notin [\theta_0 - 2\delta, \theta_0 + 2\delta]$, where $M = \sup\{u(\zeta) : |\zeta - a| = R\} - u(z_0)$. According to Theorem 4.2.3, M is finite. Additionally, $u(z)$ is subharmonic in Ω, so when $z_0 \in \partial D(a, R)$,

$$\limsup_{z \to z_0} v(z) \leq v(z_0).$$

Therefore, (4.2.1) holds everywhere in Ω with respect to v, and thus, $v(z)$ is subharmonic in Ω and $u(z) \leq v(z)$ for all $z \in \Omega$. If $f \in C(\overline{D(a, R)})$ is a harmonic majorant function for u in the disk $D(a, R)$, then

$$P_{R,u}(z) \leq \frac{1}{2\pi} \int_0^{2\pi} f(a + Re^{i\varphi}) \operatorname{Re}\left(\frac{Re^{i\varphi} + z - a}{Re^{i\varphi} - (z - a)}\right) d\varphi = f(z)$$

for all $z \in D(a, R)$. This proves the theorem. $\square$

Theorem 4.3.13. *Suppose that $R_0 \geq 0, 0 < \theta_2 - \theta_1 = \frac{\pi}{\rho} \leq 2\pi$, If u is subharmonic in the region $\Omega = \{z = Re^{i\theta} : r > R_0, \theta_1 < \theta < \theta_2\}$ such that*

$$\limsup_{z \in \Omega, z \to \zeta} u(z) \leq a$$

for every $l \zeta \in \partial \Omega$ and there is $\lambda \in (0, \rho)$ such that

$$\liminf_{R \to \infty} \frac{1}{R^\lambda} \sup\{u(z) : |z| = R, z \in \Omega\} < \infty,$$

then $u(z) \leq a$ throughout Ω.

Proof. First, we assume that $\Omega = \{z = Re^{i\theta} : r > R_0, |\theta| < \frac{\pi}{2\rho}\}$. Let $\rho_1 \in (\lambda, \rho)$, $\delta > 0$, consider subharmonic functions

$$w_\delta(z) = u(z) - a - \delta|z|^{\rho_1} \cos \rho_1 \theta, \quad z = |z|e^{i\theta} \in \Omega.$$

Then there exist R_0 and a sequence of positive numbers $\{R_n\}$ ($R_n > R_0$) tending to infinity such that for any $\zeta \in \partial \Omega_n$,

$$\limsup_{z \in \Omega_n, z \to \zeta} w_\delta(z) \leq \delta,$$

where $\Omega_n = \{z = Re^{i\theta} : R_0 < r < R_n, |\theta| < \frac{\pi}{2\rho}\}$. Theorem 4.3.11 implies that $w_\delta(z) \leq \delta$ for all $z \in \Omega_n$. We see that $u(z) \leq a$ for all $z \in \Omega$ by letting $\delta \to 0$.

In general cases, we can consider the subharmonic $\tilde{u}(z) = u(z\exp\{\frac{i}{2}(\theta_1+\theta_2)\})$ in the domain $\{z = Re^{i\theta} : r > R_0, |\theta| < \frac{\pi}{2\rho}\}$. □

Theorem 4.3.14 (Phragmén–Lindelöf Theorem). *Let a, σ and σ_0 be real numbers, and $R_0 \geqslant 0, 0 < \theta_2 - \theta_1 = \frac{\pi}{\rho} \leqslant 2\pi$. Suppose that u is a subharmonic function in the region $\Omega = \{z = Re^{i\theta} : r > R_0, \theta_1 < \theta < \theta_2\}$, such that*

$$\limsup_{z \in \Omega, z \to \zeta} u(z) \leqslant a$$

for all $\zeta \in \partial\Omega$.

(a) *If*

$$\liminf_{R \to \infty} \frac{1}{R^\rho} \sup\{u(z) : |z| = R, z \in \Omega\} \leqslant \sigma,$$

then, when $\sigma \geq 0$, we have

$$u(z) \leqslant a + \frac{4\sigma}{\pi}|z|^\rho \cos\rho\left(\theta - \frac{1}{2}(\theta_1+\theta_2)\right) \qquad (4.3.11)$$

for all $z \in \Omega$.

(b) *If*

$$\limsup_{R \to \infty} \frac{1}{R^\rho} \sup\{u(z) : |z| = R, z \in \Omega\} \leqslant \sigma_0,$$

then, when $\sigma_0 \geq 0$,

$$u(z) \leqslant a + \sigma_0 |z|^\rho \cos\rho\left(\theta - \frac{1}{2}(\theta_1+\theta_2)\right) \qquad (4.3.12)$$

for all $z \in \Omega$.

Proof. (a) First, we assume that $\Omega = \{z = Re^{i\theta} : r > R_0, |\theta| < \frac{\pi}{2\rho}\}$. For all $\delta > 0$, There exists a sequence of increasing positive numbers $\{R_n\}$ greater than R_0 and approaching infinity such that

$$\sup\{u(z) - a : |z| = R_n, z \in \Omega\} \leqslant (\sigma+\delta)R_n^\rho.$$

Let $D_+(0,1) = \{z = Re^{i\theta} : |\theta| < \frac{\pi}{2}, 0 < r < 1\}$ be a half-disk with radius 1, then the function

$$h(z) = 2 + \frac{2}{\pi}(\arg(z-i) - \arg(z+i)) \quad (|\arg(z \pm i)| < \pi) \quad (4.3.13)$$

is continuous on $\overline{D_+(0,1)} \setminus \{i, -i\}$, non-negative harmonic in $D_+(0,1)$ and $0 < h(z) < 1$ for all $z \in D_+(0,1)$. Furthermore, $h(z)$ takes the value 1 on the arc $\{z = e^{i\theta} : |\theta| < \frac{\pi}{2}\}$ and takes the value 0 on the open diameter of the half-disk $\{z = iy : -1 < y < 1\}$. Let

$$G_n = \left\{ z = Re^{i\theta} : |\theta| < \frac{\pi}{2\rho}, 0 < r < R_n \right\}$$

be a sector with radius R_n and central angle $\frac{\pi}{\rho}$, then the function

$$h_n(z) = h\left(\frac{z^\rho}{R_n^\rho}\right)$$

is continuous in $\overline{G_n} \setminus \{iR_n, -iR_n\}$, non-negative harmonic and $0 < h_n(z) \leqslant 1$ in G_n. Furthermore, $h_n(z)$ takes the value 1 on

$$\left\{ z = Re^{i\theta} : |\theta| < \frac{\pi}{2\rho}, r = R_n \right\},$$

and $h_n(z)$ takes the same value 0 on the two radii $\{z = Re^{i\theta} : |\theta| = \frac{\pi}{2\rho}, 0 < r < R_n\}$ of the sector's boundary. Put

$$\Omega_n = \left\{ z = Re^{i\theta} : R_0 < r < R_n, |\theta| < \frac{\pi}{2\rho} \right\},$$

then $\Omega_n \subset G_n$. Consider a subharmonic function

$$g_n(z) = u(z) - a - (\sigma + \delta) R_n^\rho h_n(z)$$

for $z \in \Omega$,

$$\limsup_{z \in \Omega_n, z \to \zeta} g_n(z) \leqslant 0,$$

$\zeta \in \partial \Omega_n$. According to Theorem 4.3.11, $g_n(z) \leqslant 0$ for all $z \in \Omega_n$. Hence,

$$u(z) - a \leqslant (\sigma + \delta) R_n^\rho h_n(z) \quad (z \in \Omega_n).$$

In particular, for any $x \in (0, R_n)$, we have

$$u(x) - a \leqslant (\sigma + \delta) R_n^\rho h_n(x) = (\sigma + \delta) R_n^\rho \frac{4}{\pi} \arctan\left(\frac{x}{R_n}\right)^\rho.$$

First, let $n \to \infty$, we have

$$u(x) - a \leqslant (\sigma + \delta) \frac{4}{\pi} x^\rho,$$

furthermore, let $\delta \to 0$, we obtain that $u(x) - a \leqslant \sigma \frac{4}{\pi} x^\rho$. Applying Theorem 4.3.13 to the subharmonic function

$$u(z) - a - \frac{4\sigma}{\pi}|z|^\rho \cos(\rho\theta), \quad z = |z|e^{i\theta} \in \Omega$$

on regions $\Omega_+ = \{z = Re^{i\theta} : r > R_0, 0 < \theta < \frac{\pi}{2\rho}\}$ and $\Omega_- = \{z = Re^{i\theta} : r > R_0, -\frac{\pi}{2\rho} < \theta < 0\}$, we can conclude that $u(z) - a - \frac{4\sigma}{\pi}|z|^\rho \cos(\rho\theta) \leqslant 0$ for all $z \in \Omega$. This proves that (4.3.11) holds.

(b) First, we assume that $\Omega = \{z = Re^{i\theta} : r > R_0, |\theta| < \frac{\pi}{2\rho}\}$. For all $\delta > 0$,

$$\omega_\delta(z) = u(z) - a - (\sigma_0 + \delta)|z|^\rho \cos\rho\theta, \quad z = |z|e^{i\theta} \in \Omega$$

is subharmonic, and with ζ on the boundary of the regions $\Omega_+ = \{z = Re^{i\theta} : r > R_0, 0 < \theta < \frac{\pi}{2\rho}\}$ and on the boundary of $\Omega_- = \{z = Re^{i\theta} : r > R_0, -\frac{\pi}{2\rho} < \theta < 0\}$, we have

$$\limsup_{z \in \Omega_\pm, z \to \zeta} \omega_\delta(z) \leqslant 0.$$

Applying Theorem 4.3.13 to the subharmonic function $\omega_\delta(z)$ in the regions Ω_+ and Ω_-, we conclude that $u(z) \leqslant 0$. Applying Theorem 4.3.13 again to the subharmonic function $\omega_\delta(z)$ in the region Ω, we conclude again that $\omega_\delta(z) \leqslant 0$. Letting $\delta \to 0$, then $u(z) \leqslant a + \sigma_0 |z|^\rho \cos\rho\theta$ for $z \in \Omega$. This proves (4.3.12).

In the general case, we can consider a subharmonic function $\tilde{u}(z) = u(z\exp\{\frac{i}{2}(\theta_1 + \theta_2)\})$ in region $\{z = Re^{i\theta} : r > R_0, |\theta| < \frac{\pi}{2\rho}\}$. □

Theorem 4.3.15 (Phragmén–Lindelöf Theorem). *Let a be a real number, $0 < \theta_2 - \theta_1 = \frac{\pi}{\rho} \leqslant 2\pi$, u a subharmonic function*

in a region $\Omega = \{z = Re^{i\theta} : \theta_1 < \theta < \theta_2\}$, satisfying
$$\limsup_{z \in \Omega, z \to \zeta} u(z) \leqslant a$$
for all $\zeta \in \partial\Omega$. If there is $\lambda \in (0, \rho)$ such that
$$\liminf_{R \to \infty} \frac{1}{R^\lambda} \sup\{u(Re^{i\theta}) : \theta_1 < \theta < \theta_2\} < \infty$$
and
$$\liminf_{\varepsilon \to 0} \varepsilon^\lambda \sup\{u(\varepsilon e^{i\theta}) : \theta_1 < \theta < \theta_2\} < \infty,$$
then $u(z) \leqslant a$ for all $z \in \Omega$.

Proof. First, we assume that $\Omega = \{z = Re^{i\theta} : |\theta| < \frac{\pi}{2\rho}\}$. Consider for $\rho_1 \in (\lambda, \rho)$, $\delta > 0$, the auxiliary subharmonic function in Ω,
$$\omega_\delta(z) = u(z) - a - \delta(|z|^{\rho_1} + |z|^{-\rho_1})\cos\rho_1\theta,$$
$$z = Re^{i\theta}, \quad |\theta| < \tfrac{\pi}{2\rho}, \quad r > 0.$$

We claim that $u(z) \leqslant a$ for all $z \in \Omega$. We note that there exist a sequence of positive numbers $\{R_n\}$ which increases to ∞ and a sequence of positive numbers $\{\varepsilon_n\}$ which decreases to 0 with $R_1 > \varepsilon_1$ such that
$$\limsup_{z \in D_n, z \to \zeta} \omega_\delta(z) \leqslant \delta$$
for any $\zeta \in \partial D_n$, where $D_n = \{z = Re^{i\theta} : \varepsilon_n < r < R_n, |\theta| < \frac{\pi}{2\rho}\}$. By Theorem 4.3.11, we have $\omega_\delta(z) \leqslant \delta$ for all $z \in D_n$. As n was an arbitrary big quantity > 1 and $\delta > 0$ was an arbitrary small quantity, we can conclude that the equality $u(z) \leqslant a$ in $z\Omega$ holds. In the general case, we can consider the subharmonic function $\widetilde{u}(z) = u(z \exp\{\frac{i}{2}(\theta_1 + \theta_2)\})$ in the region $\{z = Re^{i\theta} : r > 0, |\theta| < \frac{\pi}{2\rho}\}$. □

Theorem 4.3.16 (Phragmén–Lindelöf Theorem). *Let a and b be real numbers, σ and σ_0 two non-negative real numbers, $0 < \theta_2 - \theta_1 = \frac{\pi}{\rho} \leqslant 2\pi$. Suppose that u is subharmonic in the region $\Omega = \{z = Re^{i\theta} : \theta_1 < \theta < \theta_2\}$, satisfying*
$$\limsup_{z \in \Omega, z \to Re^{i\theta_1}} u(z) \leqslant a, \quad \limsup_{z \in \Omega, z \to Re^{i\theta_2}} u(z) \leqslant b$$
for all $r > 0$.

(a) *If*

$$\liminf_{R \to \infty} \frac{1}{R^\rho} \sup\{u(\text{Re}^{i\theta}) : \theta_1 < \theta < \theta_2\} \leqslant \sigma$$

and

$$\liminf_{\varepsilon \to 0} \varepsilon^\rho \sup\{u(\varepsilon e^{i\theta}) : \theta_1 < \theta < \theta_2\} \leqslant \sigma,$$

then

$$u(z) \leqslant \frac{\rho}{\pi}((b-a)\theta + \theta_2 a - \theta_1 b)$$
$$+ \frac{4\sigma}{\pi}(r^\rho + r^{-\rho})\cos\left(\rho\theta - \frac{\rho}{2}(\theta_1 + \theta_2)\right) \quad (4.3.14)$$

for all $z = \text{Re}^{i\theta} \in \Omega, \theta_1 < \theta < \theta_2$.

(b) *If*

$$\limsup_{R \to \infty} \frac{1}{R^\rho} \sup\{u(\text{Re}^{i\theta}) : \theta_1 < \theta < \theta_2\} \leqslant \sigma_0$$

and

$$\limsup_{\varepsilon \to 0} \varepsilon^\rho \sup\{u(\varepsilon e^{i\theta}) : \theta_1 < \theta < \theta_2\} \leqslant \sigma_0,$$

then

$$u(z) \leqslant \frac{\rho}{\pi}((b-a)\theta + \theta_2 a - \theta_1 b)$$
$$+ \sigma_0(r^\rho + r^{-\rho})\cos\left(\rho\theta - \frac{\rho}{2}(\theta_1 + \theta_2)\right) \quad (4.3.15)$$

for all $z = \text{Re}^{i\theta} \in \Omega, \theta_1 < \theta < \theta_2$.

Proof. (a) First, we assume that $\Omega = \{z = \text{Re}^{i\theta} : |\theta| < \frac{\pi}{2\rho}\}$, $a = 0$ and $b = 0$. For all $\delta > 0$, there exist a sequence of increasing positive numbers $\{R_n\}$ to infinity and a sequence of decreasing positive numbers $\{\varepsilon_n\}$ to 0 such that $R_1 > \varepsilon_1$, and

$$\sup\{u(z) : |z| = R_n, z \in \Omega\} \leqslant (\sigma + \delta)R_n^\rho,$$
$$\sup\{u(z) : |z| = \varepsilon_n, z \in \Omega\} \leqslant (\sigma + \delta)\varepsilon_n^{-\rho}.$$

Suppose that $D_+(0,1) = \{z = \text{Re}^{i\theta} : |\theta| < \frac{\pi}{2}, 0 < r < 1\}$ is a half-disk with radius 1. The function $h(z)$ defined by (4.3.13) is continuous on $\overline{D_+(0,1)} \setminus \{i, -i\}$, non-negative subharmonic on the half-disk

$D_+(0,1)$ B, and satisfies $0 < h(z) < 1$. On the open semicircular arc $\{z = e^{i\theta} : |\theta| < \frac{\pi}{2}\}$, $h(z)$ takes the value 1, while on the open diameter $\{z = iy : |y| < 1\}$ of the half-disk, $h(z)$ takes the value 0. Let

$$D_n = \left\{ z = \mathrm{Re}^{i\theta} : |\theta| < \frac{\pi}{2\rho}, \varepsilon_n < r < R_n \right\}$$

be a sector with a central angle of $\frac{\pi}{\rho}$. The function $\alpha_n(z)$ defined by

$$\alpha_n(z) = R_n^\rho h\left(\frac{z^\rho}{R_n^\rho}\right) + \varepsilon_n^{-\rho} h\left(\frac{\varepsilon_n^\rho}{z^\rho}\right)$$

is continuous on $\overline{D_n} \setminus \{iR_n, -iR_n, i\varepsilon_n, -i\varepsilon_n\}$ and positive harmonic in D_n. The value of function $\alpha_n(z)$ is greater than R_n^ρ on the open arc $\{z = \mathrm{Re}^{i\theta} : |\theta| < \frac{\pi}{2\rho}, r = R_n\}$ of the sector's boundary ∂D_n; the value of $\alpha_n(z)$ is greater than $\varepsilon_n^{-\rho}$ on the open arc $\{z = \mathrm{Re}^{i\theta} : |\theta| < \frac{\pi}{2\rho}, r = \varepsilon_n\}$ of the sector's boundary ∂D_n; the value of $\alpha_n(z)$ is 0 on the two line segments $\{z = \mathrm{Re}^{i\theta} : |\theta| = \frac{\pi}{2\rho}, \varepsilon_n < r < R_n\}$ of the sector's boundary ∂D_n. The subharmonic function

$$\beta_n(z) = u(z) - (\sigma + \delta)\alpha_n(z)$$

in D_n satisfies

$$\limsup_{z \in D_n, z \to \zeta} \beta_n(z) \leq 0$$

for all $\zeta \in \partial D_n$. Theorem 4.3.11 implies that $\beta_n(z) \leq 0$ for all $z \in D_n$. Thus, $u(z) \leq (\sigma + \delta)\alpha_n(z)$ for all $z \in D_n$. In particular,

$$u(x) \leq (\sigma + \delta)\alpha_n(x)$$
$$= (\sigma + \delta)\frac{4}{\pi}\left(R_n^\rho \arctan\left(\frac{x}{R_n}\right)^\rho + \varepsilon_n^{-\rho} \arctan\left(\frac{\varepsilon_n}{x}\right)^\rho\right)$$

for all $x \in (\varepsilon_n, R_n)$. We obtain

$$u(x) \leq (\sigma + \delta)\frac{4}{\pi}(x^\rho + x^{-\rho})$$

for all $x \in (0, \infty)$ by letting $n \to \infty$, and we also obtain

$$u(x) \leq \sigma\frac{4}{\pi}(x^\rho + x^{-\rho})$$

for all $x \in (0, \infty)$ by letting $\delta \to 0$. Applying Theorem 4.3.15 to the subharmonic function

$$u(z) - \frac{4\sigma}{\pi}(|z|^\rho + |z|^{-\rho})\cos(\rho\theta), \quad z = |z|e^{i\theta} \in \Omega,$$

to the regions $\Omega_+ = \{z = Re^{i\theta} : 0 < \theta < \frac{\pi}{2\rho}\}$ and $\Omega_- = \{z = Re^{i\theta} : -\frac{\pi}{2\rho} < \theta < 0\}$, we can conclude that the inequality (4.3.14) holds in Ω.

(b) First, we assume that $\Omega = \{z = Re^{i\theta} : |\theta| < \frac{\pi}{2\rho}\}$, $a = 0$ and $b = 0$. Let $\Omega_+ = \{z = Re^{i\theta} : 0 < \theta < \frac{\pi}{2\rho}\}$ and $\Omega_- = \{z = Re^{i\theta} : -\frac{\pi}{2\rho} < \theta < 0\}$. Given $\delta > 0$, the function

$$\omega_\delta(z) = u(z) - (\sigma_0 + \delta)(|z|^\rho + |z|^{-\rho})\cos\rho\theta, \quad z = |z|e^{i\theta} \in \Omega,$$

is subharmonic in Ω, and there is $M > \delta$ such that

$$\limsup_{z \in \Omega_\pm, z \to \zeta} (\omega_\delta(z) - M) \leqslant 0$$

for all $\zeta \in \partial\Omega_+ \cap \Omega_-$. Applying Theorem 4.3.13 to the functions $\omega_\delta(z)$ for $z \in \Omega_+$ and $z \in \Omega_-$, we obtain that $u(z) \leqslant M$ for all $z \in \Omega$. Applying Theorem 4.3.13 to the functions $\omega_\delta(z)$ again for $z \in Omega$, we obtain $\omega_\delta(z) \leqslant \delta$. Thus, $u(z) \leqslant \sigma_0(|z|^\rho + |z|^{-\rho})\cos\rho\theta$ by letting $\delta \to 0$. This proves (4.3.15).

In the general case, we can consider a subharmonic function

$$\widetilde{u}(z) = u(z\exp\{(\theta_1 + \theta_2)/2\}) - \frac{\rho(b-a)}{\pi}\arg z + \frac{a+b}{2}$$

in the region $\{z = Re^{i\theta} : |\theta| < \frac{\pi}{2\rho}\}$. □

Theorem 4.3.17 (Phragmén–Lindelöf Theorem). *Let a be real number, σ and σ_0 two non-negative numbers, $y_2 > y_1, \rho = \frac{\pi}{y_2 - y_1}$. Suppose that u is subharmonic in the half-strip region $\Omega = \{z = x + iy : y_1 < y < y_2, x > x_0\}$, satisfying*

$$\limsup_{z \in \Omega, z \to \zeta} u(z) \leqslant a$$

for all $\zeta \in \partial\Omega$.

(a) If
$$\liminf_{x\to\infty} e^{-\rho x} \sup\{u(x+iy) : y_1 < y < y_2\} \leqslant \sigma,$$
then
$$u(z) \leqslant a + \frac{4\sigma}{\pi} e^{\rho x} \sin(\rho y - \rho y_1)$$
for all $z \in \Omega$.

(b) If
$$\limsup_{x\to\infty} e^{-\rho x} \sup\{u(x+iy) : y_1 < y < y_2\} \leqslant \sigma_0,$$
then
$$u(z) \leqslant a + \sigma_0 e^{\rho x} \sin(\rho y - \rho y_1)$$
for all $z \in \Omega$.

Proof. Since the function $w = g(z) = \exp\{\rho(z - iy_1)\}$ is a conformal mapping of the half-strip domain Ω onto the region $\widetilde{\Omega} = \{w : |w| > e^{\rho x_0}, 0 < \arg w < \pi\}$ and its inverse function $z = g^{-1}(w) = iy_1 + \frac{\log w}{\rho}$ is a conformal mapping of $\widetilde{\Omega}$ onto the region Ω, the function $\widetilde{u}(w) = u(iy_1 + \frac{\log w}{\rho})$ is subharmonic in the region $\widetilde{\Omega}$ and
$$\limsup_{w\in\widetilde{\Omega}, w\to\zeta} \widetilde{u}(w) \leqslant a,$$
for all $\zeta \in \partial\widetilde{\Omega}$, and
$$\sup\{\widetilde{u}(Re^{i\theta}) : 0 < \theta < \pi\} = \sup\{u(\rho^{-1}\log R + iy) : y_1 < y < y_2\}.$$
Theorem 4.3.14 implies that the conclusion of Theorem 4.3.17 holds. $\square$

Theorem 4.3.18. *Let a and b be two real numbers, σ and σ_0 two positive real numbers, and $y_2 > y_1$, $\rho = \frac{\pi}{y_2 - y_1}$. Suppose that u is subharmonic in the region $\Omega = \{z = x + iy : y_1 < y < y_2\}$, satisfying*
$$\limsup_{z\in\Omega, z\to\xi+iy_1} u(z) \leqslant a; \quad \limsup_{z\in\Omega, z\to\xi+iy_2} u(z) \leqslant b$$
for all real number ξ.

(a) *If*
$$\liminf_{|x|\to\infty} e^{-\rho|x|} \sup\{u(x+iy) : y_1 < y < y_2\} \leq \sigma,$$
then
$$u(x+iy) \leq \frac{y-y_1}{y_2-y_1}b + \frac{y_2-y}{y_2-y_1}a + \frac{4\sigma}{\pi}(e^{\rho x} + e^{-\rho x})\sin(\rho y - \rho y_1)$$
for all $x + iy \in \Omega$.

(b) *If*
$$\limsup_{|x|\to\infty} e^{-\rho|x|} \sup\{u(x+iy) : y_1 < y < y_2\} \leq \sigma_0,$$
then
$$u(x+iy) \leq \frac{y-y_1}{y_2-y_1}b + \frac{y_2-y}{y_2-y_1}a + \sigma_0(e^{\rho x} + e^{-\rho x})\sin(\rho y - \rho y_1)$$
for all $x + iy \in \Omega$.

Proof. The function $w = g(z) = \exp\{\rho(z - iy_1)\}$ is a conformal mapping from the strip region Ω onto the upper half-plane $\mathbb{C}_+ = \{w : 0 < \arg w < \pi\}$, its inverse function $z = g^{-1}(w) = iy_1 + \frac{\log w}{\rho}$ is a conformal mapping from the upper half-plane $\mathbb{C}_+$ onto the strip region Ω. So, the function $\tilde{u}(w) = u(iy_1 + \frac{\log w}{\rho})$ is subharmonic in the upper plane $\mathbb{C}_+$

$$\limsup_{\mathrm{Im}\,w>0, w\to r} \tilde{u}(w) \leq a, \qquad \limsup_{\mathrm{Im}\,w>0, w\to -r} \tilde{u}(w) \leq b$$

for all $r > 0$, and

$$\sup\{\tilde{u}(Re^{i\theta}) : 0 < \theta < \pi\} = \sup\{u(\rho^{-1}\log R + iy) : y_1 < y < y_2\}.$$

Theorem 4.3.16 implies that the conclusion of Theorem 4.3.18 holds. □

Theorem 4.3.19. *Suppose that $f(z)$ is analytic in the half-plane $\mathbb{C}_+ = \{z = x+iy : y > 0\}$, continuous in the closed the half-plane*

$\overline{\mathbb{C}}_+ = \{z = x+iy : y \geq 0\}$ and satisfying

$$\limsup_{z\in\mathbb{C}_+, |z|\to\infty} \frac{\log|f(z)|}{|z|} \leq \sigma$$

and

$$\int_{-\infty}^{\infty} |f(x)|^p dx = M < \infty, \quad p > 0.$$

Then

$$\int_{-\infty}^{\infty} |f(x+iy)|^p dx \leq M e^{p\sigma y} \tag{4.3.16}$$

for all $y > 0$.

Proof. For any positive integer N, the sequence of the functions $g_{n,N}(z)$ defined by

$$g_{n,N}(z) = \sum_{k=-n}^{n} \left| f\left(z + \frac{kN}{n}\right) \right|^p \frac{N}{n} \quad (n = 1, 2, \ldots)$$

converges uniformly on compact subsets of $\mathbb{C}_+$ to the function

$$h_N(z) = \int_{-N}^{N} |f(z+t)|^p dt.$$

The function

$$\log\left(|f(z + \frac{kN}{n})|^p \frac{N}{n} \right)$$

is subharmonic in $\mathbb{C}_+$, Theorem 4.3.6(f) implies that $\log g_{n,N}(z)$ is subharmonic in $\mathbb{C}_+$. Theorem 4.3.4(c) shows that $\log|h_N(z)|$ is subharmonic in $\mathbb{C}_+$, satisfying

$$\limsup_{z\in\mathbb{C}_+, |z|\to\infty} \frac{\log|h_N(z)|}{|z|} \leq p\sigma$$

and

$$|h_N(x)| \leq \int_{-\infty}^{\infty} |f(x)|^p dx = M < \infty.$$

Theorem 4.3.14(b) shows that $\log|h_N(x+iy)| \leq \log M + p\sigma y$. Therefore,

$$\int_{-N}^{N} |f(x+iy)|^p dx \leq Me^{p\sigma y}$$

for all $y > 0$. (4.3.16) holds by letting $N \to \infty$. □

Theorem 4.3.20. *Let a be a complex number, $R_0 \geq 0, 0 < \theta_2 - \theta_1 = \frac{\pi}{\rho} \leq 2\pi$, $f(z)$ an analytic function in the region $\Omega = \{z = Re^{i\theta} : r > R_0, \theta_1 < \theta < \theta_2\}$, satisfying*

$$\liminf_{R \to \infty} \frac{1}{R^\rho} \sup\{\log|f(z)| : |z| = R, z \in \Omega\} \leq 0.$$

If for all $\varepsilon > 0$, there exists $M_\varepsilon > 0, R_\varepsilon > R_0$ such that

$$\limsup_{z \in \Omega, z \to Re^{i\theta_1}} |f(z) - a| \leq \varepsilon, \quad \limsup_{z \in \Omega, z \to Re^{i\theta_2}} |f(z)| \leq M_\varepsilon$$

for all $R \geq R_\varepsilon$. Then, for all $\delta \in (0, \theta_2 - \theta_1)$, we have

$$\lim_{R \to \infty} \sup\{|f(Re^{i\theta}) - a| : \theta_1 < \theta \leq \theta_2 - \delta\} = 0. \quad (4.3.17)$$

Proof. First, we assume that $\Omega = \{z = Re^{i\theta} : r > R_0, 0 < \theta < \pi\}$. Since $\log|f(z) - a|$ is subharmonic in Ω, $f(z) - a$ is bounded in Ω by Theorem 4.3.14. Without loss of generality, we can assume $a = 0$ and $|f(z)| \leq 1$. For any $0 < \varepsilon < 1$, there exists $R_\varepsilon > 2R_0 + 1$ such that for any $\theta \in [0, \pi]$, we have

$$0 \leq \pi - \arg(R_0 e^{i\theta} - R_\varepsilon) \leq \arcsin\left(\frac{R_0}{R_\varepsilon}\right) \leq \frac{\pi R_0}{2R_\varepsilon} < \frac{\varepsilon}{-\log \varepsilon},$$

and when $R \geq R_\varepsilon$, we have

$$\limsup_{z \in \Omega, z \to R} |f(z)| \leq \varepsilon.$$

Since

$$u(z) = \log|f(z)| - \varepsilon - \left(1 - \frac{1}{\pi} \arg(z - R_\varepsilon)\right) \log \varepsilon$$

is subharmonic and has a upper bound in Ω

$$\limsup_{z\in\Omega, z\to\zeta} u(z) \leq 0$$

for all $\zeta \in \partial\Omega \setminus \{R_\varepsilon\}$. According to Theorem 4.3.11,

$$\log|f(z)| \leq \varepsilon + \left(1 - \frac{1}{\pi}\arg(z - R_\varepsilon)\right)\log\varepsilon$$

for all $z \in \Omega$. So, for all $\delta \in (0, \pi)$, when $R \geq R_\varepsilon, \theta \in (0, \pi - \delta]$, $\pi - \arg(Re^{i\theta} - R_\varepsilon) \geq \frac{\delta}{2}$, then

$$\sup\{\log|f(Re^{i\theta})| : 0 < \theta \leq \pi - \delta\} \leq \varepsilon + \frac{\delta}{2\pi}\log\varepsilon.$$

Thus, (4.3.17) holds. In the general case, we can consider analytic functions $\widetilde{f}(z) = f(z^{\frac{1}{\rho}}e^{i\theta_1})$ in the set $\{z = Re^{i\theta} : r > R_0^\rho, 0 < \theta < \pi\}$. □

Theorem 4.3.21. *Suppose a and b are complex numbers, $R_0 \geq 0, 0 < \theta_2 - \theta_1 = \frac{\pi}{\rho} \leq 2\pi$. Let $f(z)$ be an analytic function in the region and continuous on the closure of Ω, denoted as $\overline{\Omega}$. Furthermore, assume that $f(z)$ satisfies*

$$\liminf_{R\to\infty} \frac{1}{R^\rho} \sup\{\log|f(z)| : |z| = R, z \in \Omega\} \leq 0.$$

If

$$\lim_{R\to\infty} f(Re^{i\theta_1}) = a, \quad \lim_{R\to\infty} f(Re^{i\theta_2}) = b,$$

then $a = b$, and

$$\limsup_{R\to\infty}\sup\{|f(Re^{i\theta}) - a| : \theta_1 \leq \theta \leq \theta_2\} = 0.$$

Proof. $\delta \in (0, (\theta_2 - \theta_1)/2)$, $f(z) - a$. According to Theorem 4.3.20, for any $\delta \in (0, (\theta_2 - \theta_1)/2)$, $f(z) - a$ uniformly converges to 0 in $\{z = Re^{i\theta} : r > R_0, \theta_1 < \theta < \theta_2 - \delta\}$, and $f(z) - b$ uniformly converges to 0 in $\{z = Re^{i\theta} : r > R_0, \theta_1 + \delta < \theta < \theta_2\}$. Therefore, $f(z)$ uniformly converges to a and b in $\{z = Re^{i\theta} : r > R_0, \theta_1 + \delta < \theta < \theta_2 - \delta\}$. Hence, we have $a = b$, and $f(z)$ uniformly converges to a in $\{z = Re^{i\theta} : r > R_0, \theta_1 < \theta < \theta_2\}$. □

Theorem 4.3.22. *Suppose that a is a complex number, $R_0 \geqslant 0, 0 < \theta_2 - \theta_1 = \frac{\pi}{\rho} \leqslant 2\pi$, let $f(z)$ be a bounded analytic function in a region $\Omega = \{z = Re^{i\theta} : r > R_0, \theta_1 < \theta < \theta_2\}$, $\gamma \in C([0,\infty))$, $\gamma^* = \gamma([0,\infty)) \subset \Omega$ and $\lim_{t \to \infty} \gamma(t) = \infty$. If*

$$\lim_{t \to \infty} f(\gamma(t)) = a,$$

then

$$\lim_{R \to \infty} \sup\{|f(Re^{i\theta}) - a| : \theta_1 + \delta \leqslant \theta \leqslant \theta_2 - \delta\} = 0$$

for all $\delta \in (0, (\theta_2 - \theta_1)/2)$.

Proof. We first assume that

$$\Omega = \{z = Re^{i\theta} : r > R_0, 0 < \theta < \pi\}, a = 0, \quad |f(z)| \leqslant 1$$

and that γ is one-to-one on $[0,\infty)$. Put $f_1(z) = \overline{f(-\overline{z})}$, then $f_1(z) \in H(\Omega)$. Suppose that the curve $\gamma_1(t)$ is symmetric with respect to the imaginary axis, i.e., $\gamma_1(t) = -\overline{\gamma(t)}$ for all $t \in [0,\infty)$, then the union of the images of γ and γ_1, denoted as γ^* and γ_1^*, respectively, i.e., $S = \gamma([0,\infty)) \cup \gamma_1([0,\infty))$ is symmetric with respect to the imaginary axis and lies in the region Ω. $\gamma_1(t) = -\overline{\gamma(t)}, t \in [0,\infty)$. Hence, there exists a curve $\gamma_2 \in C([0,\infty))$ such that γ_2 is one-to-one in $[0,\infty)$, $\lim_{t \to \infty} \gamma_2(t) = \infty$, the images $\gamma_2^* = \gamma_2([0,\infty))$ of γ_2 consists of points located in the first quadrant and belonging to S. In other words, $\gamma_2^* = \{z \in S : \operatorname{Re} z \geqslant 0\}$. Put $g(z) = f(z)f_1(z)$ and $\gamma_2(0) = R_1 e^{i\theta_0}$, where $R_1 = |\gamma_2(0)| > R_0, 0 < \theta_0 \leqslant \frac{\pi}{2}$, then $g(z) \in H(\Omega), g(iy) = |f(iy)|^2$ and $|g(z)| \leqslant 1$. For all $0 < \varepsilon < 1$, there is $R_\varepsilon > 2R_1 + 1$ such that

$$0 \leqslant \pi - \arg(R_1 e^{i\theta} - R_\varepsilon) \leqslant \arcsin\left(\frac{R_1}{R_\varepsilon}\right) \leqslant \frac{\pi R_1}{2R_\varepsilon} < \frac{\varepsilon}{-\log \varepsilon}$$

for all $\theta \in [0, \pi]$ and

$$\limsup_{z \in \Omega, z \to \zeta} |g(z)| = |g(\zeta)| \leqslant \varepsilon$$

for all $\zeta \in \gamma_2^*, |\zeta| \geqslant R_\varepsilon$. The function $u(z) = \log|g(z)| - \varepsilon - (1 - \frac{1}{\pi}\arg(z - R_\varepsilon))\log \varepsilon$ is subharmonic, having a upper bound in Ω_0 and

$$\limsup_{z \in \Omega, z \to \zeta} u(z) \leqslant 0$$

for all $\zeta \in \partial\Omega_0 \setminus \{R_\varepsilon\}$, where $\Omega_0 \subset \Omega$, the boundary of Ω_0 is composed of the ray $(-\infty, -R_1]$, the circular arc $C_1 = \{R_1 e^{i\theta} : \theta_0 \leqslant \theta \leqslant \pi\}$ with radius R_1, and γ_2^*. According to Theorem 4.3.11, $\log|g(z)| \leqslant \varepsilon + (1 - \frac{1}{\pi}\arg(z - R_\varepsilon))\log\varepsilon$ for all $z \in \Omega_0$. In particular,

$$\log|g(iy)| \leqslant \varepsilon + \left(1 - \frac{1}{\pi}\arg(iy - R_\varepsilon)\right)\log\varepsilon$$

$$= \varepsilon + \left(\frac{1}{2} - \frac{1}{\pi}\arctan\frac{R_\varepsilon}{y}\right)\log\varepsilon$$

for all $y > R_\varepsilon$. Thus,

$$\limsup_{y\to\infty}|g(iy)|^2 \leqslant \varepsilon e^\varepsilon.$$

Taking the limit $\varepsilon \to 0$, one obtains

$$\lim_{y\to\infty}|f(iy)| = 0.$$

According to Theorem 4.3.20, for all $\delta \in (0, \frac{\pi}{2})$, we have

$$\limsup_{R\to\infty}\sup\{|f(Re^{i\theta})| : \delta \leqslant \theta \leqslant \pi - \delta\} = 0.$$

In general, we can consider the analytic function $\widetilde{f}(z) = f(z^{\frac{1}{\rho}}e^{i\theta_1})$ in the region

$$\{z = Re^{i\theta} : r > R_0^\rho, 0 < \theta < \pi\}. \qquad \square$$

Theorem 4.3.23. *Let u be subharmonic in the unit disk U. Then u has a harmonic majorant if and only if*

$$\sup\left\{\frac{1}{2\pi}\int_0^{2\pi} u(Re^{i\varphi})d\varphi : 0 < r < 1\right\} < \infty. \qquad (4.3.18)$$

In this case, u has the least harmonic majorant given by

$$h(z) = \lim_{r\to 1}\frac{1}{2\pi}\int_0^{2\pi} u(Re^{i\varphi})\operatorname{Re}\left(\frac{Re^{i\varphi} + z}{Re^{i\varphi} - z}\right)d\varphi \qquad (4.3.19)$$

for all $z \in U$.

Proof. If h is a harmonic majorant function of u in the unit disk U, then for any $r \in (0,1)$, we have

$$\frac{1}{2\pi}\int_0^{2\pi} u(\mathrm{Re}^{i\varphi})d\varphi \leqslant \frac{1}{2\pi}\int_0^{2\pi} h(\mathrm{Re}^{i\varphi})d\varphi = h(0).$$

Conversely, if (4.3.18) holds, then for any $R \in (0,1)$, the function

$$P_{R,u}(z) = \frac{1}{2\pi}\int_0^{2\pi} u(\mathrm{Re}^{i\varphi})\mathrm{Re}\left(\frac{\mathrm{Re}^{i\varphi}+z}{\mathrm{Re}^{i\varphi}-z}\right)d\varphi$$

is harmonic in the disk $D(0,R)$ and, by Theorem 4.3.12, we have $P_{R_1,u}(z) \leqslant P_{R_2,u}(z)$ ($|z| < R_1 < R_2 < 1$). Therefore, by the Harnack theorem, (4.3.19) converges uniformly on compact subsets of U, and hence, $h(z)$ is a harmonic function in U. By Theorem 4.3.12, the least harmonic majorant function of u in the unit disk U is $h(z)$. □

4.4 The Dirichlet Problem and Green's Function

In the harmonic function theory, one of the most important problems is to find a harmonic function with given boundary values. This is known as the Dirichlet problem. More precisely, given a real function $f(\zeta)$ on the boundary $\partial\Omega$ of a region Ω, the goal is to find a harmonic function $u(z)$ in Ω that satisfies $f(\zeta) = u(\zeta)$ for all $\zeta \in \partial\Omega$. According to the maximum principle, if such a function $u(z)$ exists, it must be unique.

Let $f(\zeta)$ be a bounded function on the boundary $\partial\Omega$ of a bounded region Ω. The Perron family of function of f is the family $\mathfrak{F}(f)$ of subharmonic functions u in Ω such that for every $\zeta \in \partial\Omega$, we have

$$\limsup_{z\to\zeta} u(z) \leqslant f(\zeta) \leq M, \qquad (4.4.1)$$

where $M = \sup\{|u(z)| : z \in \Omega\}$.

It is easy to see that the function set $\mathfrak{F}(f)$ has the following properties:

Theorem 4.4.1. $\mathfrak{F}(f)$ *is not empty.*

Proof. In fact, $u(z) = -M \in \mathfrak{F}(f)$, where $M = \sup\{|u(z)| : z \in \Omega\}$. □

Theorem 4.4.2. If $u_1, u_2, \ldots, u_n \in \mathfrak{F}(f)$, then $u = \max\{u_1, u_2, \ldots, u_n\} \in \mathfrak{F}(f)$.

Proof. By Theorem 4.3.4(a). □

Theorem 4.4.3. If $u \in \mathfrak{F}(f)$, the closed disk $\overline{D(a,R)} \subset \Omega$, $P_u(z) = P_{R,u}(z)$ is the Poisson integral defined by (4.3.9), then the function $v(z)$ defined by (4.3.10) also belongs to $\mathfrak{F}(f)$.

Proof. From Theorem 4.3.12, we know that Theorem 4.4.3 holds. □

Theorem 4.4.4. Let $\varphi(z) = \sup\{u(z) : u \in \mathfrak{F}(f)\}$, then the function $\varphi(z)$ is harmonic in the region Ω (the function φ is called the Perron function of f).

Proof. From the properties of the family $\mathfrak{F}(f)$, we have $\varphi(z) \leqslant M$. To prove that it is harmonic in Ω, it suffices to show that for any $z_0 \in \Omega$, there exists a closed disk $\overline{D(z_0, R)} \subset \Omega$ such that $\varphi(z)$ is harmonic in $D(z_0, R)$.

By the definition of $u(z_0)$, there exists a sequence $\{v_n\} \subset \mathfrak{F}(f)$ such that $\lim_{n\to\infty} v_n(z_0) = \varphi(z_0)$. Let $V_n = \max\{v_1, v_2, \ldots, v_n\}$, $n = 1, 2, \ldots$. According to Theorem 4.4.2, $V_n \in \mathfrak{F}(f)$ and the sequence $\{V_n\}$ is increasing in Ω. Let

$$U_n(z) = \begin{cases} P_{V_n}(z), & |z - z_0| < R, \\ V_n(z), & |z - z_0| \geqslant R, \ z \in \Omega. \end{cases}$$

According to Theorem 4.4.3, $U_n \in \mathfrak{F}(f)$, the sequence $\{U_n\}$ is increasing in Ω and harmonic in $D(z_0, R)$. The Harnack principle implies that the sequence $\{U_n\}$ converges uniformly on compact subsets of $D(z_0, R)$ to a harmonic function $U(z)$ in $D(z_0, R)$. We also have, for $z \in D(z_0, R)$,

$$v_n(z) \leqslant V_n(z) \leqslant U_n(z) \leqslant U(z).$$

Thus, $U(z_0) = u(z_0)$, $U(z) \leqslant \varphi(z)$, $z \in D(z_0, R)$. To prove that $\varphi(z)$ is harmonic in $D(z_0, R)$, it suffices to show that for any point $z_1 \in D(z_0, R)$, we have $U(z_1) = \varphi(z_1)$.

According to the definition of $\varphi(z_1)$, there is a sequence $\tilde{v}_n \in \mathfrak{F}(f)$ such that $\lim_{n\to\infty} \tilde{v}_n(z_1) = \varphi(z_1)$. Let

$$\tilde{V}_n = \max\{\tilde{v}_1, \tilde{v}_2, \ldots, \tilde{v}_n, V_n\}, \quad n = 1, 2, \ldots.$$

According to Theorem 4.4.2, $\tilde{V}_n \in \mathfrak{F}(f)$, and the sequence $\{\tilde{V}_n\}$ is increasing in Ω. Let

$$\tilde{U}_n(z) = \begin{cases} P_{\tilde{V}_n}(z), & |z - z_0| < R, \\ \tilde{V}_n(z), & |z - z_0| \geqslant R, \ z \in \Omega. \end{cases}$$

According to Theorem 4.4.3, $\tilde{U}_n \in \mathfrak{F}(f)$, the sequence $\{\tilde{U}_n\}$ is increasing in Ω, harmonic in $D(z_0, R)$. According to the Harnack principle, the sequence $\{\tilde{U}_n\}$ converges uniformly on compact subsets of $D(z_0, R)$ to a harmonic function $\tilde{U}(z)$. For all $z \in \Omega$,

$$\tilde{v}_n(z) \leqslant \hat{V}_n(z) \leqslant \tilde{U}_n(z) \leqslant \tilde{U}(z),$$

thus

$$\tilde{U}(z_1) = \varphi(z_1), \quad U(z) \leqslant \tilde{U}(z) \leqslant \varphi(z), z \in D(z_0, R).$$

So, $U(z_0) = \varphi(z_0) = \tilde{U}(z_0)$. Since the harmonic function $U(z) - \tilde{U}(z)$ attains its maximum value of 0 at $z_0 \in D(z_0, R)$, we can conclude that the harmonic function $U(z) - \tilde{U}(z)$ is identically zero in $D(z_0, R)$. Therefore, we have $U(z) \equiv \tilde{U}(z)$ in $D(z_0, R)$. Hence, for any point $z_1 \in D(z_0, R)$, we have $U(z_1) = \varphi(z_1)$. $\square$

Suppose a continuous real function f is given on the boundary $\partial\Omega$ of a bounded region Ω. According to Theorem 4.4.4, the Perron function $\varphi(z)$ associated with f is harmonic in Ω. Therefore, if the Dirichlet problem for Ω is solvable, its solution $u_0(z)$ must belong to the Perron family $\mathfrak{F}(f)$. Thus, we have $u_0(z) \leqslant \varphi(z)$. On the other hand, for any $u(z) \in \mathfrak{F}(f)$ and for every point ζ on $\partial\Omega$, we have

$$\limsup_{z\to\zeta}(u(z) - u_0(z)) \leqslant 0.$$

According to Theorem 4.3.11, $u_0(z)$ is a harmonic majorant function of $u(z)$, which means $u(z) \leqslant u_0(z)$. Therefore, $\varphi(z) \leqslant u_0(z)$. Thus, $u_0(z) = \varphi(z)$. In conclusion, if the Dirichlet problem has a solution, then its solution $u_0(z)$ must be the Perron function $\varphi(z)$ associated with f. The next step in solving the Dirichlet problem is to prove that for every point ζ on $\partial\Omega$, $\lim_{z\to\zeta}\varphi(z)$ exists and equals $f(\zeta)$. However, this is not always true and depends on the geometric properties of the domain Ω. Here is an example. Let $\Omega = U \setminus 0$ be the unit disk with the origin removed. The boundary of Ω, denoted as $\partial\Omega$, consists of the unit circle $|\zeta| = 1$ and the point at the origin $\zeta = 0$. The boundary value function is

$$f(\zeta) = \begin{cases} 0, & |\zeta| = 1, \\ 1, & \zeta = 0. \end{cases}$$

If $u(z) \in \mathfrak{F}(f)$, then since $0 \leqslant f(\zeta) \leqslant 1$, for any $z \in \Omega$, we must have $u(z) \leqslant 1$. From (4.4.1) and Theorem 4.3.11, it follows that $u(z) \leqslant 0$ for all $z \in \Omega$. Therefore, the Perron function with respect to f is identically zero, and the Dirichlet problem on Ω is not solvable. To remedy this, we introduce the concept of a barrier function.

Definition 4.4.5. Let Ω be a bounded domain, and let $\zeta_0 \in \partial\Omega$. If there exists a function $\omega(\zeta)$ that is harmonic in Ω, continuous on $\overline{\Omega}$, satisfies $\omega(\zeta_0) = 0$, and for any $z \in \overline{\Omega} \setminus \{\zeta_0\}$, we have $\omega(z) > 0$, then we call $\omega(z)$ the barrier function of Ω at ζ_0.

Theorem 4.4.6. *Let Ω be a bounded domain and $\zeta_0 \in \partial\Omega$. If the barrier function $\omega(z)$ of Ω at ζ_0 exists and $f(\zeta)$ is bounded on $\partial\Omega$ with $|f(\zeta)| \leqslant M$, and $f(\zeta)$ is continuous at ζ_0, then the harmonic function $\varphi(z)$ in Theorem 4.4.4 satisfies*

$$\lim_{z\to\zeta_0} \varphi(z) = f(\zeta_0). \tag{4.4.2}$$

Proof. It suffices to prove that

$$\limsup_{z\to\zeta_0} \varphi(z) \leqslant f(\zeta_0), \quad \liminf_{z\to\zeta_0} \varphi(z) \geqslant f(\zeta_0). \tag{4.4.3}$$

To prove the first inequality in (4.4.3), for any $\varepsilon > 0$, there exists a neighborhood $D(\zeta_0, \delta)$ of ζ_0 such that when $\zeta \in D(\zeta_0, \delta) \cap (\partial\Omega)$, we have

$$|f(\zeta) - f(\zeta_0)| < \varepsilon.$$

Let $\omega(z)$ be the barrier function of Ω at ζ_0, which has a positive lower bound ω_0 on $\overline{\Omega} \setminus D(\zeta_0, \delta)$. Consider the auxiliary function

$$W(z) = f(\zeta_0) + \varepsilon + \frac{\omega(z)}{\omega_0}(M - f(\zeta_0)).$$

Then

$$W(\zeta) \geqslant f(\zeta_0) + \varepsilon \geqslant f(\zeta),$$

everywhere $\zeta \in D(\zeta_0, \delta) \cap \partial\Omega$. If $\zeta \in \partial\Omega \setminus D(\zeta_0, \delta)$, then we have

$$W(\zeta) \geqslant f(\zeta_0) + \varepsilon + M - f(\zeta_0) = M + \varepsilon \geqslant f(\zeta).$$

Hence, if $\zeta \in \partial\Omega$, then we have

$$W(\zeta) \geqslant f(\zeta).$$

For all $v \in \mathfrak{F}(f), \zeta \in \partial\Omega$, we have

$$\limsup_{z \to \zeta}(v(z) - W(z)) \leqslant f(\zeta) - W(\zeta) \leqslant 0.$$

According to the maximum principle for subharmonic functions in Ω, we have

$$v(z) \leqslant W(z).$$

On the other word, we have

$$\varphi(z) = \sup\{v(z) : v \in \mathfrak{F}(f)\} \leqslant W(z), \quad z \in \Omega.$$

As a consequence, we obtain

$$\limsup_{z \to \zeta_0} u(z) \leqslant \limsup_{z \to \zeta} W(z) = f(\zeta_0) + \varepsilon$$

for all $\varepsilon > 0$, so

$$\limsup_{z \to \zeta_0} u(z) \leqslant f(\zeta_0).$$

To prove the second inequality in (4.4.3), we consider another auxiliary function

$$V(z) = f(\zeta_0) - \varepsilon - \frac{\omega(z)}{\omega_0}(M + f(\zeta_0)).$$

Then
$$V(\zeta) \leqslant f(\zeta_0) - \varepsilon \leqslant f(\zeta),$$
everywhere $\zeta \in D(\zeta_0, \delta) \cap (\partial\Omega)$. If $\zeta \in \partial\Omega \setminus D(\zeta_0, \delta)$, then we have
$$V(\zeta) \leqslant f(\zeta_0) - \varepsilon - (M + f(\zeta_0)) = -M - \varepsilon \leqslant f(\zeta).$$
So, when $\zeta \in \partial\Omega$, $V(\zeta) \geqslant f(\zeta)$. According to the definition of $\mathfrak{F}(f)$, we have $V \in \mathfrak{F}(f)$, we can conclude that
$$\varphi(z) \geqslant W(z)$$
for all $z \in \Omega$, and
$$\liminf_{z \to \zeta_0} u(z) \geqslant \liminf_{z \to \zeta} V(z) = f(\zeta_0) - \varepsilon.$$
$\varepsilon > 0$ is arbitrary,
$$\liminf_{z \to \zeta_0} u(z) \geqslant f(\zeta_0).$$
This proves (4.4.2). □

Lemma 4.4.7. *Let Ω be a bounded region and $\zeta_0 \in \partial\Omega$. If there is a simply connected region D such that every point on the boundary of D is a simple boundary point, and $\Omega \subset D$ with $\zeta_0 \in \partial D \cap \partial\Omega$, then the barrier function $\omega(z)$ of Ω at ζ_0 exists.*

Proof. By the Riemann mapping theorem, there exists a conformal map $f(z)$ from the simply connected domain D to the unit disk U, which can be extended to a homeomorphism from $\overline{D}$ to the closed unit disk $\overline{U}$ such that $f(\zeta_0) = -1$. Therefore, the function $\omega(z) = \operatorname{Re} f(z) + 1$ is harmonic in Ω, continuous on $\overline{\Omega}$, and satisfies $\omega(\zeta_0) = 0$ and $\omega(z) > 0$ for all $z \in \overline{\Omega} \setminus \{\zeta_0\}$. Thus, $\omega(z)$ is a barrier function of Ω at ζ_0. □

Definition 4.4.8. *If the Dirichlet problem has a solution in a region Ω, then Ω is referred to as a Dirichlet region.*

Theorem 4.4.9. *If Ω is a bounded region with its boundary consisting of a finite number of Jordan curves, then Ω is a Dirichlet region.*

Definition 4.4.10. Let Ω be a region and $a \in \Omega$. The Green function of Ω with the singularity at a is harmonic function in $\Omega \setminus \{a\}$, which is continuous on $\overline{\Omega} \setminus \{a\}$ such that

(a) if $a \neq \infty$, $g(z,a) + \log|z-a|$, where is harmonic in Ω;
(b) if $a = \infty$, $g(z,\infty) - \log|z|$ is harmonic in Ω;
(c) for any $\zeta \in \partial \Omega$, $\lim_{z \to \zeta} g(z,a) = 0$.

There is at most one Green function $g(z,a)$ with the singularity at a for a given region Ω and a point a in Ω. For example, if Ω is the extended complex plane $\overline{\mathbb{C}}, a = \infty$, then the Green's function with the singularity at a does not exist. In fact, if $g(z,\infty)$ exists, then it is harmonic in the complex plane $\mathbb{C}$ and can be expressed as $g(z,\infty) = \log|z| + G(z)$, where $G(z)$ is harmonic in a neighborhood of ∞. Therefore, there exists an entire function

$$f(z) = \sum_{n=0}^{\infty} a_n z^n,$$

such that $g(z,\infty) = \mathrm{Re} f(z)$. Let $A(r)$ and $M(r)$ denote the maximum values of $|f(z) - f(0)|$ and $\mathrm{Re}(f(z) - f(0))$ on $|z| = r$, respectively. When $0 < r < R$, by the Carathé inequality, we have

$$M(r) \leqslant \frac{2r}{R-r} A(R). \tag{4.4.4}$$

However, $A(R) - \log R$ is bounded for $R \in [1, \infty)$. Taking the limit as $R \to \infty$ in (4.4.4), we have $f(z) \equiv f(0)$. Therefore, $g(z, \infty) = \mathrm{Re} f(z)$ is a constant, which leads to a contradiction.

If a Green's function $g(z, a)$ with the singularity point $a \in \Omega$, it is unique. In fact, if $g_1(z, a)$ is also a Green's function for the same point a within Ω, then $g(z, a) - g_1(z, a)$ is harmonic within Ω and vanishes on $\partial \Omega$. By the maximum principle, $g(z, a) - g_1(z, a)$ is identically zero in Ω, i.e., $g_1(z, a) = g(z, a)$.

Theorem 4.4.11. *Let Ω be a bounded Dirichlet region. Then, for any point a in Ω, the Green function $g(z, a)$ with the singularity at a exists.*

Proof. We can solve the Dirichlet problem using $f(\zeta) = \log|z - a|$ as the boundary value. Let its solution be denoted as $G(z)$. Then, the function $g(z, a) = -\log|z - a| + G(z)$ is Green's function with the singularity at the point $a \in \Omega$. □

Theorem 4.4.12. *If Ω is a bounded region with its boundary consisting of a finite number of piecewise smooth Jordan curves, $g(z, a) = -\log|z - a| + G(z)$, which is Green's function with the singularity at the point $a \in \Omega$, then for any given continuous real function $f(\zeta)$ on $\partial\Omega$, the solution to the Dirichlet problem can be expressed as*

$$u(z) = \frac{1}{2\pi} \int_{\partial\Omega} f(\zeta) \frac{\partial g(\zeta, z)}{\partial n} ds, \qquad (4.4.5)$$

where $\frac{\partial g(\zeta,z)}{\partial n}$ is the outward normal derivative of $g(\zeta, z)$ with respect to ζ on $\partial\Omega$.

Proof. Under the additional assumption that $u(z)$ and $g(z, z_0)$ have piecewise continuous first-order partial derivatives on $\partial\Omega$, we can prove this theorem. Let a be any point in Ω. Consider a small disk $D(a, \varepsilon)$, $\overline{D(a, \varepsilon)} \subset \Omega$. Let $\Omega_\varepsilon = \Omega \setminus D(a, \varepsilon)$. By taking $v = g(z, a)$ and $u = u(z)$ in Green's formula (4.1.16), and noting that $g(z, a)$ and $u(z)$ are both harmonic inside Ω_ε and $g(z, a)$ is zero on $\partial\Omega$, we obtain

$$\int_{\partial\Omega} u(\zeta) \frac{\partial g(\zeta, a)}{\partial n} ds = \int_{\partial D(a,\varepsilon)} \left(u(z) \frac{\partial g(z, a)}{\partial n} - g(z, a) \frac{\partial u(z)}{\partial n} \right) ds.$$

Since $u(z)$ and $\frac{\partial u(z)}{\partial n}$ is continuous at the point a and continuous in $\partial D(a, \varepsilon)$, and

$$\frac{\partial g(z, a)}{\partial n} = -\frac{dg(a + \varepsilon e^{i\theta}, a)}{d\varepsilon} = \frac{1}{\varepsilon} - \frac{dG(a + \varepsilon e^{i\theta})}{d\varepsilon},$$

where $G(z)$ is harmonic in the region Ω. Thus,

$$\int_{\partial\Omega} u(\zeta) \frac{\partial g(\zeta, a)}{\partial n} ds = ((u(a) + o(1)) \left[\varepsilon^{-1} + O(1)\right] + O(-\log \varepsilon)) 2\pi\varepsilon$$

$$\to 2\pi u(a) \ (\varepsilon \to 0).$$

This proves that (4.4.5) holds. □

Theorem 4.4.13 (Carleman Formula). *Let*

$$\Omega(\rho, R) = \{z : \rho < |z| < R, \operatorname{Re} z > 0\} \quad (0 < \rho < R),$$

and let $f(z)$ be holomorphic in $\overline{\Omega(\rho, R)} = \{z : \rho \leqslant |z| \leqslant R, \operatorname{Re} z \geqslant 0\}$. Let $\lambda_k = r_k e^{i\theta_k}$ ($k = 1, 2, \ldots, n$) be the zeros of $f(z)$ in $\overline{\Omega(\rho, R)}$ (m multiplicity of the zeros taken in the account m time), suppose that $f(z)$ has no zero in $\{z = \rho e^{i\theta} : |\theta| < \frac{\pi}{2}\}$. Then

$$\sum_{\rho < |\lambda_k| \leqslant R} \left(\frac{1}{|\lambda_k|} - \frac{|\lambda_k|}{R^2} \right) \cos \theta_n$$

$$= \frac{1}{\pi R} \int_{-\frac{\pi}{2}}^{\frac{\pi}{2}} \log |f(Re^{i\theta})| \cos \theta \, d\theta$$

$$+ \frac{1}{2\pi} \int_{\rho}^{R} \left(\frac{1}{y^2} - \frac{1}{R^2} \right) \log |f(iy) f(-iy)| \, dy + A_f(\rho, R), \quad (4.4.6)$$

where

$$A_f(\rho, R) = -\frac{1}{2\pi} \int_{-\frac{\pi}{2}}^{\frac{\pi}{2}} \left(\left(\frac{1}{\rho} + \frac{\rho}{R^2} \right) \log |f(\rho e^{i\theta})| \right.$$

$$\left. + \left(1 - \frac{\rho^2}{R^2} \right) \frac{\partial \log |f(\rho e^{i\theta})|}{\partial \rho} \right) \cos \theta \, d\theta. \quad (4.4.7)$$

Proof. There is $\varepsilon > 0$ such that $\overline{D(\lambda_k, \varepsilon)} \cap \overline{D(\lambda_j, \varepsilon)} = \varnothing$ for all $\lambda_k \neq \lambda_j$. Applying Green formula (4.1.16) for the functions $u(z) = \log|f(z)|$ and $v(z) = \operatorname{Re}\left(\frac{1}{z} - \frac{z}{R^2}\right)$ in the region $\Omega_\varepsilon(\rho, R) = \Omega(\rho, R) \setminus \bigcup_{k=1}^{n} \overline{D(\lambda_k, \varepsilon)}$,

$$\int_{\partial \Omega_\varepsilon(\rho, R)} \left(v \frac{\partial u}{\partial n} - u \frac{\partial v}{\partial n} \right) ds = 0,$$

where $\frac{\partial}{\partial n}$ is the derivative in the outward normal direction on $\partial \Omega_\varepsilon(\rho, R)$. The function $v(z)$ is harmonic in $\mathbb{C} \setminus \{0\}$:

$$v(Re^{i\theta}) = 0, \quad -\frac{\partial v}{\partial n} = \frac{2\cos\theta}{R^2}$$

on the semicircle $\{z = Re^{i\theta} : |\theta| < \frac{\pi}{2}\}$ of radius R;

$$v(\pm it) = 0, \quad -\frac{\partial v}{\partial n} = \frac{1}{t^2} - \frac{1}{R^2}$$

on the two vertical lines $\{z = \pm it : \rho < t < R\}$. Also,

$$\lim_{\varepsilon \to 0} \int_{\partial D(\lambda_k,\varepsilon)} \left(v\frac{\partial u}{\partial n} - u\frac{\partial v}{\partial n}\right) ds = 2\pi m_k \cos\theta_k \left(\frac{1}{r_k} - \frac{r_k}{R^2}\right)$$

for $\lambda_k \in \Omega_\varepsilon(\rho, R)$, where m_k is the multiplicity of $f(z)$ at the point λ_k,

$$\lim_{\varepsilon \to 0} \int_{\partial(D(\lambda_k,\varepsilon) \cap \Omega_\varepsilon(\rho,R))} \left(v\frac{\partial u}{\partial n} - u\frac{\partial v}{\partial n}\right) ds = 0$$

for $\lambda_k \in \partial\Omega(\rho, R)$ and $|\lambda_k| > \rho$. Thus,

$$\sum_{\rho < |\lambda_n| \leq R} \left(\frac{1}{|\lambda_n|} - \frac{|\lambda_n|}{R^2}\right) \cos\theta_n$$

$$= \frac{1}{\pi R} \int_{-\frac{\pi}{2}}^{\frac{\pi}{2}} \log|f(Re^{i\theta})| \cos\theta\, d\theta$$

$$+ \frac{1}{2\pi} \int_\rho^R \left(\frac{1}{y^2} - \frac{1}{R^2}\right) \log|f(iy)f(-iy)|\, dy + A_f(\rho, R),$$

where

$$A_f(\rho, R)$$
$$= \frac{1}{2\pi} \int_{-\frac{\pi}{2}}^{\frac{\pi}{2}} \left(\log|f(\rho e^{i\theta})| \frac{\partial v(\rho e^{i\theta})}{\partial \rho} - v(\rho e^{i\theta}) \frac{\partial \log|f(\rho e^{i\theta})|}{\partial \rho}\right) \rho\, d\theta$$

$$= -\frac{1}{2\pi} \int_{-\frac{\pi}{2}}^{\frac{\pi}{2}} \left(\left(\frac{1}{\rho} + \frac{\rho}{R^2}\right) \log|f(\rho e^{i\theta})|\right.$$

$$\left. + \left(1 - \frac{\rho^2}{R^2}\right) \frac{\partial \log|f(\rho e^{i\theta})|}{\partial \rho}\right) \cos\theta\, d\theta.$$

This proves (4.4.6). □

Theorem 4.4.14 (Carleman Formula). *Let $R > 0$, $\Omega(R) = \{z : |z| < R, \operatorname{Re} z > 0\}$, suppose $f(z)$ is holomorphic in $\overline{\Omega(R)} = \{z : |z| \leq R, \operatorname{Re} z \geq 0\}$. Let $\lambda_k = r_k e^{i\theta_k}$ $(k = 1, 2, \ldots, n)$ be the zeros of $f(z)$ in $\overline{\Omega(R)}$ (m multiplicity of the zeros taken in the account m time).*

(a) *For given $R > \rho > 0$, $A_f(\rho, R)$ in (4.4.7) can be written as*

$$A_f(\rho, R) = \frac{1}{2\pi}\left(\frac{1}{\rho^2} - \frac{1}{R^2}\right) \int_0^\rho \log|f(iy)f(-iy)|dy$$

$$- \left(\frac{1}{\rho^2} - \frac{1}{R^2}\right) \sum_{0 < |\lambda_n| \leq \rho} |\lambda_n| \cos\theta_n$$

$$- \frac{1}{\pi\rho} \int_{-\frac{\pi}{2}}^{\frac{\pi}{2}} \log|f(\rho e^{i\theta})| \cos\theta \, d\theta. \tag{4.4.8}$$

(b) *If $f(0) = 1$, then*

$$\sum_{|\lambda_n| \leq R} \left(\frac{1}{|\lambda_n|} - \frac{|\lambda_n|}{R^2}\right) \cos\theta_n$$

$$= \frac{1}{\pi R} \int_{-\frac{\pi}{2}}^{\frac{\pi}{2}} \log|f(Re^{i\theta})| \cos\theta \, d\theta$$

$$+ \frac{1}{2\pi} \int_0^R \left(\frac{1}{y^2} - \frac{1}{R^2}\right) \log|f(iy)f(-iy)| dy - \frac{1}{2}\operatorname{Re} f'(0).$$
$$\tag{4.4.9}$$

Proof. We first prove (b). Suppose that $f(0) = 1$, put $\rho \to 0$ in (4.4.7), then

$$\lim_{\rho \to 0} A_f(\rho, R)$$

$$= \lim_{\rho \to 0} \frac{-1}{2\pi} \int_{-\frac{\pi}{2}}^{\frac{\pi}{2}} \operatorname{Re}\left(\frac{1}{\rho}\log f(\rho e^{i\theta}) + \frac{\partial \log f(\rho e^{i\theta})}{\partial \rho}\right) \cos\theta \, d\theta$$

$$= -\frac{1}{2}\operatorname{Re} f'(0).$$

This proves (4.4.9). Therefore, if $f(0) = 1$, for $R > \rho > 0$, (4.4.9) holds for both R and $R = \rho$. Subtracting the two equations, we conclude that (4.4.8) holds. If 0 is a zero of $f(z)$ order k, then there exists a holomorphic function $g(z)$ on $\overline{\Omega(R)}$, and a constant $c \neq 0$ such that $f(z) = cz^k g(z)$ and $g(0) = 1$, After calculation, it is found that (4.4.8) holds for non-zero constant functions c and power functions cz^k. Therefore, it can be concluded that for $R > \rho > 0$, (4.4.8) holds for general functions $f(z)$. □

4.5 Harmonic Functions in the Unit Disk

The collection of all complex Borel measures with finite total variation on $[-\pi, \pi]$ is denoted by $\mathscr{M}([-\pi, \pi])$. For $\mu \in \mathscr{M}([-\pi, \pi])$, the function $\mu(t) = \mu([-\pi, t]))$ is called the distribution function of μ, which is a bounded variation function on $[-\pi, \pi]$. As a result, the real part $\operatorname{Re}, \mu(t)$ and the imaginary part $\operatorname{Im}, \mu(t)$ can both be expressed as the difference of two monotonically increasing functions. The Lebesgue decomposition theorem [26] states that there exists $f \in L^1([-\pi, \pi])$ such that $\mu'(t) = f(t)$ holds almost everywhere on $[-\pi, \pi]$. The singular complex measure of μ, denoted by $d\mu_s = d\mu - f(t)dt$, has a distribution function $\mu_s(t)$ such that for almost every $t \in [-\pi, \pi]), (\mu_s'(t) = 0$ holds. $\|\mu\| = \int_{[-\pi,\pi]} d|\mu|(x)$ is called the total variation norm of μ. The space $\mathscr{M}([-\pi, \pi])$ equipped with this norm becomes a Banach space. By the Riesz representation theorem [26], in an isomorphic sense, $\mathscr{M}([-\pi, \pi])$ is the dual space of $C([-\pi, \pi])$, i.e., for any bounded linear functional $T \in (C([-\pi, \pi]))^*$ on $C([-\pi, \pi])$, there exists $\mu \in \mathscr{M}([-\pi, \pi])$ such that

$$\|T\| = \|\mu\| = \sup \left\{ \left| \int_{[-\pi,\pi]} f(t)\overline{d\mu(t)} \right| : f \in C([-\pi, \pi]), \|f\|_\infty = 1 \right\}$$

and

$$T(f) = \int_{[-\pi,\pi]} f(x)\overline{d\mu(x)}, \qquad f \in C([-\pi, \pi]). \tag{4.5.1}$$

Since $C([-\pi, \pi])$ is a separable Banach space, the Banach–Alaoglu theorem [26] states that the unit ball in $\mathscr{M}([-\pi, \pi])$, denoted by $\{\mu \in \mathscr{M}([-\pi, \pi]) : \|\mu\| \leqslant 1\}$, is weak* compact. In other words, for any sequence of bounded linear functionals $T_n \in (C([-\pi, \pi]))^*$ such that $\|T_n\| \leqslant 1$ for $n = 1, 2, \ldots$, there exists a subsequence $\{T_{n_k}\}$ and

a functional $T \in (C([-\pi, \pi]))^*$ such that
$$T(f) = \lim_{k \to \infty} T_{n_k}(f) \tag{4.5.2}$$
for all $f \in C([-\pi, \pi])$. If we identify an element f in $L^1([-\pi, \pi])$ with the measure $f(t)dt \in \mathscr{M}([-\pi, \pi])$, then $L^1([-\pi, \pi])$ can be viewed as a closed subspace of $\mathscr{M}([-\pi, \pi])$. Furthermore, let $U = D(0, 1)$ be the unit disk and $\partial U = \partial D(0, 1)$ be the boundary of the unit disk. Then, $\mathscr{M}_P(\partial U)$ denotes the set of finite complex-valued Borel measures on ∂U and $C_P(\partial U)$ denotes the set of continuous function g on ∂U. Then, $\mathscr{M}_P(\partial U)$ is the dual space of $C_P(\partial U)$.

Definition 4.5.1. The function
$$P(\zeta, z) = \operatorname{Re}\left(\frac{\zeta + z}{\zeta - z}\right) = \frac{1 - |z|^2}{|\zeta - z|^2} = \frac{1 - r^2}{1 + r^2 - 2r\cos(\theta - t)}$$
for $z = \operatorname{Re}^{i\theta}$ and $\zeta = e^{it}$ is called the Poisson kernel on the unit disk U. If $f(t) \in L^1([-\pi, \pi])$ or $f(t) = g(e^{it}) \in L^1([-\pi, \pi])$, the function
$$\frac{1}{2\pi} \int_{-\pi}^{\pi} P(e^{it}, \operatorname{Re}^{i\theta}) f(t) dt$$
is called the Poisson integral of f and is denoted by $P(f)(z)$ or $P(g)(z)$, i.e.,
$$P(f)(z) = \frac{1}{2\pi} \int_{-\pi}^{\pi} P(e^{it}, \operatorname{Re}^{i\theta}) f(t) dt \tag{4.5.3}$$
for $z = \operatorname{Re}^{i\theta}$. If $\mu \in \mathscr{M}([-\pi, \pi])$ is a complex measure on $[-\pi, \pi]$, we will write the Poisson integral $u = P(d\mu)$ of μ in the form
$$P(\mu)(\operatorname{Re}^{i\theta}) = \frac{1}{2\pi} \int_{-\pi}^{\pi} P(e^{it}, \operatorname{Re}^{i\theta}) d\mu(t). \tag{4.5.4}$$

Theorem 4.5.2. *Let $\mu \in \mathscr{M}([-\pi, \pi])$, then the Poisson integral $u = P(d\mu)$ of μ,*
$$u(z) = P(\mu)(z) = \frac{1}{2\pi} \int_{-\pi}^{\pi} P(e^{it}, \operatorname{Re}^{i\theta}) d\mu(t)$$
is a harmonic function in the unit disk U.

Proof. We can assume, without loss of generality, that $\mu \in \mathcal{M}([-\pi, \pi])$ is a positive measure, the function u defined by

$$u(z) = \operatorname{Re}\left(\frac{1}{2\pi}\int_{-\pi}^{\pi}\frac{e^{it}+z}{e^{it}-z}d\mu(t)\right) \qquad (4.5.5)$$

is real part of an analytic function. Therefore, $u(z)$ is harmonic in the unit disk U. □

The region

$$\Gamma_\alpha(e^{i\theta_0}) = \{\operatorname{Re}^{i\theta} \in U : |\theta - \theta_0| < \alpha(1-r)\} \qquad (4.5.6)$$

in the unit disk U is called a non-tangential approach region, with vertex $e^{i\theta_0}$, where $\alpha > 0$.

Definition 4.5.3. Let $f(t) = g(e^{it})$ be a measurable function having a period of 2π, if for any $\alpha > 0$,

$$\lim_{\substack{z \to e^{i\theta_0} \\ z \in \Gamma_\alpha(e^{i\theta_0})}} P(f)(z) = f(\theta_0) = g(e^{i\theta_0}), \qquad (4.5.7)$$

then Poisson integral $u(z) = P(f)(z) = P(g)(z)$ is said to have non-tangential limit $f(\theta_0) = g(e^{i\theta_0})$.

Theorem 4.5.4. If $f \in L^p([-\pi, \pi])$ $(1 \leq p \leq \infty)$, then the Poisson integral $u(z) = P(f)(z)$ of f satisfies the following:

(a) For $r \in (0, 1)$, we have $u_r(t) = u(\operatorname{Re}^{it}) \in L^p([-\pi, \pi])$, and

$$\|u_r\|_p^p = \int_{-\pi}^{\pi}|u(\operatorname{Re}^{it})|^p dt \leq \|f\|_p^p = \int_{-\pi}^{\pi}|f(t)|^p dt. \qquad (4.5.8)$$

(b) For all $1 \leq p < \infty$, let $(Pf)_r(t) = P(f)(\operatorname{Re}^{it})$, then

$$\lim_{r \to 1}\|(Pf)_r - f\|_p = 0. \qquad (4.5.9)$$

If $f \in C([-\pi, \pi])$ and $f(-\pi) = f(\pi)$, then

$$\lim_{r \to 1}\sup\{|P(f)(\operatorname{Re}^{i\theta}) - f(\theta)| : |\theta| \leq \pi\} = 0. \qquad (4.5.10)$$

(c) If $f \in L^p([-\pi,\pi])$ $(1 \leqslant p < \infty)$ with respect to the period 2π, then for any Lebesgue point θ_0 of the function f, we have, for all $\alpha > 0$,

$$\lim_{\substack{z \to e^{i\theta_0} \\ z \in \Gamma_\alpha(e^{i\theta_0})}} P(|f - f(\theta_0)|)(z) = 0 \qquad (4.5.11)$$

and

$$\lim_{\substack{z \to e^{i\theta_0} \\ z \in \Gamma_\alpha(e^{i\theta_0})}} P(f)(z) = f(\theta_0), \qquad (4.5.12)$$

i.e., $P(f)(z)$ is almost everywhere non-tangential convergence to f.

Proof. (a) Since $P(e^{it}, \mathrm{Re}^{i\theta}) = P(e^{i(t-\theta)}, r)$ and

$$\frac{1}{2\pi} \int_{-\pi}^{\pi} P(e^{it}, \mathrm{Re}^{i\theta_0}) = 1, \qquad (4.5.13)$$

by applying Fubini's theorem and Minkowski's inequality, we can obtain (a).

(b) For $1 \leqslant p < \infty$, $r \in (0,1)$ and for all $\delta \in (0, \frac{\pi}{4})$, by applying Minkowski's inequality, we can obtain

$2\pi \|(P(f))_r - f\|_p$

$\leqslant \int_{|t|<\delta} \|T_t(f) - f\|_p P(e^{it}, r)dt + \int_{\delta \leqslant |t| \leqslant \pi} \|T_t(f) - f\|_p P(e^{it}, r)dt,$

where $T_t(f)(s) = f(s+t)$. Since

$$\sup\{P(e^{it}, r) : \delta \leqslant |t| \leqslant \pi\} = P(e^{i\delta}, r) \to 0 \quad (r \to 1) \qquad (4.5.14)$$

and the continuity of L^p of the integral of f, we have

$$\limsup_{r \to 1} \|(Pf)_r - f\|_p \leqslant \sup\{\|T_t(f) - f\|_p : |t| \leqslant \delta\} \to 0 \quad (\delta \to 0).$$

If $f \in C([-\pi,\pi])$ and $f(-\pi) = f(\pi)$, then for all $\delta \in (0, \frac{\pi}{4})$, we also have

$$\limsup_{r \to 1} \{|P(f)(\mathrm{Re}^{i\theta}) - f(\theta)| : |\theta| \leqslant \pi\}$$

$$\leqslant \sup\{|f(\theta - t) - f(\theta)| : |t| \leqslant \delta\} \to 0 \quad (\delta \to 0).$$

(c) When $z = Re^{i\theta}$, then

$$2\pi|P(f)(z) - f(\theta_0)| \leq 2\pi P(|f - f(\theta_0)|)(z) \leq I_1 + I_2,$$

where

$$I_1 = \int_{|t|<\eta} |f(t+\theta_0) - f(\theta_0)|P(e^{i(t+\theta_0-\theta)}, r)dt,$$

$$I_2 = \int_{\eta \leq |t| \leq \pi} |f(t+\theta_0) - f(\theta_0)|P(e^{i(t+\theta_0-\theta)}, r)dt.$$

For all $\delta > 0$, there is $\eta \in (0, \pi)$ such that if $0 < s \leq \eta$, we have

$$G(s) = \int_0^s (|f(\theta_0 - t) - f(\theta_0)| + |f(\theta_0 + t) - f(\theta_0)|)dt \leq \delta s. \tag{4.5.15}$$

Now, let $r \in (\frac{1}{2}, 1)$, for all $\alpha > 0$ and $z = Re^{i\theta} \in \Gamma_\alpha(e^{i\theta_0})$ and $\alpha(1-r) \leq \frac{\eta}{4}$, we have $|\theta - \theta_0| \leq \alpha(1-r) \leq \frac{\eta}{4}$, so if $\eta \leq |t| \leq \pi$, we have

$$P(e^{i(t-\theta+\theta_0)}, r) \leq P(e^{i\frac{\eta}{2}}, r) \leq \frac{1-r}{\sin^2 \frac{\eta}{2}},$$

thus

$$I_2 \leq 4\pi(|f(\theta_0)| + \|f\|_1)\frac{2(1-r)}{\sin^2 \frac{\eta}{2}}.$$

For I_1, the abbreviation of the expression $P(e^{i(t-\theta+\theta_0)}, r)$ as $P(t)$, then

$$I_1 \leq \int_0^\eta (P(s) + P(-s))G'(s)ds$$

$$= G(s)(P(s) + P(-s))\Big|_0^\eta - \int_0^\eta G(s)\frac{d}{ds}(P(s) + P(-s))ds$$

$$\leq \delta\left(\eta(P(\eta) + P(-\eta)) + \int_0^\eta s(|P'(s)| + |P'(-s)|)ds\right).$$

Since when $0 \leqslant t \leqslant \pi$, we have $\frac{d}{dt}P(e^{it},r) \leqslant 0$, so when $0 \leqslant s \leqslant |\theta - \theta_0| \leqslant \eta$, $(\theta - \theta_0)P'(s) \geqslant 0$, $(\theta - \theta_0)P'(-s) \geqslant 0$, therefore

$$\int_0^{|\theta-\theta_0|} s(|P'(s)| + |P'(-s)|)ds$$

$$= \left| \int_0^{|\theta-\theta_0|} s(P'(s) + P'(-s))ds \right|$$

$$= \left| s(P(s) - P(-s)) \Big|_0^{|\theta-\theta_0|} - \int_0^{|\theta-\theta_0|} (P(s) - P(-s))ds \right|$$

$$\leqslant \frac{4|\theta - \theta_0|}{1-r} + 2\pi \leqslant 4\alpha + 2\pi$$

and

$$\int_{|\theta-\theta_0|}^{\eta} s(|P'(s)| + |P'(-s)|)ds$$

$$= \left| \int_{|\theta-\theta_0|}^{\eta} s(P'(s) + P'(-s))ds \right|$$

$$\leqslant \frac{4\eta(1-r)}{\sin^2 \frac{\eta}{2}} + \frac{4|\theta - \theta_0|}{1-r} + 2\pi \leqslant \frac{\eta(1-r)}{\sin^2 \frac{\eta}{2}} + 4\alpha + 2\pi.$$

Therefore, for all $\delta > 0$, there exists $\eta \in (0, \pi)$ such that, for all $\alpha > 0$ and $z = \text{Re}^{i\theta} \in \Gamma_\alpha(e^{i\theta_0})$, when $r \in (\frac{1}{2}, 1)$ and $\alpha(1-r) \leqslant \frac{\eta}{4}$, we have

$$I_1 + I_2 \leqslant \delta \left(\frac{3\eta(1-r)}{\sin^2 \frac{\eta}{2}} + 8\alpha + 4\pi \right) + 2\pi(|f(\theta_0)| + \|f\|_1) \frac{1-r}{\sin^2 \frac{\eta}{2}}.$$

Due to the arbitrary of $\delta > 0$, (c) holds. □

If we strengthen the conditions on f, then Theorem 4.5.4 yields the following result for $p = \infty$.

Corollary 4.5.5. *Let $f \in L^\infty([-\pi, \pi])$, then the Poisson integral $u(z) = P(f)(z)$ of f satisfies the following:*

(a) *If $f \in C([-\pi, \pi])$ and $f(-\pi) = f(\pi)$, as $r \to 1$, the function $u(\text{Re}^{i\theta})$ uniformly converges to $f()$ on the interval $[-\pi, \pi]$ and*

the function $u(z)$ defined by
$$u(z) = \begin{cases} P(f)(z), & |z| < 1, \\ f(\theta), & z = e^{i\theta}, \ |\theta| \leqslant \pi \end{cases}$$
is harmonic in U and continuous on $\overline{U}$.
(b) If $f \in L^\infty([-\pi,\pi])$, as $r \to 1$, $u(Re^{i\theta})$ is weak* convergence to f, i.e., for all $\varphi \in C([-\pi,\pi])$ with $\varphi(-\pi) = \varphi(\pi)$, we have
$$\lim_{r \to 1} \int_{-\pi}^{\pi} u(Re^{i\theta})\varphi(\theta)d\theta = \int_{-\pi}^{\pi} f(t)\varphi(t)dt.$$

Proof. (a) If $f \in C([-\pi,\pi])$ and $f(-\pi) = f(\pi)$, then f can be extended to a continuous function on $\mathbb{R}$ with period 2π. According to Theorem 4.5.4(c), $\lim_{r \to 1} u(Re^{i\theta}) = f(\theta)$ holds everywhere on $\mathbb{R}$, and f is uniformly continuous on $\mathbb{R}$. In other words, for any $\varepsilon > 0$, there exists $\delta \in (0,\pi)$ such that for any θ, whenever $|\theta' - \theta| < 2\delta$,
$$|f(\theta') - f(\theta)| < \varepsilon/2. \tag{4.5.16}$$
On the other hand, there exists $\eta > 0$ such that when $0 < 1 - r < \eta$, we have
$$2\|f\|_\infty \cdot P(e^{i\delta}, r) < \varepsilon/2. \tag{4.5.17}$$
Therefore, combining (4.5.16) and (4.5.17), we conclude that for $|\theta' - \theta| < \delta$ and $0 < 1 - r < \eta$,
$$2\pi|u(Re^{i\theta'}) - f(\theta)| \leqslant \int_{-\pi}^{\pi} |f(\theta' - t) - f(\theta)| P(e^{it}, r)dt$$
$$= \int_{|t|<\delta} + \int_{\delta \leqslant |t| \leqslant \pi} \leqslant \varepsilon/2 + 2\|f\|_\infty \cdot P(e^{i\delta}, r) < \varepsilon.$$
Therefore, as $z \in U$ approaches $e^{i\theta}$, $u(z)$ converges uniformly to f. In other words, $u(z)$ is continuous at the point $e^{i\theta}$.
(b) For any $\varphi \in C([-\pi,\pi])$ and $\varphi(-\pi) = \varphi(\pi)$, let $v(z) = P(\varphi)(z)$, then
$$\int_{-\pi}^{\pi} u(Re^{i\theta})\varphi(\theta)d\theta = \frac{1}{2\pi}\int_{-\pi}^{\pi}\int_{-\pi}^{\pi} P(e^{i\theta}, Re^{it})f(t)dt\varphi(\theta)d\theta$$
$$= \int_{-\pi}^{\pi} v(Re^{it})f(t)dt.$$

Let $v_r(t) = v(\mathrm{Re}^{it})$, by Corollary 4.5.5(a),

$$\left| \int_{-\pi}^{\pi} u(\mathrm{Re}^{i\theta})\varphi(\theta)d\theta - \int_{-\pi}^{\pi} f(t)\varphi(t)dt \right|$$

$$\leqslant \int_{-\pi}^{\pi} |v(\mathrm{Re}^{it}) - \varphi(t)||f(t)|dt$$

$$\leqslant \|f\|_\infty \|v_r - \varphi\|_1 \to 0 \quad (r \to 1).$$

This proves the theorem. □

Following are several properties of the Poisson integral $P(\mu)$ for a measure μ.

Theorem 4.5.6. Let $\mu \in \mathscr{M}([-\pi, \pi])$, $u(z) = P(\mu)(z)$ and $u_r(t) = u(\mathrm{Re}^{it})$, then

(a) for all $r \in (0, 1)$, $\|u_r\|_1 \leqslant \|\mu\|$
(b) as $r \to 1$, u_r weak* converges to μ, i.e., for $\varphi \in C([-\pi, \pi])$ with $\varphi(-\pi) = \varphi(\pi)$,

$$\lim_{r \to 1} \int_{-\pi}^{\pi} u(\mathrm{Re}^{i\theta})\varphi(\theta)d\theta = \int_{-\pi}^{\pi} \varphi(\theta)d\mu(\theta). \qquad (4.5.18)$$

Proof. Using (4.5.3) and Fubini's theorem, we can obtain the following conclusion (a). For any $\varphi \in C([-\pi, \pi])$ such that $\varphi(-\pi) = \varphi(\pi)$, let $v(z) = P(\varphi)(z)$, applying Fubini's theorem, we have

$$\int_{-\pi}^{\pi} u(\mathrm{Re}^{i\theta})\varphi(\theta)d\theta = \int_{-\pi}^{\pi} v(\mathrm{Re}^{it})d\mu(t).$$

So, according to Corollary 4.5.5(a),

$$\left| \int_{-\pi}^{\pi} u(\mathrm{Re}^{i\theta})\varphi(\theta)d\theta - \int_{-\pi}^{\pi} \varphi(t)d\mu(t) \right|$$

$$\leqslant \int_{-\pi}^{\pi} |v(\mathrm{Re}^{it}) - \varphi(t)|d|\mu|(t)$$

$$\leqslant \|\mu\| \|v_r - \varphi\|_\infty \to 0 \quad (r \to 1).$$

This proves the theorem. □

According to Theorem 4.5.2, if $f \in L^p([-\pi, \pi])$ $(1 \leq p \leq \infty)$, then the Poisson integral $P(f)(z)$ of f is harmonic in the unit disk U. Now, we will discuss the converse problem, namely, for a harmonic function u in the unit disk U, under what conditions is u the Poisson integral of an L^p function?

Let us associate to any function u in U a family of functions u_r on $[-\pi, \pi]$, defined by

$$u_r(\theta) = u(\mathrm{Re}^{i\theta}) \quad (0 \leq r < 1).$$

Theorem 4.5.7. *Suppose that $u(z)$ is harmonic in the unit disk U, $1 \leq p \leq \infty$,*

$$C = \sup_{0 < r < 1} \|u_r\|_p < \infty. \tag{4.5.19}$$

(a) *If $1 < p \leq \infty$, it follows that there is a unique $f \in L^p([-\pi, \pi])$ so that $u = P(f)$ is the Poisson integral of f.*

(b) *If $p = 1$, it follows that there is a unique complex Borel measure $\mu \in \mathcal{M}([-\pi, \pi])$ so that $u = P(\mu)$ is the Poisson integral of μ.*

(c) *If $p = 1$, and*

$$\lim_{r_1, r_2 \to 1} \|u_{r_1} - u_{r_2}\|_1 = 0,$$

it follows that there is a unique $f \in L^1([-\pi, \pi])$ so that $u = P(f)$ is the Poisson integral of f.

Proof. (a) If $1 < p \leq \infty$, let p' be the conjugate exponents to p (i.e., $1 \leq p' < \infty$, $\frac{1}{p} + \frac{1}{p'} = 1$). From (4.5.19), the L^p norm of u_r is uniformly bounded with respect to r and $L^{p'}([-\pi, \pi])$ is separable for any sequence $\{r_k\}$ satisfying $1 > r_k > 0$ and $r_k \to 1$ as $k \to \infty$), the sequence $\{u_{r_k}\}$, as a bounded linear functional on $L^{p'}([-\pi, \pi])$, is weak* sequentially compact. Therefore, there exists a subsequence $\{u(r_{k_j} e^{i\theta})\}$ of $\{u(r_k e^{i\theta})\}$ and $f \in L^p([-\pi, \pi])$ such that $\{u(r_{k_j} e^{i\theta})\}$ weak* converges to f, i.e., for all $g \in L^{p'}([-\pi, \pi])$, we have

$$\lim_{j \to \infty} \int_{-\pi}^{\pi} u(r_{k_j} e^{it}) g(t) dt = \int_{-\pi}^{\pi} f(t) g(t) dt.$$

Let $g(t) = P(e^{it}, \mathrm{Re}^{i\theta})$, then

$$\lim_{j \to \infty} \int_{-\pi}^{\pi} u(r_{k_j} e^{it}) P(e^{it}, \mathrm{Re}^{i\theta}) dt = \int_{-\pi}^{\pi} f(t) P(e^{it}, \mathrm{Re}^{i\theta}) dt. \tag{4.5.20}$$

On the other hand, since $u_j(z) = u(r_{k_j} z)$ is harmonic and bounded in U and continuous in $\overline{U}$, by applying Theorem 4.1.3, we have

$$\lim_{j\to\infty} \int_{-\pi}^{\pi} u(r_{k_j} e^{it}) P(e^{it}, \mathrm{R}e^{i\theta}) dt = 2\pi \lim_{j\to\infty} u(r_{k_j} \mathrm{R}e^{i\theta}) = 2\pi u(\mathrm{R}e^{i\theta}). \tag{4.5.21}$$

Using (4.5.20) and (4.5.21), we can conclude that u is the Poisson integral of f.

(b) If $p = 1$, for any sequence $\{r_k\}$ satisfying $1 > r_k > 0$ and $r_k \to 1$ as $k \to \infty$), by (4.5.19), the sequence $\{u_r(\theta) = u(\mathrm{R}e^{i\theta})\}$, as a bounded linear functional on $C([-\pi, \pi])$, applying the Banach–Alaoglu theorem to the sequence $\{u_r(\theta) = u(\mathrm{R}e^{i\theta})\}$, there exists a subsequence $\{u(r_{k_j} e^{i\theta})\}$ of $\{u(r_k e^{i\theta})\}$ and $\mu \in \mathscr{M}([-\pi, \pi])$ such that $\{u(r_{k_j} e^{i\theta})\}$ weak* converges to μ. Let $g(t) = P(e^{it}, \mathrm{R}e^{i\theta})$, then

$$\lim_{j\to\infty} \int_{-\pi}^{\pi} u(r_{k_j} e^{it}) P(e^{it}, \mathrm{R}e^{i\theta}) dt = \int_{-\pi}^{\pi} P(e^{it}, \mathrm{R}e^{i\theta}) d\mu(t). \tag{4.5.22}$$

Since $u_j(z) = u(r_{k_j} z)$ is harmonic and bounded in U and continuous in $\overline{U}$, by applying Theorem 4.1.3, we have

$$\lim_{j\to\infty} \frac{1}{2\pi} \int_{-\pi}^{\pi} u(r_{k_j} e^{it}) P(e^{it}, \mathrm{R}e^{i\theta}) dt = \lim_{j\to\infty} u(r_{k_j} \mathrm{R}e^{i\theta}) = u(\mathrm{R}e^{i\theta}). \tag{4.5.23}$$

By (4.5.22) and (4.5.23),

$$u(\mathrm{R}e^{i\theta}) = \frac{1}{2\pi} \int_{-\pi}^{\pi} P(e^{it}, \mathrm{R}e^{i\theta}) d\mu(t).$$

(c) If u_r satisfies the Cauchy condition in the $f \in L^1([-\pi, \pi])$ norm as $r \to 1$, then there exists $f \in L^1([-\pi, \pi])$ such that

$$\lim_{r'\to 1} \|u_{r'} - f\|_1 = 0. \tag{4.5.24}$$

Thus, for all $g \in L^\infty([-\pi, \pi])$,

$$\int_{-\pi}^{\pi} u(r' e^{it}) g(t) dt \to \int_{-\pi}^{\pi} f(t) g(t) dt \quad (r' \to 1). \tag{4.5.25}$$

Let $g(t) = P(e^{it}, \mathrm{Re}^{i\theta})$, then

$$\lim_{r' \to 1} \int_{-\pi}^{\pi} u(r'e^{it})P(e^{it}, \mathrm{Re}^{i\theta})dt = \int_{-\pi}^{\pi} P(e^{it}, \mathrm{Re}^{i\theta})f(t)dt.$$

since $u_{r'}(z) = u(r'z))$ is harmonic and bounded in U and continuous in $\overline{U}$, by applying Theorem 4.1.3, we have that

$$\lim_{r' \to 1} \frac{1}{2\pi} \int_{-\pi}^{\pi} u(r'e^{it})P(e^{it}, \mathrm{Re}^{i\theta})dt = \lim_{r' \to 1} u(r'\mathrm{Re}^{i\theta}) = u(\mathrm{Re}^{i\theta}).$$
(4.5.26)

Using (4.5.25) and (4.5.26), we can conclude that u is the Poisson integral of f. □

Theorem 4.5.8 (Fatou Theorem). *Suppose that $u(z)$ is harmonic in U, $1 \leqslant p \leqslant \infty$, and*

$$C = \sup_{0 < r < 1} \|u_r\|_p < \infty,$$

then u almost everywhere has non-tangential limit $f(e^{i\theta})$ at ∂U, i.e., for all $\alpha > 0$ and and almost everywhere $\theta_0 \in \mathbb{R}$, we have

$$\lim_{\substack{z \in \Gamma_\alpha(e^{i\theta_0}) \\ z \to e^{i\theta_0}}} u(z) = f(e^{i\theta_0}). \qquad (4.5.27)$$

where

(a) *if $1 < p \leqslant \infty$, u is the Poisson integral $P(f)$ of f,*
(b) *if $p = 1$, $f(e^{i\theta})d\theta$ is the absolutely continuous part of the measure $\mathscr{M}([-\pi, \pi])$ and u is the Poisson–Stieltjes integral $P(\mu)$ of μ.*

Proof. Since u is harmonic in U and satisfying (4.5.19), by applying Theorem 4.5.7, we see that, if $1 < p \leqslant \infty$, u is the Poisson integral $P(f)$ of $g(\theta) = f(e^{i\theta}) \in L^p([-\pi, \pi])$. By Theorem 4.5.4(c), we see that (4.5.27) holds.

If $p = 1$, applying Theorem 4.5.7 to u, it follows that there is unique complex Borel measure $\mu \in \mathscr{M}([-\pi, \pi])$ so that $u = P(\mu)$ is the Poisson integral of μ. According to the Lebesgue decomposition theorem [26], there exists $g(\theta) = f(e^{i\theta}) \in L^1([-\pi, \pi])$ such that

$d\mu(\theta) = f(e^{i\theta})d\theta + d\mu_s(\theta)$, where μ_s is mutually singular with respect to the Lebesgue measure. Therefore, we have
$$u(z) = P(\mu)(z) = P(f)(z) + P(\mu_s)(z).$$
On the one hand, according to Theorem 4.5.4(d), the non-tangential limit of $P(f)$ on ∂U is almost everywhere equal to f. On the other hand, the non-tangential limit of $P(\mu_s)$ on ∂U is almost everywhere zero. This can be proven as follows: Since the distribution function $\mu_s(t)$ of μ_s is such that $\mu_s'(t) = 0$ holds for almost all $t \in [-\pi, \pi]$. Let $\theta_0 \in (-\pi, \pi)$ be such that $\mu_s'(\theta_0) = 0$, then for all $\delta > 0$, there is $\eta > 0$
$$|\mu_s(\theta_0 + t) - \mu_s(\theta_0)| \leqslant \delta|t|$$
for all $|t| \leqslant \eta$. We can assume that $\mu_s(\theta_0) = 0$, so for all $\alpha > 0$ and $z = \text{Re}^{i\theta} \in \Gamma_\alpha(e^{i\theta_0})$, let $r \in (\frac{1}{2}, 1)$ and $\alpha(1-r) \leqslant \frac{\eta}{4}$, then we have
$$|\theta - \theta_0| \leqslant \alpha(1-r) \leqslant \frac{\eta}{4},$$
so (4.5.15) is satisfied. Thus,
$$|P(\mu_s)(\text{Re}^{i\theta})| = \left|\int_{-\pi}^{\pi} P(e^{i(t-\theta+\theta_0)}, r) d\mu_s(t+\theta_0)\right|$$
$$\leqslant \left|\int_{|t|<\eta}\right| + \left|\int_{\eta \leqslant |t| \leqslant \pi}\right| = \tilde{I}_1 + \tilde{I}_2.$$
For $\tilde{I}_1$, let's denote $P(e^{i(t-\theta+\theta_0)}, r)$ as $P(t)$. Using integration by parts, we have
$$\tilde{I}_1 = \left|\mu(t_0+t)P(t)\Big|_{-\eta}^{\eta} - \int_{-\eta}^{\eta} \mu_s(t+t_0)P'(t)dt\right|$$
$$\leqslant \delta\left(\eta(P(\eta) + P(-\eta)) + \int_0^{\eta} s(|P'(s)| + |P'(-s)|)ds\right).$$
Using the estimation provided in Theorem 4.5.4(c), we can conclude that
$$\tilde{I}_1 + \tilde{I}_2 \leqslant \delta\left(\frac{3\eta(1-r)}{\sin^2 \frac{\eta}{2}} + 8\alpha + 4\pi\right) + 2\pi\|\mu_s\|\frac{1-r}{\sin^2 \frac{\eta}{2}}.$$

By the arbitrariness of $\delta > 0$, we know that the non-tangential limits of $P(\mu_s)$ on ∂U are almost everywhere zero. Hence, conclusion (b) is established. □

4.6 Harmonic Functions in the Upper Half-Plane

Let $\mathscr{M}(\mathbb{R})$ be the set of finite complex-valued Borel measures on $\mathbb{R}$. The function $\mu(t) = \mu((-\infty, t])$ is a bounded variation function on $\mathbb{R} = (-\infty, \infty)$ and is called the distribution function of μ. Both the real part $\operatorname{Re}, \mu(t)$ and the imaginary part $\operatorname{Im} \mu(t)$ can be expressed as the difference of two monotone increasing functions. The Lebesgue decomposition theorem [26] states that there exists a function $f \in L^1(\mathbb{R})$ such that $\mu'(t) = f(t)$ holds almost everywhere on $\mathbb{R}$. The measure $d\mu_s = d\mu - f(t)dt$ is called the singular complex measure associated with μ, and its distribution function satisfies $\mu_s'(t) = 0$ for almost everywhere $t \in \mathbb{R}$.

For $\mu \in \mathscr{M}(\mathbb{R})$, the total variation norm of μ, denoted by $\|\mu\|_{\mathscr{M}}$, is defined as $\int_{\mathbb{R}} d|\mu|(x)$. This norm makes $\mathscr{M}$ a Banach space. By the Riesz representation theorem, in an isomorphic sense, $\mathscr{M}(\mathbb{R})$ is the dual space of $C_0(\mathbb{R}) = \{f \in C(\mathbb{R}) : \lim_{|t| \to \infty} f(t) = 0\}$ [26]. In other words, for any bounded linear functional $T \in (C_0(\mathbb{R}))^*$ on $C_0(\mathbb{R})$, there exists $\mu \in \mathscr{M}(\mathbb{R})$ such that $\|T\| = \|\mu\|_{\mathscr{M}} = \int_{\mathbb{R}} d|\mu|(x)$ and

$$T(f) = \int_{\mathbb{R}} f(x)\overline{d\mu(x)} \qquad (f \in C_0(\mathbb{R})). \tag{4.6.1}$$

The Banach–Alaoglu theorem states that the unit ball in $\mathscr{M}(\mathbb{R})$, given by $\{\mu \in \mathscr{M}(\mathbb{R}) : \|\mu\|_{\mathscr{M}} \leq 1\}$, is weak* compact, i.e., for any sequence of bounded linear functionals $\{\}T_n\}$ on $(C_0(\mathbb{R}))^*$ with $\|T_n\| \leq 1$ for $n = 1, 2, \ldots$, there exists $T \in (C_0(\mathbb{R}))^*$ and a subsequence $\{T_{n_k}\}$ of $\{T_n\}$ such that the sequence converges weak* to T, i.e.,

$$T(f) = \lim_{k \to \infty} T_{n_k}(f) \qquad (f \in C_0(\mathbb{R})). \tag{4.6.2}$$

If we identify elements f in $L^1(\mathbb{R})$ with $f(t)dt \in \mathscr{M}(\mathbb{R})$, then $L^1(\mathbb{R})$ can be considered as a closed subspace of $\mathscr{M}(\mathbb{R})$.

Definition 4.6.1. Let f, g be two measurable functions on $\mathbb{R}$. If the integral
$$\int_{\mathbb{R}} f(x-t)g(t)dt$$
exists for almost every $x \in \mathbb{R}$, then it is called the convolution of f and g, denoted as $f * g$, i.e.,
$$(f * g)(x) = \int_{\mathbb{R}} f(x-t)g(t)dt, \quad x \in \mathbb{R}.$$
Let $\mu \in \mathscr{M}(\mathbb{R})$ and $f \in L^p(\mathbb{R})$ $(1 \leqslant p \leqslant \infty)$, the convolution of the function f and the measure μ is defined by
$$(f * \mu)(x) = \int_{\mathbb{R}} f(x-t)d\mu(t), \quad x \in \mathbb{R}.$$
By applying Fubini's theorem and Minkowski's inequality, we can obtain the following result regarding the convolution of measures and L^p functions.

Theorem 4.6.2. Let $\mu \in \mathscr{M}(\mathbb{R})$ and $f \in L^p(\mathbb{R})$ $(1 \leqslant p \leqslant \infty)$, then $f * \mu \in L^p(\mathbb{R})$, and
$$\|f * \mu\|_p \leqslant \|f\|_p \|\mu\|_{\mathscr{M}}.$$

Theorem 4.6.3. Let $\mu \in \mathscr{M}(\mathbb{R})$, the Poisson–Stieltjes integral of measure μ
$$u(x, y) = (P_y * \mu)(x) = \int_{\mathbb{R}} P_y(x - t)d\mu(t) \quad (y > 0)$$
is harmonic in the upper half-plane $\mathbb{C}_+ = \{z = x+iy : y > 0\}$, where $P_y(x) = P(x, y) = \frac{y}{\pi(x^2+y^2)}$ is called the Poisson kernel with respect to $\mathbb{C}_+$, $P_y * \mu$ is called the Poisson–Stieltjes integral of measure μ.

Proof. Let $\mu \in \mathscr{M}(\mathbb{R})$ be a real measure. Let $F(z)$ be the Cauchy-type integral on $\mathbb{R}$ defined by
$$F(z) = \frac{1}{\pi i} \int_{\mathbb{R}} \frac{d\mu(t)}{t - z}, \tag{4.6.3}$$
where $z = x+iy$, $y \neq 0$, according to Theorem 1.2.13, $F(z)$ is analytic in $\mathbb{C}_+$, so $\operatorname{Re} F(z) = (P_y * \mu)(x) = u(x, y)$ is harmonic in $\mathbb{C}_+$. □

Let $f \in L^p(\mathbb{R})$ $(1 \leqslant p \leqslant \infty)$, the function

$$(f * P_y)(x) = \int_{\mathbb{R}} f(x-t) P_y(t) dt \quad (y > 0)$$

is called the Poisson integral of f.

It is easy to show that $P_y \in L(\mathbb{R})$, $P_{y_1} * P_{y_2} = P_{y_1+y_2}$ and $\int_{\mathbb{R}} P_y(x) dx = 1$, the Poisson integral of f $u(x,y) = (f * P_y)(x)$ is harmonic in the upper half-plane $\mathbb{C}_+ = \{z = x + iy : y > 0\}$.

The region in the upper half-plane $\mathbb{C}+$ given by

$$\Gamma_\alpha(x_0) = \{x + iy \in \mathbb{C}_+ : |x - x_0| < \alpha y\},$$

where $\alpha > 0$, is called a wedge-shaped region with vertex x_0 in the boundary of $\mathbb{C}_+$.

Definition 4.6.4. Let $u(z)$ be defined in the upper half-plane $\mathbb{C}+$. If for any $\alpha > 0$, we have

$$\lim_{\substack{z \to x_0 \\ z \in \Gamma_\alpha(x_0)}} u(z) = f(x_0),$$

then we say that $u(z)$ converges non-tangentially to $f(x_0)$.

Theorem 4.6.5. *Let $f \in L^p(\mathbb{R})$ $(1 \leqslant p \leqslant \infty)$, then the Poisson integral u of f defined by $u(x,y) = (f * P_y)(x)$ satisfies the following properties:*

(a) *For fixed $y > 0$, $u_y(x) = u(x,y) \in L^p(\mathbb{R})$, and $\|u_y\|_p \leqslant \|f\|_p$.*
(b) *If $1 \leqslant p < \infty$, then*

$$\lim_{y \to 0} \|u_y - f\|_p = 0.$$

(c) *If $f \in L^p(\mathbb{R})$, $1 \leqslant p < \infty$, then at the Lebesgue point x_0 of f, for any $\alpha > 0$, we have*

$$\lim_{\substack{x+iy \to x_0 \\ x+iy \in \Gamma_\alpha(x_0)}} \int_{\mathbb{R}} |f(x-t) - f(x_0)| P_y(t) dt = 0$$

and

$$\lim_{\substack{z \to x_0 \\ z \in \Gamma_\alpha(x_0)}} (f * P_y)(x) = f(x_0),$$

which means that $f * P_y$ converges non-tangentially to f almost everywhere.

Proof. (a) Since $\int_{\mathbb{R}} P_y(x-t)dt = 1$ $(y > 0)$, by applying Fubini's theorem and the Minkowski inequality, we can obtain (a).

(b) For $1 \leq p < \infty$, $y > 0$, by applying Minkowski's inequality, we can obtain

$$\|f * P_y - f\|_p \leq \int_{\mathbb{R}} \left(\int_{\mathbb{R}} |f(x-yt) - f(x)|^p dx \right)^{1/p} P_1(t) dt.$$

By using the continuity of L^p integrability of f and the Lebesgue-dominated convergence theorem, we can obtain the following result:

$$\lim_{y \to 0} \|(f * P_y) - f\|_p \leq \int_{\mathbb{R}} \lim_{y \to 0} \left(\int_{\mathbb{R}} |f(x-yt) - f(x)|^p dx \right)^{1/p} P_1(t) dt = 0.$$

This shows that as y approaches 0, the L^p norm of the convolution $f * P_y$ with kernel P_y converges non-tangentially to f, which proves (b). Moreover, if $f \in C_0(\mathbb{R})$, then

$$\lim_{y \to 0} |(f * P_y)(x) - f(x)| \leq \int_{\mathbb{R}} \lim_{y \to 0} |f(x-yt) - f(x)| P_1(t) dt = 0$$

uniformly with respect to $x\mathbb{R}$.

(c) For all $\delta > 0$, there exists $\eta > 0$ such that, when $0 < s \leq \eta$,

$$G(s) = \int_0^s (|f(x_0 - t) - f(x_0)| + |f(x_0 + t) - f(x_0)|) dt \leq \delta s. \tag{4.6.4}$$

Now, take $y > 0$, then

$$|(f * P_y)(x) - f(x_0)| \leq \int_{\mathbb{R}} |f(x_0 - t) - f(x_0)| P_y(t + x - x_0) dt$$

$$= I_1 + I_2,$$

where

$$I_1 = \int_{|t| < \eta} |f(x_0 - t) - f(x_0)| P_y(t + x - x_0) dt,$$

$$I_2 = \int_{|t| \geq \eta} |f(x_0 - t) - f(x_0)| P_y(t + x - x_0) dt.$$

For all $\alpha > 0$ and $z = x + iy \in \Gamma_\alpha(x_0), 2\alpha y < \eta$, we first estimate I_1. Let $Q(t) = P_y(t + x - x_0)$, then $Q(-t) = P_y(t - (x - x_0))$. For $0 \leqslant s \leqslant \eta$, we have $G(s) \leqslant \delta s$ and $\frac{d}{ds}P_y(s) \leqslant 0$, thus

$$I_1 \leqslant \int_0^\eta (Q(s) + Q(-s))G'(s)ds$$

$$= G(s)(Q(s) + Q(-s))\Big|_0^\eta - \int_0^\eta G(s)\frac{d}{ds}(Q(s) + Q(-s))ds$$

$$\leqslant \delta\left(\eta(Q(\eta) + Q(-\eta)) + \int_0^\eta s(|Q'(s)| + |Q'(-s)|)ds\right).$$

When $t \geqslant 0$, $\frac{d}{dt}P_y(t) \leqslant 0$, so when $0 \leqslant s \leqslant |x - x_0| \leqslant \eta$, $(x - x_0)Q'(s) \leqslant 0, (x - x_0)Q'(-s) \leqslant 0$, we have

$$\int_0^{|x-x_0|} s(|Q'(s)| + |Q'(-s)|)ds$$

$$= \left|\int_0^{|x-x_0|} s(Q'(s) + Q'(-s))ds\right|$$

$$= \left|s(Q(s) - Q(-s))\Big|_0^{|x-x_0|} - \int_0^{|x-x_0|}(Q(s) - Q(-s))ds\right|$$

$$\leqslant \frac{4|x - x_0|}{y} + 2 \leqslant 4\alpha + 2$$

and

$$\int_{|x-x_0|}^\eta s(|Q'(s)| + |Q'(-s)|)ds$$

$$= \left|\int_{|x-x_0|}^\eta s(Q'(s) + Q'(-s))ds\right|$$

$$= \left|s(Q(s) - Q(-s))\Big|_{|x-x_0|}^\eta - \int_{|x-x_0|}^\eta (Q(s) - Q(-s))ds\right|$$

$$\leqslant \frac{8y}{\pi\eta} + \frac{2|x - x_0|}{\pi y} + 2 \leqslant \frac{4y}{\eta} + \alpha + 2.$$

Next, let us estimate I_2. Let p' satisfy $\frac{1}{p} + \frac{1}{p'} = 1$, according to the Hölder inequality,

$$I_2 \leq \int_{|t| \geq \eta} |f(x_0 - t)| \frac{2y}{t^2} dt + \int_{|t| \geq \eta} |f(x_0)| \frac{2y}{t^2} dt$$

$$\leq \frac{2y\|f\|_p}{((2p'-1)\eta^{2p'-1})^{\frac{1}{p'}}} + \frac{4y|f(x_0)|}{\eta}.$$

Therefore, for all $\delta > 0$, there exists $\eta > 0$ such that, for all $\alpha > 0$ and $z = x + iy \in \Gamma_\alpha(e^{i\theta_0})$, when $\alpha y \leq \frac{\eta}{4}$, we have

$$I_1 + I_2 \leq \delta \left(\frac{10y}{\eta} + 8\alpha + 4 \right) + \frac{2y\|f\|_p}{((2p'-1)\eta^{2p'-1})^{\frac{1}{p'}}} + \frac{4y|f(x_0)|}{\eta}.$$

Since $\delta > 0$ is arbitrary, we conclude that statement (c) holds. □

If we strengthen the conditions on f, then Theorem 4.6.5(c) has the following conclusion for $p = \infty$.

Corollary 4.6.6. *Let $f \in L^\infty(\mathbb{R})$, then the Poisson integral u of f defined by $u(x, y) = (f * P_y)(x)$ satisfies the following properties:*

(a) *If $f \in C(\mathbb{R}) \cap L^\infty(\mathbb{R})$, then, as $y \to 0$, the function $u(x + iy)$ uniformly converges to $f(x)$ on any compact set $F \subset \mathbb{R} R$. Therefore, the function*

$$u(z) = \begin{cases} (f * P_y)(x), & z = x + iy \in \mathbb{C}_+, \\ f(x), & z = x, \ y = 0 \end{cases}$$

is harmonic in $\mathbb{C}_+$ and continuous in $\overline{\mathbb{C}}_+$.

(b) *If $f \in C_0(\mathbb{R}) \subset L^\infty(\mathbb{R})$, then, as $y \to 0$, the function $u(x + iy)$ uniformly converges to $f(x)$ on $\mathbb{R}$.*

(c) *If $f \in L^\infty(\mathbb{R})$, as $y \to 0$, $u(x, y)$ weakly* converges to $f(x)$ on $\mathbb{R}$, i.e., for all $\varphi \in L^1(\mathbb{R})$,*

$$\lim_{y \to 0} \int_\mathbb{R} u(x, y) \varphi(x) dx = \int_\mathbb{R} f(x) \varphi(x) dx.$$

Proof. (a) Since f is continuous on $\mathbb{R}$, by Theorem 4.6.5(c), we know that $\lim_{y \to 0} u(x, y) = f(x)$ holds everywhere on $\mathbb{R}$. Let F be any compact set in $\mathbb{R}$. Then, f is uniformly continuous on F. This means that for any $\delta > 0$, there exists $\eta_1 > 0$ such that for any $x \in F$ and $|t| < \eta_1$, we have

$$|f(x-t) - f(x)| < \delta/2. \tag{4.6.5}$$

On the other hand, there exists $\eta_2 > 0$ such that for $0 < y < \eta_2$, we have

$$2y\|f\|_\infty \int_{|t| \geq \eta_1} \frac{dt}{|t|^2} < \delta. \tag{4.6.6}$$

Therefore, combining (4.6.5) and (4.6.6), we have

$$|u(x,y) - f(x)| \leq \int_{\mathbb{R}} |f(x-t) - f(x)| P_y(t) dt$$

$$\leq \int_{|t| < \eta_1} |f(x-t) - f(x)| P_y(t) dt$$

$$+ \int_{|t| \geq \eta_1} |f(x-t) - f(x)| P_y(t) dt$$

$$< \delta/2 + 2y\|f\|_\infty \int_{|t| \geq \eta_1} \frac{dt}{\pi |t|^2} < \delta.$$

Therefore, as $y \to 0$, u uniformly converges to f on F. For any $x_0 \in \mathbb{R}$, since f is continuous at x_0, for any $\varepsilon > 0$, there exists $\eta_1 > 0$ such that for $|t| < \eta_1$, we have $|f(x_0 - t) - f(x_0)| < \varepsilon/2$. Let $F = \{x \in \mathbb{R} : |x - x_0| \leq \eta_1/2\}$, then F is a compact in $\mathbb{R}$. For the above ε, there exists $\eta_2 > 0$ such that for $0 < y < \eta_2$ and any $x \in F$, we have

$$|u(x,y) - f(x)| < \varepsilon/2. \tag{4.6.7}$$

Thus, for the given ε, we can choose $0 < \delta < \min\{\eta_1/2, \eta_2\}$ such that for any $z = x + iy \in \mathbb{C}_+$ with $|z - x_0| < \delta$, we have $|x - x_0| < \delta < \eta_1/2$ and $y < \delta < \eta_2$. Using (4.6.7), we obtain

$$|u(z) - f(x_0)| \leq |u(z) - f(x)| + |f(x) - f(x_0)| < \varepsilon.$$

This shows that u is continuous at x_0, and since x_0 was arbitrary, u is continuous on $\mathbb{R}$.

(b) Since $f \in C_0(\mathbb{R})$, it is uniformly continuous on $\mathbb{R}$. Hence, for any $\delta > 0$, there exists $\eta > 0$ such that for any $x \in \mathbb{R}$ and $|t| < \eta$, (4.6.5) still holds.

From the proof of (a), we know that as $y \to 0$, u uniformly converges to f on $\mathbb{R}$.

(c) For all $\varphi \in L^1(\mathbb{R})$, let $v_y(x) = v(x+iy) = (P_y * \varphi)(x)$, then

$$\int_{\mathbb{R}} u(x+iy)\varphi(x)dx = \int_{\mathbb{R}}\int_{\mathbb{R}} P_y(x-t)f(t)dt\varphi(x)dx$$
$$= \int_{\mathbb{R}} v(t+iy)f(t)dt.$$

According to Theorem 4.6.5(b),

$$\left|\int_{\mathbb{R}} u(x+iy)\varphi(x)dx - \int_{\mathbb{R}} f(x)\varphi(x)dx\right|$$
$$\leqslant \int_{\mathbb{R}} |v(t+iy) - \varphi(t)||f(t)|dt$$
$$\leqslant \|f\|_\infty \|v_y - \varphi\|_1 \to 0 \quad (y \to 0).$$

This proves Corollary 4.6.6. □

Here are several properties of the Poisson–Stieltjes integral with respect to μ.

Theorem 4.6.7. *Let $\mu \in \mathscr{M}(\mathbb{R})$, $u(x+iy)$ the Poisson–Stieltjes integral with respect to μ, $u_y(x) = u(x+iy)$ satisfies the following properties:*

(a) *For all $y > 0$, $\|u_y\|_1 \leqslant \|\mu\|_\mathscr{M}$.*
(b) *As $y \to 0$, $u(x,y)$ weakly* converges to μ on $\mathbb{R}$, i.e., for all $\varphi \in C_0(\mathbb{R})$,*

$$\lim_{y \to 0} \int_{\mathbb{R}} u(x,y)\varphi(x)dx = \int_{\mathbb{R}} \varphi(x)d\mu(x).$$

Proof. Since $\int_{\mathbb{R}} P_y(x-t)dt = 1$ $(y > 0)$, by applying Fubini's theorem, we can obtain (a). Since $\mathscr{M}(\mathbb{R}) = (C_0(\mathbb{R}))^*$ and for all

$\varphi \in C_0(\mathbb{R})$, let $v_y(x) = v(x+iy) = (P_y * \varphi)(x)$, then

$$\int_{\mathbb{R}} u(x+iy)\varphi(x)dx = \int_{\mathbb{R}}\int_{\mathbb{R}} P_y(x-t)d\mu(t)\varphi(x)dx = \int_{\mathbb{R}} v(t+iy)d\mu(t).$$

By applying Corollary 4.6.6(b),

$$\left|\int_{\mathbb{R}} u(x+iy)\varphi(x)dx - \int_{\mathbb{R}} \varphi(x)d\mu(x)\right|$$

$$\leqslant \int_{\mathbb{R}} |v(t+iy) - \varphi(t)||d\mu(t)| \leqslant \|\mu\|_{\mathscr{M}}\|v_y - \varphi\|_\infty \to 0 \quad (y \to 0).$$

This proves Theorem 4.6.7. □

If $f \in L^p(\mathbb{R})$ ($1 \leqslant p \leqslant \infty$), the Poisson integral u of f defined by $u(x,y) = (P_y * f)(x)$ is harmonic in $\mathbb{C}_+$. Now, we will discuss the converse problem, namely, for a harmonic function u in the upper half-plane $\mathbb{C}_+$, under what conditions is u the Poisson integral of an L^p function?

Let us associate to any function u in $\mathbb{C}_+$ a family of functions u_y on $y \in (0, \infty)$, defined by $u_y(x) = u(x+iy)$.

Theorem 4.6.8. Let $u(z)$ be harmonic in $\mathbb{C}_+$, $0 < p \leqslant \infty$ and

$$C = \sup_{y>0} \|u_y\|_p < \infty, \qquad (4.6.8)$$

then

$$|u(x+iy)| \leqslant CC_p \left(\frac{2}{\pi y}\right)^{\frac{1}{p}}, \qquad (4.6.9)$$

where, when $1 \leqslant p \leqslant \infty$, $C_p = 1$, and when $0 < p < 1$,

$$C_p = \exp\left\{\frac{2^{\frac{4}{1-p}}}{p^2(1-p)}\right\}.$$

In particular, for all $y_0 > 0$, $u(z+iy_0)$ is harmonic in $\mathbb{C}_+$, bounded and continuous in $\overline{\mathbb{C}}_+$.

Proof. When $p = \infty$, the inequality (4.6.9) holds trivially. When $1 \leqslant p < \infty$, for any chosen $z_0 = x_0 + iy_0 \in \mathbb{C}_+$, we consider a disk centered at z_0 with radius y_0, denoted by $D(z_0, y_0)$. According to the mean value property of harmonic functions (as stated in (4.1.7)), when $1 \leqslant p < \infty$, we have the following result:

$$|u(z_0)| = \left| \int_{D(z_0,y_0)} u(\zeta) \frac{d\lambda(\zeta)}{\pi y_0^2} \right| \leqslant \left(\int_{D(z_0,y_0)} |u(\zeta)|^p \frac{d\lambda(\zeta)}{\pi y_0^2} \right)^{\frac{1}{p}}$$

$$\leqslant \frac{1}{(\pi y_0^2)^{1/p}} \left(\int_0^{2y_0} \int_{\mathbb{R}} |u(\xi + i\eta)|^p d\xi\, d\eta \right)^{1/p} \leqslant C \left(\frac{2}{\pi y_0} \right)^{\frac{1}{p}}.$$

When $0 < p < 1$, for all $z_0 = x_0 + iy_0, y_0 > 0, 0 < r < y_0$, let

$$M(r) = \max\{|u(z_0 + Re^{i\theta})| : \theta \in [-\pi, \pi]\},$$

$$I(r) = \int_{-\pi}^{\pi} |u(z_0 + Re^{i\theta})|^p d\theta.$$

In order to prove (4.6.9), we first prove that

$$|u(x_0 + iy_0)| \leqslant C_p \left(\frac{1}{\pi R^2} \int_0^R I(r) r\, dr \right)^{\frac{1}{p}} \quad (0 < R < y_0). \quad (4.6.10)$$

Assuming that $R = 1$ and $\int_0^1 I(r) r\, dr = \pi$, we need to prove that $|u(x_0 + iy_0)| \leqslant C_p$. If $|u(x_0 + iy_0)| > 1$, by the maximum modulus principle, we have $M(r) > 1$ Then, by the geometric and arithmetic mean inequalities, we can write

$$\exp\left\{ \frac{1}{\log 2} \int_{\frac{1}{2}}^{1} \log I(r) \frac{dr}{r} \right\} \leqslant \frac{1}{\log 2} \int_{\frac{1}{2}}^{1} I(r) \frac{dr}{r} \leqslant \frac{4\pi}{\log 2} \leqslant \exp\left\{ \frac{4}{\log 2} \right\}.$$

Let $\alpha > 1$ be such that $\alpha(1 - p) < 1$, for example, we can take $\alpha = \alpha(p) = \frac{2-p}{2-2p}$. By using the inequalities $1 > r > r^\alpha > 0$, (4.1.4), and Theorem 4.1.7(c), we have

$$M(r^\alpha) \leqslant \frac{r}{\pi(r - r^\alpha)} \int_{-\pi}^{\pi} |u(z_0 + Re^{i\theta})| d\theta \leqslant \frac{r}{\pi(r - r^\alpha)} (M(r))^{1-p} I(r),$$

so
$$\log M(r^\alpha) \leqslant (1-p)\log M(r) + \log I(r) + \log\left(\frac{r}{\pi(r-r^\alpha)}\right).$$

Integrate from $\frac{1}{2}$ to 1, and since

$$\int_{\frac{1}{2}}^1 \log I(r)\frac{dr}{r} \leqslant 4 \quad \text{and} \quad \int_{\frac{1}{2}}^1 \log\left(\frac{r}{\pi(r-r^\alpha)}\right)\frac{dr}{r} \leqslant \frac{2^{\alpha-1}}{\alpha-1},$$

we have

$$\int_{\frac{1}{2}}^1 \log M(r^\alpha)\frac{dr}{r} = \frac{1}{\alpha}\int_{(\frac{1}{2})^\alpha}^1 \log M(s)\frac{ds}{s}$$
$$\leqslant (1-p)\int_{\frac{1}{2}}^1 \log M(r)\frac{dr}{r} + 4 + \frac{2^{\alpha-1}}{\alpha-1}.$$

Since $\alpha > 1$ and $M(r) \geqslant 1$, then

$$\left(\frac{1}{\alpha} - (1-p)\right)\int_{\frac{1}{2}}^1 \log M(r)\frac{dr}{r} \leqslant 4 + \frac{2^{\alpha-1}}{\alpha-1}.$$

Let $\alpha = \alpha(p) = \frac{2-p}{2-2p}$, then

$$\log\min\left\{M(r) : \frac{1}{2} \leqslant r \leqslant 1\right\}$$
$$\leqslant \frac{8 + \frac{2^{\alpha(p)}}{\alpha(p)-1}}{\left(\frac{1}{\alpha(p)} - (1-p)\right)} \leqslant \frac{2^{\frac{4}{1-p}}}{p^2(1-p)} = \log C_p.$$

Thus, $|u(x_0 + iy_0)| \leqslant \min\{M(r) : \frac{1}{2} \leqslant r \leqslant 1\} \leqslant C_p$.

If $|u(x_0 + iy_0)| \leqslant 1$, it is obvious that $|u(x_0 + iy_0)| \leqslant 1 \leqslant C_p$. Thus, if $R = 1$ and $\int_0^1 I(r)rdr = \pi$, (4.6.10) holds. If either $R = 1$ or $\int_0^1 I(r)rdr = \pi$ does not hold, then consider the harmonic function

$$\tilde{u}(z) = \frac{u(z_0 + zR)}{\int_0^R I(r)rdr}.$$

It can be seen that (4.6.10) still holds. Therefore, we have for all $R \in (0, y_0)$,

$$|u(z_0)| \leqslant C_p \left(\int_{D(z_0,R)} |u(\zeta)|^p \frac{d\lambda(\zeta)}{\pi R^2} \right)^{\frac{1}{p}}$$

$$\leqslant C_p \frac{1}{(\pi R^2)^{1/p}} \left(\int_{y_0-R}^{y_0+R} \int_{\mathbb{R}} |u(\xi + i\eta)|^p d\xi\, d\eta \right)^{1/p}$$

$$\leqslant CC_p \left(\frac{2}{\pi R} \right)^{\frac{1}{p}}.$$

Let $R \to y_0$, we see that (4.6.9) holds. In particular, consider the harmonic function $u(z + iy_0)$ in the upper half-plane $\mathbb{C}_+$, we have $|u(z + iy_0)| \leqslant C \left(\frac{2}{\pi y_0} \right)^{\frac{1}{p}}$. $\square$

Lemma 4.6.9. *Suppose that u is a bounded harmonic in $\mathbb{C}_+$ and continuous in $\overline{\mathbb{C}}_+$, then for all $y > 0$,*

$$u(z) = \int_{\mathbb{R}} u(t) P_y(x-t) dt. \qquad (4.6.11)$$

Proof. Let

$$u_1(z) = \begin{cases} \int_{\mathbb{R}} u(t) P_y(x-t) dt, & y > 0, \\ u(x), & y = 0. \end{cases}$$

According to Theorem 4.6.5(a), u_1 is a bounded harmonic function in $\mathbb{C}_+$. By Corollary 4.6.6(a), u_1 is continuous in $\overline{\mathbb{C}}_+$. Let $h(z) = u(z) - u_1(z)$, then h is a bounded harmonic in $\mathbb{C}_+$, continuous in $\overline{\mathbb{C}}_+$, and $h(x) = u(x) - u_1(x)$ is identically zero on $\mathbb{R}$, by Corollary 4.1.19, h is identically zero on $\mathbb{C}$, i.e., for any $y > 0$, (4.6.11) holds. $\square$

Theorem 4.6.10. *Let u be harmonic in $\mathbb{C}_+$, $1 \leqslant p \leqslant \infty$ and (4.6.8) hold for some constant $C \geqslant 0$.*

(a) *If $1 < p \leqslant \infty$, then there is $f \in L^p(\mathbb{R})$ such that u is the Poisson integral of f.*
(b) *If $p = 1$, then there is $\mu \in \mathscr{M}(\mathbb{R})$ such that u is the Poisson–Stieltjes integral of μ.*

(c) If $p = 1$ and u_y satisfies the Cauchy condition in the L^1 norm as $y \to 0$, i.e.,
$$\lim_{y_1,y_2 \to 0} \|u_{y_1} - u_{y_2}\|_1 = 0,$$
then there is $f \in L^p(\mathbb{R})$ such that u is the Poisson integral of f.

Proof.
(a) If $1 < p \leq \infty$, by the condition (4.6.8), The L^p norm of u is uniformly bounded with respect to $y > 0$. Since $L^{p'}(\mathbb{R})$ ($1 \leq p' < \infty$, $p^{-1} + p'^{-1} = 1$) is a separable Banach space, for any sequence $\{y_k\}$ such that $y_k > 0$ and $y_k \to 0$ as $k \to \infty$ and the sequence, $\{u_{y_k}\}$, viewed as a bounded linear functional on $L^{p'}(\mathbb{R})$ is weak*-sequentially compact. Hence, there exists a subsequence $\{u_{y_{k_j}}\}$ of $\{u_{y_k}\}$ and $f \in L^p(\mathbb{R})$ such that $\{u_{y_{k_j}}\}$ weak* converges to f, i.e., for any $g \in L^{p'}(\mathbb{R})$,
$$\lim_{j \to \infty} \int_{\mathbb{R}} u(t + iy_{k_j})g(t)dt = \int_{\mathbb{R}} f(t)g(t)dt.$$
Now, let $g(t) = P_y(x - t)$, then
$$\lim_{j \to \infty} \int_{\mathbb{R}} u(t + iy_{k_j})P_y(x - t)dt = \int_{\mathbb{R}} f(t)P_y(x - t)dt. \quad (4.6.12)$$
On the other hand, according to Lemma 4.6.8, $u(z + y_{k_j})$ is bounded and harmonic in $\mathbb{C}_+$ and continuous in $\overline{\mathbb{C}_+}$. By applying Lemma 4.6.9, we have
$$\lim_{j \to \infty} \int_{\mathbb{R}} u(t + iy_{k_j})P_y(x - t)dt = \lim_{j \to \infty} u(x + iy_{k_j} + y) = u(x + iy).$$
From (4.6.12), we conclude that u is the Poisson integral of f.

(b) If $p = 1$, for any sequence $\{y_k\}$ such that $y_k > 0$ and $y_k \to 0$ as $k \to \infty$), we can deduce from (4.6.8) that $\{u_{y_k}\}$, viewed as a bounded linear functional on $C_0(\mathbb{R})$, is uniformly bounded. By applying the Banach–Alaoglu theorem, there exists a subsequence $\{u(x + iy_{k_j})\}$ and $\mu \in \mathcal{M}(\mathbb{R})$ such that $\{u_{y_{k_j}}\}$ weak* converges to μ. Let $P_{y,x}(t) = P_y(x - t)$, then $P_{y,x} \in C_0(\mathbb{R})$ and
$$\lim_{j \to \infty} \int_{\mathbb{R}} u(t + iy_{k_j})P_y(x - t)dt = \int_{\mathbb{R}} P_y(x - t)d\mu(t). \quad (4.6.13)$$
By (4.6.12) and (4.6.13), u is the Poisson–Stieltjes integral of μ.

(c) If $u_y(x) = u(x+iy)$ satisfies the Cauchy condition in the L^1 norm as $y \to 0$, then there exists $f \in L^1(\mathbb{R})$ such that
$$\lim_{y' \to 0} \|u_{y'} - f\|_1 = 0.$$
Therefore, for any $g \in L^\infty(\mathbb{R})$,
$$\int_\mathbb{R} u(t+iy')g(t)dt \to \int_\mathbb{R} f(t)g(t)dt \quad (y' \to 0). \tag{4.6.14}$$
Let $g(t) = P_y(x-t)$, this is expressed by (4.6.14),
$$\int_\mathbb{R} u(t+iy')P_y(x-t)dt \to \int_\mathbb{R} f(t)P_y(x-t)dt \quad (y' \to 0). \tag{4.6.15}$$
On the one hand, from (4.6.11) and (4.6.12), we have
$$\int_\mathbb{R} u(t+iy')P_y(x-t)dt = u(x+i(y'+y)) \to u(x+iy) \quad (y' \to 0).$$
Finally, combining this with (4.6.15), we conclude that u is the Poisson integral of f. $\square$

Theorem 4.6.11 (Fatou's Theorem). *Suppose u is harmonic in $\mathbb{C}_+$. If there exists $1 \leq p \leq \infty$ such that (4.6.8) holds, then u has non-tangential limits almost everywhere on $\mathbb{R}$. In other words, there exists a function f on $\mathbb{R}$ such that for any $\alpha > 0$ and almost every $x_0 \in \mathbb{R}$,*
$$\lim_{\substack{z \in \Gamma_\alpha(x_0) \\ z \to x_0}} u(z) = f(x_0), \tag{4.6.16}$$
where

(a) *when $1 < p \leq \infty$, u is the Poisson integral of f;*
(b) *when $p = 1$, $f(x)dx$ is the absolutely continuous part of the measure μ in $\mathscr{M}(\mathbb{R})$, and u is the Poisson–Stieltjes integral of μ.*

Proof. Since u is harmonic in $\mathbb{C}_+$ and satisfies (4.6.8), when $1 < p \leq \infty$, by Theorem 4.6.10(a), we know that u is the Poisson integral of some $L^p(\mathbb{R})$ function. Furthermore, by Theorem 4.6.5(d), we conclude that (4.6.16) holds for almost every $x_0 \in \mathbb{R}$.

When $p = 1$, by Theorem 4.6.10(b), u is the Poisson–Stieltjes integral of a measure μ in $\mathcal{M}(\mathbb{R})$. By the Lebesgue decomposition theorem [26], there exists $f \in L^1(\mathbb{R})$ such that $d\mu(x) = f(x)dx + d\mu_s(x)$, where μ_s is singular with respect to the Lebesgue measure. Therefore,
$$u(z) = P_y * \mu(x) = P_y * f(x) + P_y * \mu_s(x).$$
On the one hand, by Theorem 4.6.5(d), the non-tangential limits of $P_y * f$ on $\mathbb{R}$ almost everywhere are equal to f. On the other hand, the non-tangential limits of $P_y * \mu_s$ on $\mathbb{R}$ almost everywhere are equal to zero. The proof is as follows: Since the distribution function of μ_s satisfies $\mu_s'(t) = 0$ for almost every $t \in \mathbb{R}$, let $x_0 \in \mathbb{R}$ be such that $\mu_s'(x_0) = 0$. For any $\delta > 0$, there exists $\eta > 0$ such that for $|t| \leqslant \eta$, we have $|\mu_s(x_0 - t) - \mu_s(x_0)| \leqslant \delta |t|$. Without loss of generality, assume $\mu_s(x_0) = 0$. For any $\alpha > 0$ and $z = x + iy \in \Gamma_\alpha(x_0)$, if $0 < \alpha y \leqslant \frac{\eta}{4}$, then $|x - x_0| \leqslant \alpha y \leqslant \frac{\eta}{4}$. Hence,
$$|(P_y * \mu_s)(x + iy)| \leqslant I_1 + I_2,$$
where
$$I_1 = \left| \int_{|t| < \eta} P_y(t + x - x_0) d\mu_s(x_0 - t) \right|,$$
$$I_2 = \left| \int_{|t| \geqslant \eta} P_y(t + x - x_0) d\mu_s(x_0 - t) \right|.$$

Let's estimate I_1. Let $Q(t) = P_y(t + x - x_0)$, then $Q(-t) = P_y(t - (x - x_0))$. Since $|\mu_s(x_0 - t)| \leqslant \delta|t|$ and $\frac{d}{ds}P_y(s) \leqslant 0$ for $0 \leqslant s \leqslant \eta$, we have
$$I_1 = \left| \mu_s(x_0 - t)Q(t) \Big|_{-\eta}^{\eta} - \int_{-\eta}^{\eta} \mu_s(x_0 - t)Q'(t)dt \right|$$
$$\leqslant \delta \left(\eta(Q(\eta) + Q(-\eta)) + \int_0^\eta s(|Q'(s)| + |Q'(-s)|)ds \right).$$

Similar to the proof of Theorem 4.6.5(c), we can derive the following estimation for I_1:
$$I_1 \leqslant \delta \left(\frac{10y}{\eta} + 8\alpha + 4 \right).$$

For I_2, we have
$$I_2 \leq \|\mu_s\| P_y\left(\frac{\eta}{2}\right) \leq \|\mu_s\| \frac{4y}{\eta^2}.$$

Thus,
$$\lim_{\substack{x+iy \in \Gamma_\alpha(x_0) \\ x+iy \to x_0}} |(P_y * \mu_s)(x+iy)| = 0.$$

Therefore, conclusion (b) holds. □

Theorem 4.6.12. *If $v(z)$ is subharmonic in $\mathbb{C}_+$ and*
$$C = \sup_{y>0} \int_\mathbb{R} |v(x+iy)| dx < \infty, \tag{4.6.17}$$

then
$$v(x+iy) \leq \frac{2C}{\pi y}.$$

Proof. For any $z = x + iy \in \mathbb{C}_+$, consider the disk $D(z, y)$ centered at z with radius y. By the mean value property of subharmonic functions (as shown in (4.3.1)), we have
$$v(z) \leq \int_{D(z,y)} v(\zeta) \frac{d\lambda(\zeta)}{\pi y^2} \leq \int_{D(z,y)} |v(\zeta)| \frac{d\lambda(\zeta)}{\pi y^2}$$
$$\leq \frac{1}{\pi y^2} \int_0^{2y} \int_\mathbb{R} |v(\xi, \eta)| d\xi \, d\eta \leq \frac{2C}{\pi y}.$$

Thus, the conclusion holds. □

Theorem 4.6.13. (a) *Let $b > a$. Suppose that $v(z)$ is a non-negative subharmonic in $\Omega_{a,b} = z = x + iy : x \in \mathbb{R}, y \in (a,b)$ and*
$$\sup_{c \leq y \leq d} \int_{-\infty}^{+\infty} v(x+iy) dx = M_{c,d} < \infty$$

for any $c, d \in (a,b)$ with $c < d$, then the function
$$\varphi(y) = \int_\mathbb{R} v(x+iy) dx \tag{4.6.18}$$

is a convex function on (a,b).

(b) *Suppose that $v(z)$ is a non-negative subharmonic in $\mathbb{C}_+$ and satisfies (4.6.17), then the function $\varphi(y)$ defined by (4.6.18) is decreasing and convex on $(0, \infty)$ and there exists a positive measure $\mu \in \mathscr{M}(\mathbb{R})$ such that u defined by $u(x + iy) = P_y * \mu$ is the Poisson–Stieltjes integral of μ and satisfies $v(z) \leq u(z)$ for all $z \in \mathbb{C}_+$, and $C = \|\mu\|$.*

Proof. (a) For any $c, d \in (a, b)$ with $c < d$, we have

$$\int_c^d \int_{-\infty}^{+\infty} v(x+iy)\,dx\,dy \leq (d-c)M_{c,d} < \infty.$$

Let $y_0 \in (c, d)$, and choose $\delta_0 > 0$ such that $[y_0 - 2\delta_0, y_0 + 2\delta_0] \subset (c, d)$. For any $y \in (y_0 - \delta_0, y_0 + \delta_0)$ and any $k > 1 + 2\delta_0$, we define

$$\varphi(y) = \int_{|x| \geq k} v(x+iy)\,dx + \int_{|x| < k} v(x+iy)\,dx.$$

Let $M_k(y) = \int_{|x| \geq k} v(x+iy)\,dx$, since

$$M_k(y) \leq \int_{|x| \geq k} \frac{1}{\pi\delta_0^2} \int_{(\xi-x)^2+(\eta-y)^2 \leq \delta_0^2} v(\xi+i\eta)\,d\xi\,d\eta\,dx$$

$$\leq \frac{1}{\pi\delta_0^2} \int_{|x| \geq k} \int_{-\delta_0}^{\delta_0} \int_{-\delta_0}^{\delta_0} v(\xi + x + i(\eta+y))\,d\xi\,d\eta\,dx$$

$$\leq \frac{1}{\pi\delta_0^2} \int_{|x| \geq k-\delta_0} \int_{-\delta_0}^{\delta_0} v(x + i(\eta+y))\,d\eta(2\delta_0)\,dx$$

$$\leq \frac{2}{\pi\delta_0} \int_{|x| \geq k-\delta_0} \int_c^d v(x+iy)\,dx\,dy,$$

and we have

$$\int_{-\infty}^{+\infty} \int_c^d v(x+iy)\,dx\,dy < \infty,$$

by the Lebesgue-dominated convergence theorem, we know that

$$\lim_{k \to +\infty} \int_{|x| \geq k-\delta_0} \int_c^d v(x+iy)\,dx\,dy = 0.$$

Therefore, for any $\epsilon > 0$, there exists $k_\epsilon > 1 + 2\delta_0$ such that

$$\int_{|x| \geq k_\epsilon - \delta_0} \int_c^d v(x+iy) dx dy < \frac{\epsilon}{2}.$$

Moreover, since $v(x+iy)$ is upper semicontinuous in Ω, there exists an upper bound A_ϵ for $v(x+iy)$ over the compact set $[-k_\epsilon, k_\epsilon] \times [y_0 - \delta_0, y_0 + \delta_0]$. By Fatou's lemma, we have

$$\int_{|x| \leq k_\epsilon} \liminf_{y \to y_0} (A_\epsilon - v(x+iy)) dx \leq \liminf_{y \to y_0} \int_{|x| \leq k_\epsilon} (A_\epsilon - v(x+iy)) dx,$$

which implies that

$$\limsup_{y \to y_0} \int_{|x| \leq k_\epsilon} v(x+iy) dx \leq \int_{|x| \leq k_\epsilon} \limsup_{y \to y_0} v(x+iy) dx$$

$$\leq \int_{|x| \leq k_\epsilon} v(x+iy_0) dx.$$

Therefore, there exists $\delta \in (0, \delta_0)$ such that for $|y - y_0| < \delta$, we have

$$\int_{|x| \leq k_\epsilon} v(x+iy) dx \leq \int_{|x| \leq k_\epsilon} v(x+iy_0) dx + \frac{\epsilon}{2},$$

which implies that $\varphi(y) < \varphi(y_0) + \epsilon$. Thus, $\varphi(y)$ is upper semicontinuous in (a, b). Let

$$\widetilde{\varphi}(z) = \int_{\mathbb{R}} v(t+z) \, dt, \quad z \in \Omega_{a,b},$$

then

$$\widetilde{\varphi}(z) = \varphi(y), \quad z = x + iy \in \Omega_{a,b}.$$

For any $z_0 \in \Omega_{a,b}$, there exists $\delta_0 > 0$ such that $D(z_0, \delta_0) = \{z : |z - z_0| < \delta\} \subset \Omega_{a,b}$. Therefore, for $0 < r < \delta$, we have

$$\widetilde{\varphi}(z_0) \leq \int_{\mathbb{R}} \left(\frac{1}{2\pi} \int_0^{2\pi} v(t + z_0 + Re^{i\theta}) \, d\theta \right) dt$$

$$= \frac{1}{2\pi} \int_0^{2\pi} \widetilde{\varphi}(z_0 + Re^{i\theta}) \, d\theta.$$

Thus, the function $\widetilde{\varphi}(z)$ is subharmonic in $D(z_0, \delta_0)$. This shows that $\widetilde{\varphi}(z)$ is subharmonic in $\Omega_{a,b}$. Let $\alpha(t)$ be a non-negative infinitely differentiable function on $\mathbb{R}$ with $\operatorname{supp}\alpha \subset [0,1]$ and with

$$2\pi \int_0^1 t\alpha(t^2)dt = 1.$$

Let $\alpha_\delta(z) = \frac{1}{\delta^2}\alpha(\frac{|z|^2}{\delta^2})$, according to Theorem 4.3.9, the function

$$u_\delta(z) = \int_{D(z,\delta)} \widetilde{\varphi}(\zeta)\alpha_\delta(z-\zeta)d\lambda(\zeta)$$

is infinitely differentiable and subharmonic on $\Omega_\delta = \{z : d(z, \Omega_{a,b}^c) > \delta\} = \{z = x+iy : a+\delta < x < b-\delta, y \in \mathbb{R}\}$ $(0 < \delta < 2^{-1}(a+b))$ and as δ tends to zero monotonically, $u_\delta(z)$ on $\Omega_\delta = \{z : d(z, \Omega^c) > \delta\}$ decreases pointwise toward $\widetilde{\varphi}(z) = \varphi(y)$, where $d\lambda(\zeta)$ is the Lebesgue measure. By Theorem 4.3.10(b),

$$\Delta u_\delta(z) = \frac{\partial^2}{\partial x^2}u_\delta + \frac{\partial^2}{\partial y^2}u_\delta \geq 0.$$

Since $u_\delta(x+iy) = u_\delta(iy)$, it follows that

$$\Delta u_\delta = \frac{d^2}{dy^2}u_\delta(iy) \geq 0.$$

Thus, the function $u_\delta(iy)$ is convex in $(a+\delta, b-\delta)$. As δ tends to zero monotonically, $u_\delta(z)$ decreases pointwise on $\Omega_\delta = \{z : d(z, \Omega^c) > \delta\}$ toward $\widetilde{\varphi}(z) = \varphi(y)$. Therefore, the function $\varphi(y)$ is convex in (a,b).

(b) From (a), we know that the function $\varphi(y)$ is convex in $(0, \infty)$, which implies that $\varphi(y)$ is a monotonically decreasing convex function in $(0, \infty)$. Now, consider a fixed $y_0 > 0$. The function

$$u_{y_0}(z) = \int_{\mathbb{R}} v(t+iy_0)P_y(x-t)dt$$

is harmonic in $\mathbb{C}_+$. Let $\varepsilon > 0$ and $y_0 > 0$ be fixed, and set $\delta_0 = \frac{y_0}{2}$, from Theorem 4.6.12, there exists $y_1 > 2y_0$ such that for all $y \geq y_1$, and any x, we have

$$v(x+i(y_0+y)) < \frac{\varepsilon}{2}.$$

Moreover, from the proof of (a), there exists $A > 1 + 4\delta_0$ such that

$$\int_{|\xi| \geq A-\delta_0} \int_{\delta_0}^{y_1+4\delta_0} v(\xi+i\eta)d\xi d\eta < \frac{\epsilon}{2}.$$

Therefore, for $0 \leq y \leq y_1$ and $|x| \geq A$, we have

$$v(z+iy_0) \leq \frac{1}{\pi\delta_0^2} \int_{(\xi-x)^2+(\eta-y)^2 \leq \delta_0^2} v(\xi+i\eta+iy_0)d\xi d\eta$$

$$\leq \frac{1}{\pi\delta_0^2} \int_{-\delta_0}^{\delta_0} \int_{-\delta_0}^{\delta_0} v(\xi+x+i(\eta+y+y_0))d\xi d\eta$$

$$\leq \frac{1}{\pi\delta_0^2} \int_{|\xi| \geq |x|-\delta_0} \int_{\delta_0}^{4\delta_0+y_1} v(\xi+i\eta)d\xi d\eta$$

$$\leq \frac{2}{\pi\delta_0^2} \int_{|\xi| \geq |x|-\delta_0} \int_{\delta_0}^{y_1+4\delta_0} v(\xi+i\eta)d\xi d\eta < \frac{\varepsilon}{2}.$$

Since there exists a sequence of monotonically decreasing functions $\{u_n(t)\}$ on $[-A, A]$ that converge uniformly to $v(t+iy_0)$ and are continuous, we have

$$U_n(z) = \int_{-A}^{A} P_y(x-t)u_n(t)dt,$$

which is harmonic in $\mathbb{C}_+$, and the function

$$V_n(z) = v(z+iy_0) - U_n(z)$$

is subharmonic in $\mathbb{C}_+$. Now, consider the set $G = \{z = x+iy : |x| < A, 0 < y < y_1\}$. For any $z = x+iy \notin G$ with $y > 0$, we have

$$V_n(z) \leq v(z+iy_0) < \varepsilon.$$

For all $x_0 \in \mathbb{R}$ with $|x_0| \leq A$, we have

$$\limsup_{z \to x_0, y > 0} V_n(z) \leq v(x_0+iy_0) - u_n(x_0) \leq 0.$$

Therefore, for any boundary point $z_0 \in \partial G$, where $G = \{z = x+iy : |x| < A, 0 < y < y_1\}$ is a bounded region, we have

$$\limsup_{z \to z_0, z \in G} V_n(z) \leq \varepsilon.$$

Hence, for any $z \in G = \{z = x+iy : |x| < A, 0 < y < y_1\}$, we have $V_n(z) \leq \varepsilon$. Consequently, for any $z \in \mathbb{C}_+$, we have $V_n(z) \leq \varepsilon$. Taking the limit as $n \to \infty$, we obtain

$$v(z+iy_0) \leq \varepsilon + \int_{-A}^{A} P_y(x-t)v(t+iy_0)dt \quad (y>0).$$

Therefore,

$$v(z+iy_0) \leq \varepsilon + \int_{-\infty}^{\infty} P_y(x-t)v(t+iy_0)dt \quad (y>0).$$

As $\varepsilon \to 0$,

$$v(z+iy_0) \leq \int_{-\infty}^{\infty} P_y(x-t)v(t+iy_0)dt \quad (y>0).$$

Therefore, for $y > y_0 > 0$, we have

$$v(z) \leq \int_{-\infty}^{\infty} P_{y-y_0}(x-t)v(t+iy_0)dt.$$

Since there exists a sequence $\{y_k\}$ such that $y_k > 0$, $y_k \to 0$ as $(k \to \infty)$ and $\|v_{y_k}\|_1 = \varphi(y_k) \to C$ $(k \to \infty)$, where $v_y(x) = v(x+iy)$. By (4.6.17), we can apply the Banach–Alaoglu theorem. There exist a subsequence $\{v_{y_{k_j}}\}$ and $\mu \in \mathscr{M}(\mathbb{R})$ such that $\{v_{y_{k_j}}\}$ weakly* converges to μ. In other words, for any $\psi \in C_0(\mathbb{R})$,

$$\lim_{j \to \infty} \int_{\mathbb{R}} v(t+iy_{k_j})\psi(t)dt = \int_{\mathbb{R}} \psi(t)d\mu(t). \qquad (4.6.19)$$

Therefore,

$$\|\mu\| = \sup\left\{\left|\int_{\mathbb{R}} \psi(t)d\mu(t)\right| : \psi \in C_0(\mathbb{R}), \|\psi\|_\infty = 1\right\}$$

$$\leq \liminf_{j \to \infty} \int_{\mathbb{R}} v(t+iy_{k_j})dt = C.$$

Since there exists a sequence $\{\psi_j\} \subset C_0(\mathbb{R})$ with $\|\psi_j\|_\infty = 1$ such that

$$\|\mu\| + \frac{1}{j} \geq \int_{\mathbb{R}} \psi_j(t)v(t+iy_{k_j})dt + \frac{1}{j} \geq \int_{\mathbb{R}} v(t+iy_{k_j})dt,$$

we have $C \leqslant \|\mu\|$. Thus, we get $C = \|\mu\|$. By taking $\psi(t) = P_y(x-t)$ $\in C_0(\mathbb{R})$ in (4.6.19), we have

$$\lim_{j \to \infty} \int_{\mathbb{R}} v(t + iy_{k_j}) P_y(x-t) dt. = \int_{\mathbb{R}} P_y(x-t) d\mu(t) = u(x+iy).$$

For $y > 2y_k$, we have

$$|P_y(x-t) - P_{y-y_k}(x-t)| \leqslant \frac{y_k}{(x-t)^2 + (y-y_k)^2} \leqslant \frac{4y_k}{y^2}.$$

Therefore,

$$\lim_{j \to \infty} \int_{\mathbb{R}} |P_{y-y_{k_j}}(x-t) - P_y(x-t)||v(t+iy_{k_j})| dt = 0.$$

Therefore,

$$v(z) \leqslant \lim_{j \to \infty} \int_{\mathbb{R}} v(t+iy_{k_j}) P_y(x-t) dt = \int_{\mathbb{R}} P_y(x-t) d\mu(t)$$

for all $y > 0$. $\square$

Theorem 4.6.14 (Poisson Representation). *Let $v(z)$ be a non-negative harmonic function on $\mathbb{C}_+$. Then, there exists $c \geqslant 0$ and a non-negative Borel measure μ on $\mathbb{R}$ such that*

$$\int_{-\infty}^{\infty} \frac{d\mu(t)}{1+t^2} < \infty \qquad (4.6.20)$$

and

$$v(z) = cy + \frac{y}{\pi} \int_{-\infty}^{\infty} \frac{d\mu(t)}{(t-x)^2 + y^2}, \quad y > 0. \qquad (4.6.21)$$

Proof. The mapping

$$\alpha(w) = i\frac{1-w}{1+w} \qquad (4.6.22)$$

maps U onto the upper half-plane $\mathbb{C}_+$, its inverse mapping is

$$\beta(z) = \frac{i-z}{i+z}. \qquad (4.6.23)$$

Then the function $u(w) = v(\alpha(w))$ is a non-negative harmonic function in the unit disk U, thus we have

$$u(0) = \frac{1}{2\pi} \int_{-\pi}^{\pi} u(Re^{i\theta}) d\theta.$$

By Theorem 4.5.7, there exists a non-negative Borel measure ν on $[-\pi, \pi)$ such that

$$u(w) = \frac{1}{2\pi} \int_{[-\pi,\pi)} \operatorname{Re} \frac{e^{i\theta} + w}{e^{i\theta} - w} d\nu(\theta) \quad (w \in U).$$

Thus, for $z = x + iy, y > 0$,

$$v(z) = \frac{1}{2\pi} \operatorname{Re} \frac{-1 + \beta(z)}{-1 - \beta(z)} \nu(\{-\pi\}) + \frac{1}{2\pi} \int_{(-\pi,\pi)} \operatorname{Re} \frac{e^{i\theta} + \beta(z)}{e^{i\theta} - \beta(z)} d\nu(\theta)$$

$$= cy + \frac{y}{2\pi} \int_{\mathbb{R}} \frac{1 + t^2}{(t-x)^2 + y^2} d\nu(\theta(t)),$$

where $c = \frac{\nu(\{-\pi\})}{2\pi} \geqslant 0$, $\theta = \theta(t) = 2\arctan t$ is the inverse function of $t = \tan \frac{\theta}{2}$. We define a non-negative Borel measure μ on $(-\infty, \infty)$ by $2d\mu(t) = (1 + t^2) d\nu(\theta(t))$. Then, we see that Theorem 4.6.14 holds. □

Theorem 4.6.15 (Nevanlinna Representation). *Let $f(z)$ be an analytic function on $\mathbb{C}_+$ and satisfy $\operatorname{Im} f(z) \geqslant 0$ for $z \in \mathbb{C}_+$. Then, there exist $b = \bar{b}, c \geqslant 0$ and a non-negative Borel measure μ on $\mathbb{R}$ such that (4.6.20) holds and*

$$f(z) = b + cz + \frac{1}{\pi} \int_{-\infty}^{\infty} \left[\frac{1}{t-z} - \frac{t}{1+t^2} \right] d\mu(t), \quad y > 0. \quad (4.6.24)$$

Proof. By Theorem 4.6.14, there exist $c \geqslant 0$ and a non-negative Borel measure μ such that (4.6.20) holds and $v(z) = \operatorname{Im} f(z)$ satisfies (4.6.21). Thus, the function

$$F(z) = f(z) - cz - \frac{1}{\pi} \int_{-\infty}^{\infty} \left[\frac{1}{t-z} - \frac{t}{1+t^2} \right] d\mu(t)$$

is analytic on $\mathbb{C}_+$ and $\operatorname{Im} F(z)$ is identically zero on $\mathbb{C}_+$. Therefore, $F(z)$ is a real constant on $\mathbb{C}_+$. □

Theorem 4.6.16 (Stieltjes Inversion Formula). *Let $v(z)$ be a non-negative harmonic function on $\mathbb{C}_+$ defined by (4.6.21), where $c \geqslant 0$, and μ is a non-negative Borel measure on $\mathbb{R}$ satisfying (4.6.20). If $-\infty < a < b < \infty$, then*

$$\lim_{y>0, y\to 0} \int_a^b v(x+iy)dx = \mu([a,b]) - \frac{1}{2}(\mu(\{a\}) + \mu(\{b\})). \quad (4.6.25)$$

Proof. Let

$$\chi(t) = \begin{cases} 1 & \text{if } a < t < b, \\ \frac{1}{2} & \text{if } t = a \text{ or } t = b, \\ 0 & \text{otherwise,} \end{cases}$$

and

$$g(t,y) = (1+t^2)\left(\frac{y}{\pi}\int_a^b \frac{dx}{(t-x)^2+y^2} - \chi(t)\right).$$

Since $|g(t,y)| \leqslant 4(b-a)$ when $|t| \geqslant 2(|a|+|b|+1), 0 < y \leqslant 1$, and $|g(t,y)| \leqslant (1+4(|a|+|b|+1)^2$ when $|t| \leqslant 2(|a|+|b|+1)$ and $0 < y \leqslant 1$, we can apply Fubini's theorem and the Lebesgue-Dominated Convergence Theorem.

$$\lim_{y>0, y\to 0} \left(\int_a^b v(x+iy)dx - \mu([a,b]) + \frac{1}{2}(\mu(\{a\}) + \mu(\{b\}))\right)$$

$$= \lim_{y>0, y\to 0} \left(c(b-a)y + \int_{-\infty}^\infty \frac{g(t,y)d\mu(t)}{1+t^2}\right) = 0.$$

This proves (4.6.24). □

Lemma 4.6.17. *Let $g(x)$ be a non-negative and non-decreasing function on $[0,1)$. Let $p(x)$ be any non-negative measurable function on $(0,1)$ such that $p \in L^1[0,a]$ for every $a \in (0,1)$ and*

$$\int_0^1 p(t)dt = \infty.$$

Then

$$\lim_{x<1, x\to 1} g(x) = \sup\left\{\frac{\int_0^1 g(t)p(\lambda t)dt}{\int_0^1 p(\lambda t)dt} : 0 < \lambda < 1\right\}.$$

Proof. Set

$$q(\lambda) = \frac{\int_0^1 g(t)p(\lambda t)dt}{\int_0^1 p(\lambda t)dt}$$

for $0 < \lambda$. If $\lim_{x<1, x\to 1} g(x) = M < \infty$, then $q(\lambda) \leqslant M$ for $\lambda \in (0,1)$, and so, $M \geqslant K$, where $K = \sup\{q(\lambda) : 0 < \lambda < 1\}$. When $0 < x < \lambda < 1$,

$$K \geqslant q(\lambda)$$
$$\geqslant \frac{\int_x^1 g(t)p(\lambda t)dt}{\int_0^1 p(\lambda t)dt} \geqslant g(x)\frac{\int_x^1 p(\lambda t)dt}{\int_0^1 p(\lambda t)dt} = g(x)\left(1 - \frac{\int_0^{\lambda x} p(t)dt}{\int_0^{\lambda} p(t)dt}\right).$$

Letting $\lambda \to 1$, we obtain $K \geqslant g(x)$, it follows that $K \geqslant M$. This proves the lemma. □

Theorem 4.6.18. *A non-negative subharmonic function $v(z)$ on the upper half-plane $\mathbb{C}_+$ has a harmonic majorant $u(z)$ on $\mathbb{C}_+$ (i.e., $v(z) \leqslant u(z)$) for all $z \in \mathbb{C}_+$) if and only if*

$$\sup_{y>0} \int_{-\infty}^{\infty} \frac{v(x+iy)}{x^2 + (y+1)^2} dx < \infty. \tag{4.6.26}$$

Proof. Suppose first that $v(z)$ has a harmonic majorant $u(z)$ on $\mathbb{C}_+$. By Theorem 4.6.14 (the Poisson representation of a positive harmonic function), $u(z)$ can be written as

$$u(z) = cy + \frac{y}{\pi}\int_{-\infty}^{\infty} \frac{d\mu(t)}{(t-x)^2 + y^2}, \quad y > 0,$$

where $c \geqslant 0$ and μ is a positive Borel measure on $\mathbb{R}$, satisfying

$$\int_{-\infty}^{\infty} \frac{d\mu(t)}{1+t^2} < \infty.$$

Let $P_y(x) = P(x,y)$ be the Poisson kernel on the upper half-plane $\mathbb{C}_+$, the function $u_a(x,y) = P(x, a+y)(a > 0)$ is harmonic on the upper half-plane $\mathbb{C}_+$, satisfying

$$P_{a+y} \in L(\mathbb{R}), \quad P_a * P_y = P_{a+y} \quad \text{and} \quad \int_{\mathbb{R}} P_y(x)dx = 1.$$

Therefore,

$$\frac{1}{\pi}\int_{\mathbb{R}} \frac{v(x+iy)dx}{x^2+(y+1)^2}$$

$$= \frac{1}{y+1}\int_{\mathbb{R}} v(x+iy)P_{y+1}(x)dx$$

$$\leqslant \int_{\mathbb{R}} \frac{cy}{y+1}P_{y+1}(x)dx + \frac{1}{y+1}\int_{\mathbb{R}} P_y * P_{y+1}(t)d\mu(t)$$

$$= \frac{cy}{y+1} + \frac{1}{y+1}\int_{\mathbb{R}} P_{2y+1}(t)d\mu(t) \leqslant c + 2\int_{\mathbb{R}} \frac{d\mu(t)}{1+t^2}$$

for all $y > 0$, and (4.6.26) follows.

Conversely, assume that (4.6.26) holds. We use the mapping $\alpha(w) = i\frac{1-w}{1+w}$ defined by (4.6.22). We must show that $v(z)$ has a harmonic majorant on $\mathbb{C}_+$, which is the same thing as $v_1(w) = v(\alpha(w))$ having a harmonic majorant on U. The condition for this (Theorem 4.3.23) is that

$$g(r) = r\int_{-\pi}^{\pi} v_1(\mathrm{Re}^{i\theta})d\theta$$

remain bounded and increasing on $r \in [\frac{1}{2}, 1)$. By Lemma 4.6.17, it is sufficient to show that there is a positive constant $A > 0$ such that

$$\int_0^1 \frac{g(r)}{1-\lambda^2 r^2}dr \leqslant A\int_0^1 \frac{1}{1-\lambda^2 t^2}dt$$

for all $\lambda \in (0,1)$. Calculate as follows:

$$\int_0^1 \frac{g(r)}{1-\lambda^2 r^2}dr$$

$$= \int_0^1 \int_{-\pi}^{\pi} \frac{v_1(\mathrm{Re}^{i\theta})}{1-\lambda^2 r^2}d\theta r dr$$

$$= \frac{1}{2\pi}\int_U \frac{v_1(w)}{1-\lambda^2|w|^2}d\lambda(w) = \frac{1}{2\pi}\int_{\mathbb{C}_+} \frac{v(z)}{1-\lambda^2|\beta(z)|^2}|\beta'(z)|d\lambda(z)$$

$$= \frac{2}{\pi}\int_{\mathbb{C}_+} \frac{v(z)}{|z+i|^2}\frac{d\lambda(z)}{|z+i|^2 - \lambda^2|z-i|^2}$$

$$= \frac{2}{\pi} \int_{\mathbb{C}_+} \frac{v(x+iy)}{x^2+(y+1)^2} \frac{dxdy}{(1-\lambda^2)x^2+(y+1)^2-\lambda^2(y-1)^2}$$

$$\leq \frac{2}{\pi} \int_0^\infty \left(\int_{\mathbb{R}} \frac{v(x+iy)dx}{x^2+(y+1)^2} \right) \frac{dy}{(y+1)^2-\lambda^2(y-1)^2}.$$

By (4.6.24),

$$\int_0^1 \frac{g(r)}{1-\lambda^2 r^2} dr \leq A \int_0^\infty \frac{dy}{(y+1)^2-\lambda^2(y-1)^2}$$

$$= \frac{A}{2} \int_{-1}^1 \frac{dt}{1-\lambda^2 t^2} = A \int_0^1 \frac{1}{1-\lambda^2 t^2} dt.$$

The change of variables is made with the substitution $t = \frac{y-1}{y+1}$. The theorem follows. □

Exercises IV

1. Prove the minimum principle of harmonic functions: Let u be a harmonic function in the domain Ω, satisfying $B = \inf_{x \in \Omega} u(x) > -\infty$. If u is not constant, then for any $z \in \Omega$, we have $u(z) > B$.
2. Let u_1 and u_2 be both harmonic in a bounded region Ω and continuous on the closure $\overline{\Omega}$. Show the following:

    (a) If u_1 is not a constant function, then its maximum (minimum) value is attained only on the boundary $\partial \Omega = \overline{\Omega} \setminus \Omega$.
    (b) If $u_1 = u_2$ on the boundary $\partial \Omega$, then for all $z \in \overline{\Omega}$, we have $u_1(z) = u_2(z)$.

3. **The Hadamard Three-Circle Theorem:** If $f(z)$ is analytic in an annular region $R_1 < |z| < R_2$, and if $M(r) = \max\{|f(z)| : |z| = r\}$, $R_1 < r < R_2$, then $\log M(r)$ is a convex function of $\log r$ on the interval (R_1, R_2), in other words, when $R_1 < a < r < b < R_2$, we have

$$\log M(r) \leq \frac{\log(b/r)}{\log(b/a)} \log M(a) + \frac{\log(r/a)}{\log(b/a)} \log M(b).$$

4. **The Harnack Inequality:** Suppose Ω is a region, $K \subset \Omega$ is a compact, and $z_0 \in \Omega$. Prove that there exist positive constants α and β, which depend only on z_0, K and Ω such that

$$\alpha u(z_0) \leqslant u(z) \leqslant \beta u(z_0)$$

for every positive harmonic function $u(z)$ in Ω and for all $z \in K$.

5. Let $I = [a, b]$ be a closed interval on the real axis, $b > a$. Suppose that φ is continuous on I and that $\Omega = \mathbb{C} \setminus I$ is the complement of I, relative to the plane, and define

$$f(z) = \frac{1}{2\pi i} \int_a^b \frac{\varphi(t)}{t - z} dt \quad (z \notin I).$$

Prove that $f \in H(\Omega)$ and

$$\lim_{\epsilon > 0, \epsilon \to 0} [f(x + \epsilon) - f(x - i\epsilon)]$$

exists for every real x and express this limit using the φ. How would the above result be stated if we only assume that $\varphi \in L^1([a, b])$? If we further assume that $\varphi \in L^1([a, b])$ and has left and right limits at point x, how would the above result be stated?

6. Let μ be a positive measure on the unit disk U, satisfying

$$\int_U (1 - |\zeta|) d\mu(\zeta) < \infty.$$

Show that the Green's potential

$$G_\mu(z) = \frac{1}{2\pi} \int_U \log \left| \frac{\zeta - z}{1 - \bar{\zeta} z} \right| d\mu(\zeta)$$

satisfies the following properties:

(a) $G_\mu(z) \in L^1(U)$;
(b) $G_\mu(z)$ is a negative subharmonic function on U;
(c) $G_\mu(z)$ is the solution of the equation $\Delta u = \mu$ in the distribution sense. This means that for any test function $\phi \in C_0^2(U)$, we have

$$\int_U G_\mu(z) \Delta \phi(z) d\lambda(z) = \int_U \phi(z) d\mu(z);$$

(d) In the following sense,
$$\lim_{r<1, r\to 1} \int_0^{2\pi} G_\mu(\mathrm{Re}^{i\theta}) d\theta = 0,$$
$G_\mu(z)$ is zero on the boundary.

7. Suppose that u is continuous in the region Ω and $a \in \Omega$. Prove that if u is subharmonic in the region $\Omega \setminus \{a\}$, then u is subharmonic in the region Ω.
8. Let $f(z)$ be an analytic function in the domain Ω, and let it not be identically zero. Prove that $u(z) = \log|f(z)|$ and $v(z) = |f(z)|^\alpha$ ($\alpha > 0$) are subharmonic in the region Ω.
9. Let u be a continuous function in the domain Ω. If both u and $-u$ are subharmonic in Ω, then u is harmonic in Ω.
10. Let $0 < R_1 < R_2 < \infty$, $L_1: |z| = R_1$, $L_2: |z| = R_2$. Find a function $u(z)$ that is constant A on L_1, constant B on L_2, continuous on $R_1 \leqslant |z| \leqslant R_2$, and harmonic on $R_1 < |z| < R_2$.
11. Let H be the interior of the ellipse $3x^2 + 4y^2 = 12$, and L be the line passing through its foci. Find a function $u(z)$ that is constant A on L, constant B on $3x^2 + 4y^2 = 12$, continuous on the closure $\overline{H}$ of H, and harmonic on $H \setminus L$.
12. Let u be Lebesgue-measurable in the domain Ω, $u \in L^1_{loc}(\Omega)$ and
$$u(a) = \frac{1}{\pi r^2} \int_{D(a,r)} u(z) d\lambda(z)$$
for all $D(a, r) \subset\subset \Omega$. Prove that u is harmonic in Ω.
13. If $u(z)$ is harmonic in the annular region $R_1 < |z| < R_2$, then $A(r) = \max\{u(z) : |z| = r\}$, $R_1 < r < R_2$ is a convex function of $\log r$ in $(\log R_1, \log R_2)$.
14. Prove that the Green's function $g(z, z_0)$ for the domain Ω has symmetry
$$g(z, z_0) = g(z_0, z).$$
15. Show that the Green's function $g(z, z_0)$ for the domain Ω is positive on Ω.
16. Suppose that $u(z)$ is a subharmonic function in the domain Ω with $m < u(z) < M$ and $g(t)$ is a non-decreasing convex function on $[m, M]$. Show that $g(u(z))$ is a subharmonic function in Ω.

17. Suppose $f(z)$ is analytic in the half-plane $\Pi = \{z = x + iy : \operatorname{Re} z = x > 0\}$, and for all ζ in $\partial \Pi = \{z = x + iy : x = 0, y \in \mathbb{R}\}$,

$$\varlimsup_{z \in \Pi, z \to \zeta} |f(z)| \leqslant 1, \quad \text{and} \quad \varlimsup_{r \to \infty} \frac{\log M(r)}{r^\alpha} < \infty,$$

where $M(r) = \sup\{|f(z)| : |z| \leqslant r, z \in \Pi\}, 0 < \alpha < 1$. Show that

(a) $|f(z)| \leqslant 1$ for all $z \in \Pi$;

(b) if

$$\lim_{r \to \infty} \frac{\log |f(r)|}{r} = -\infty,$$

then $f(z) \equiv 0$.

18. Let $u \not\equiv -\infty$ be a subharmonic in the unit disk U. Prove that

$$\sup \left\{ \int_{-\pi}^{\pi} u^+(Re^{i\theta}) d\theta : 0 < r < 1 \right\} < \infty$$

if and only if

$$\sup \left\{ \int_{-\pi}^{\pi} |u(Re^{i\theta})| d\theta : \frac{1}{2} \leqslant r < 1 \right\} < \infty.$$

19. Suppose that $u \not\equiv -\infty$ is subharmonic in the unit disk U. Prove that the function

$$m(r) = \int_{-\pi}^{\pi} u(Re^{i\theta}) d\theta$$

is continuous in $(0, 1)$.

20. Find all subharmonic functions $u(z)$ in the unit disk U that satisfy $u(z) = u(|z|)$.

21. Let $\Omega \neq \mathbb{C}$ be a region, Prove that the function $u(z) = -\log d(z, \Omega^c)$ is subharmonic in Ω, where $d(z, \Omega^c)$ is the distance from z to Ω^c.

Chapter 5

H^p Spaces

5.1 H^p Spaces in the Unit Disk

Theorem 5.1.1 (Jensen's Formula). *Let $F \in H(D(0,R))$, $F(z) \not\equiv 0$, $0 < r < R$ and suppose that $F(z)$ has a zero at $z = 0$ of multiplicity k and $z_1, z_2, \ldots, z_n$ are the other zeros of $F(z)$ in $\overline{D(0,r)}$ repeated according to the multiplicity, then*

$$\log\left|\frac{F^{(k)}(0)}{k!}\right| + \sum_{j=1}^{n} \log\frac{r}{|z_j|} = \frac{1}{2\pi}\int_0^{2\pi} \log|F(re^{i\theta})|d\theta - k\log r.$$
(5.1.1)

Proof. Order the points z_j so that $z_1, z_2, \ldots, z_m$ are in $D(0, r)$, and $|z_{m+1}| = \cdots = |z_n| = r$ (of course, we may have $m = n$ or $m = 0$). Put

$$G(z) = \frac{F(z)}{z^k} \prod_{j=1}^{m} \frac{r^2 - z\bar{z}_j}{r(z - z_j)} \prod_{j=m+1}^{n} \frac{z_j}{z - z_j},$$

Then $G(z) \in H(D(0, r + \varepsilon))$ for some $\varepsilon \in (0, R - r)$, and $G(z) \neq 0$ for any $z \in D(0, r + \varepsilon)$, hence $\log|G(z)|$ is harmonic in $D(0, r + \varepsilon)$, and so,

$$\log|G(0)| = \frac{1}{2\pi}\int_0^{2\pi} \log|G(re^{i\theta})|d\theta,$$

i.e.,

$$\log\left|\frac{F^{(k)}(0)}{k!}\right| + \sum_{j=1}^{m} \log\frac{r}{|z_j|} = \log\left|\frac{F^{(k)}(0)}{k!}\right| + \sum_{j=1}^{n} \log\frac{r}{|z_j|}$$

$$= \frac{1}{2\pi}\int_0^{2\pi} \log|F(re^{i\theta})|d\theta - k\log r$$

$$- \sum_{j=m+1}^{n} \frac{1}{2\pi}\int_{-\pi}^{\pi} \log|1 - e^{i(\theta-\theta_j)}|d\theta.$$

Therefore, it suffices to prove that

$$\int_{-\pi}^{\pi} \log|1 - e^{i\theta}|d\theta = 0. \tag{5.1.2}$$

In fact, $\log|1-z|$ is harmonic in U, thus

$$\int_{-\pi}^{\pi} \log|1 - re^{i\theta}|d\theta = 0 \quad (0 < r < 1). \tag{5.1.3}$$

When $|\theta| \leqslant \frac{\pi}{3}$ and $r \geqslant \frac{1}{2}$, we have $\frac{\theta^2}{\pi^2} \leqslant 4r\sin^2\frac{\theta}{2} \leqslant |1-re^{i\theta}|^2 = 1 - 2r\cos\theta + r^2 \leqslant 1$, hence

$$|\log|1 - re^{i\theta}|| \leqslant -\log|\theta| + \log\pi \in L[-\frac{\pi}{3}, \frac{\pi}{3}],$$

and when $|\theta| \geqslant \frac{\pi}{3}$ and $r \geqslant \frac{1}{2}$, we have $\frac{1}{2} \leqslant |1-re^{i\theta}| \leqslant 2$, thus $|\log|1-re^{i\theta}|| \leqslant \log 2$. By taking the limit as $r \to 1$ in (5.1.3) and using the Lebesgue-dominated convergence theorem, we obtain (5.1.2). $\square$

Definition 5.1.2. Let $0 < p \leqslant \infty$. The space $H^p(U) = \{f \in H(U) : \|f\|_{H^p} < \infty\}$ is called H^p space in the unit disk in U, where

$$\|f\|_{H^p} = \sup\{m_p(f,r) : 0 < r < 1\},$$

$$m_p(f,r) = \left(\frac{1}{2\pi}\int_{-\pi}^{\pi} |f(re^{i\theta})|^p d\theta\right)^{\frac{1}{p}},$$

$$m_\infty(f,r) = \sup\{|f(re^{i\theta})| : \theta \in [-\pi,\pi]\}.$$

The class $H^0(U) = \{f \in H(U) : \|f\|_{H^0} < \infty\}$ is called the Nevanlinna class, where

$$\|f\|_{H^0} = \sup\{m_0(f,r) : 0 < r < 1\},$$

$$m_0(f,r) = \exp\left\{\frac{1}{2\pi}\int_{-\pi}^{\pi}\log^+|f(re^{i\theta})|d\theta\right\}.$$

Since $\log^+|f|, |f|^p$ is subharmonic in the unit disk U, Theorem 4.3.8 shows that $m_p(f,r)$ is increasing and converges to $\|f\|_{H^p}$ as $r \to 1$. H^p has the following property:

(a) $H^\infty(U) \subset H^q(U) \subset H^p(U) \subset H^0(U), 0 < p < q < \infty$.

(b) $f(z) = \sum_{n=0}^{\infty} a_n z^n \in H^2(U)$ if and only if $\|f\|_{H^2}^2 = \sum_{n=0}^{\infty} |a_n|^2 < \infty$.

(c) For $p \geqslant 1$, $\|\cdot\|_{H^p}$ is a normed linear space, $\|\cdot\|_{H^p}$ satisfies the triangle inequality

$$\|f+g\|_{H^p} \leqslant \|f\|_{H^p} + \|g\|_{H^p}.$$

For $0 < p < 1$, $\rho(f,g) = \|f-g\|_{H^p}^p$ is a metric in $H^p(U)$, $H^p(U)$ is a linear metric space.

(d) If $f \in H^0(U)$, $f(z) \not\equiv 0$, then

$$\sup\left\{\frac{1}{2\pi}\int_{-\pi}^{\pi}|\log|f(re^{i\theta})||d\theta : \varepsilon < r < 1\right\} < \infty \qquad (5.1.4)$$

for all $\varepsilon \in (0,1)$.

In fact, applying Jensen's formula to $f \in H^0(U)$, $f(z) \not\equiv 0$, 0 is a zero of multiplicity $k \geqslant 0$ of $f(z)$, then

$$\log\left|\frac{f^{(k)}(0)}{k!}\right| + k\log r \leqslant \frac{1}{2\pi}\int_{-\pi}^{\pi}\log|f(re^{i\theta})|d\theta$$

$$= \frac{1}{2\pi}\int_{-\pi}^{\pi}\log^+|f(re^{i\theta})|d\theta - \frac{1}{2\pi}\int_{-\pi}^{\pi}\log^-|f(re^{i\theta})|d\theta,$$

where $\log^- t = \max\{-\log t, 0\}$. Therefore,

$$\frac{1}{2\pi}\int_{-\pi}^{\pi}\log^-|f(re^{i\theta})|d\theta \leqslant \frac{1}{2\pi}\int_{-\pi}^{\pi}\log^+|f(re^{i\theta})|d\theta$$

$$- \log\left|\frac{f^{(k)}(0)}{k!}\right| - k\log r,$$

so

$$\frac{1}{2\pi}\int_{-\pi}^{\pi}|\log|f(re^{i\theta})||d\theta \leqslant \frac{1}{\pi}\int_{-\pi}^{\pi}\log^+|f(re^{i\theta})|d\theta$$

$$-\log\left|\frac{f^{(k)}(0)}{k!}\right| - k\log r.$$

Theorem 5.1.3. *If $z_1, z_2, \ldots, z_n, \ldots$ is a sequence in U such that $z_j \neq 0$ and*

$$\sum_{j=1}^{\infty}(1-|z_j|) < \infty. \tag{5.1.5}$$

If k is a non-negative integer, and if

$$B(z) = z^k \prod_{j=1}^{\infty}\frac{|z_j|(z_j-z)}{z_j(1-z\bar{z}_j)} \quad (z \in U), \tag{5.1.6}$$

then $B(z) \in H^{\infty}(U)$ converges uniformly on $\overline{D(0,r)}$ $(0 < r < 1)$ and B has no zeros except at the points z_j (and at the origin, if $k > 0$).

We call this function B a Blaschke product. Note that some of z_j may be repeated, in which case B has multiple zeros at those points. Note also that each factor in (5.1.2) has an absolute value 1 on the boundary of the unit disk U.

Proof. The jth term in the product (5.1.6) satisfies that

$$\left|1 - \frac{|z_j|(z_j-z)}{z_j(1-z\bar{z}_j)}\right| = (1-|z_j|)\left|\frac{z_j + z|z_j|}{z_j - z|z_j|^2}\right| \leqslant (1-|z_j|)\frac{1+r}{1-r}$$

if $|z| \leqslant r < 1$. Thus, $B(z)$ converges uniform on $\overline{D(0,r)}$ $(0 < r < 1)$, so $B \in H(U)$ and $|B(z)| \leqslant 1$ ($|z| < 1$), $B(z)$ has no zeros except at the points z_j (and at the origin if $k > 0$). $\square$

Theorem 5.1.4. *Suppose $f \in H^0(U)$, $f(z)$ is not identically 0 in U, and $z_1, z_2, \ldots, z_n, \ldots$ are the zeros of $f(z)$ in U, listed according to their multiplicities. Then (5.1.5) holds.*

Proof. If f has a zero of order k at the origin, and $g(z) = z^k f(z)$, then $g \in H^0(U)$, and g has the same zeros as f except at

the origin. Hence, we may assume, without loss of generality, that $f(0) \neq 0$ and
$$0 < |z_1| \leq |z_2| \leq \cdots \leq |z_n| \leq |z_{n+1}| \leq \cdots.$$
Take $r \in (\frac{1}{2}, 1)$, then Jensen's formula
$$\sum_{|z_j| \leq r} \log \frac{r}{|z_j|} = \frac{1}{2\pi} \int_0^{2\pi} \log |f(re^{i\theta})| d\theta$$
$$\leq \frac{1}{2\pi} \int_0^{2\pi} \log^+ |f(re^{i\theta})| d\theta \leq M < \infty,$$
where M is independent of $r \in (\frac{1}{2}, 1)$. Thus,
$$\sum_{j=1}^n \log \frac{r}{|z_j|} \leq M$$
for all n. The inequality persists, for every n, as $r \to 1$,
$$\sum_{j=1}^n \log \frac{1}{|z_j|} \leq M - \log \left| \frac{f^{(k)}(0)}{k!} \right|.$$
Hence,
$$\sum_{j=1}^\infty \log \frac{1}{|z_j|} \leq M.$$
This implies (5.1.5). □

By Fatou's theorem, for any $\alpha > 0$, for almost all $t \in [-\pi, \pi]$, the non-tangential limit
$$B^*(e^{it}) = \lim_{z \in \Gamma_\alpha(e^{it}), z \to e^{it}} B(z)$$
exists.

Theorem 5.1.5. *If $z_1, z_2, \ldots, z_n, \ldots$ is a sequence in U such that $z_j \neq 0$ and satisfies (5.1.5), $B(z)$ is a Blaschke Product defined by (5.1.6). Then, for all $\alpha > 0$,*
$$\lim_{z \in \Gamma_\alpha(e^{it}), z \to e^{it}} |B(z)| = 1, \qquad (5.1.7)$$

for almost everywhere $t \in [-\pi, \pi]$, and

$$\lim_{r \to 1} \int_{-\pi}^{\pi} |\log|B(re^{i\theta})||d\theta = 0. \tag{5.1.8}$$

Proof. Since $\log|B(z)| \leqslant 0$ is subharmonic, by Theorem 4.3.8, the integral on the left side of (5.1.8) is monotonically decreasing, indicating that the limit on the left side of (5.1.8) exists. By Fatou's lemma, we have

$$0 \leqslant \int_{-\pi}^{\pi} (-\log|B^*(e^{i\theta})|)d\theta \leqslant \lim_{r \to 1} \int_{-\pi}^{\pi} (-\log|B(re^{i\theta})|)d\theta.$$

For all n, let

$$B_n(z) = z^k \prod_{j=1}^{n} \frac{|z_j|(z_j - z)}{z_j(1 - z\bar{z}_j)}$$

($B_n(z)$ is called a finite Blaschke Product), then $|B_n(e^{it})| = 1$. As $r \to 1$, $\log|B_n(re^{it})| \to 0$ uniformly for $t \in [-\pi, \pi]$, thus

$$\lim_{r \to 1} \int_{-\pi}^{\pi} \log|B(re^{i\theta})|d\theta = \lim_{r \to 1} \int_{-\pi}^{\pi} \log \frac{|B(re^{i\theta})|}{|B_n(re^{i\theta})|} d\theta \leqslant 0.$$

On the other hand,

$$\frac{1}{2\pi} \int_{-\pi}^{\pi} \log \frac{|B(re^{i\theta})|}{|B_n(re^{i\theta})|} d\theta \geqslant \log \frac{|B(0)|}{|B_n(0)|} = \sum_{j=n+1}^{\infty} \log|z_j|.$$

From (5.1.5), we know that the series $\sum_{j=1}^{\infty} \log|z_j|$ converges, which implies that $\sum_{j=n+1}^{\infty} \log|z_j| \to 0$ as $n \to \infty$. Therefore,

$$\lim_{r \to 1} \int_{-\pi}^{\pi} \log|B(re^{i\theta})|d\theta \geqslant 0.$$

This completes the proof of Theorem 5.1.5. □

Theorem 5.1.6 ($H^p(U)$ Factorization Theorems). *For any $f \in H^p(U)(0 \leqslant p \leqslant \infty)$, the following decomposition holds:*

$$f(z) = G(z)B(z), \tag{5.1.9}$$

where $B(z)$ is the Blaschke product formed by the zeros of $f(z)$, $G(z) \neq 0$, $G(z) \in H^p(U)$, and $\|G\|_{H^p} = \|f\|_{H^p}$.

Proof. Suppose that $f(z)$ has a zero of multiplicity $k \geqslant 0$ at the origin and $z_1, z_2, \ldots, z_n$ are the other zeros of $f(z)$ in $D(0,1)$ repeated according to the multiplicity. Let

$$B_n(z) = z^k \prod_{j=1}^{n} \frac{|z_j|(z_j - z)}{z_j(1 - z\bar{z}_j)}, \quad B(z) = z^k \prod_{j=1}^{\infty} \frac{|z_j|(z_j - z)}{z_j(1 - z\bar{z}_j)},$$

$G(z) = f(z)/B(z)$. As $r \to 1$, $\log|B_n(re^{it})| \to 0$ uniformly for $t \in [-\pi, \pi]$, thus

$$\lim_{r \to 1} m_p(f, r) = \lim_{r \to 1} m_p\left(\frac{f}{B_n}, r\right) \geqslant m_p\left(\frac{f}{B_n}, r\right).$$

As $n \to \infty$, $\frac{|f(z)|}{|B_n(z)|}$ monotonically increases toward $|G(z)|$. When $p > 0$, by the Lebesgue monotone convergence theorem, we know that

$$\|f\|_{H^p} \geqslant m_p(G, r).$$

This implies that $\|f\|_{H^p} \geqslant \|G\|_{H^p}$. When $p = 0$, since

$$\log^+ |G| \leqslant \log^+ |f| + \log \frac{1}{|B|},$$

we can use Theorem 5.1.5 to show that $\|G\|_{H^0} \leqslant \|f\|_{H^0}$. On the other hand, it is clear that $|f(z)| \leqslant |G(z)|$, hence $\|f\|_{H^p} \leqslant \|G\|_{H^p}$. This proves that $\|f\|_{H^p} = \|G\|_{H^p}$. □

Theorem 5.1.7. *For any $f \in H^p(U)$, where $0 < p < \infty$, the following conclusion holds:*

(a) *For any $\alpha > 0$, for almost every $t \in [-\pi, \pi]$, the non-tangential limit*

$$\lim_{z \in \Gamma_\alpha(e^{it}),\ z \to e^{it}} f(z) = f^*(e^{it})$$

exists and $f^*(e^{it}) \in L^p([-\pi, \pi])$,

(b) $\|f_r - f^*\|_p^p = \int_{-\pi}^{\pi} |f(re^{it}) - f^*(e^{it})|^p dt \to 0 \quad (r \to 1)$, *where* $f_r(e^{it}) = f(re^{it})$,

(c)
$$\|f\|_{H^p} = \lim_{r \to 1} \left(\frac{1}{2\pi} \int_{-\pi}^{\pi} |f(re^{i\theta})|^p d\theta\right)^{\frac{1}{p}} = \left(\frac{1}{2\pi} \int_{-\pi}^{\pi} |f^*(e^{i\theta})|^p d\theta\right)^{\frac{1}{p}}.$$

Proof. Since analytic functions are harmonic functions, when $p > 1$, this can be derived from Theorems 4.5.4 and 4.5.7. Therefore, we only need to consider the case when $0 < p \leqslant 1$. Let $f(z) = G(z)B(z)$, where $B(z)$ is the Blaschke product formed by the zeros of $f(z)$, $G(z) \neq 0$, and $G \in H^p(U)$ with $\|G\|_{H^p} = \|f\|_{H^p}$. Suppose n is a positive integer satisfying $pn > 1$, since $G(z)$ has no zeros in the unit disk U, there exists $h(z) \in H(U)$ such that $h^n(z) = G(z)$, then $h \in H^{np}(U)$ and $\|G\|_{H^p}^p = \|h\|_{H^{np}}^{np}$. Thus, h has non-tangential boundary values $h^*(e^{it})$ almost everywhere, and hence, G has a non-tangential limit $G^*(e^{it}) = (h^*(e^{it}))^n \in L^p([-\pi, \pi])$. Therefore, f has a non-tangential limit $f^*(e^{it}) = B^*(e^{it})G^*(e^{it}) \in L^p([-\pi, \pi])$, which proves the desired conclusion in (a). According to Theorem 4.5.4, we know that as $r \to 1$, $\|h_r - h^*\|_{np} \to 0$, where $h_r(e^{it}) = h(re^{it})$, therefore

$$\|f_r - f^*\|_p^p = \|B_r(G_r - G^*) + (B_r - B^*)G^*\|_p^p$$
$$\leqslant \|B_r(G_r - G^*)\|_p^p + \|(B_r - B^*)G^*\|_p^p.$$

Since $|B_r| \leqslant 1$ and

$$|G_r - G^*| \leqslant |h_r - h^*| \sum_{k=0}^{n-1} |h_r|^{n-1-k} |h^*|^k$$
$$\leqslant n|h_r - h^*|(|h_r| + |h^*|)^{n-1},$$

by using the Hölder's inequality $(n^{-1} + (\frac{n}{n-1})^{-1} = 1)$, we have

$$\|G_r - G^*\|_p^p \leqslant n^p \|h_r - h^*\|_{np}^p \| |h_r| + |h^*| \|_{pn}^{p(n-1)}$$
$$\leqslant 2^{p(n-1)} n^p \|h_r - h^*\|_{np}^p \|h^*\|_{pn}^{p(n-1)}.$$

Since $\|h^*\|_{pn}^{pn} = \|G^*\|_p^p$, we have

$$\|B_r(G_r - G^*)\|_p^p \leqslant \|G_r - G^*\|_p^p \leqslant 2^{p(n-1)} n^p \|h_r - h^*\|_{np}^p$$
$$\|G^*\|_p^{p(1-n^{-1})} \to 0 \quad (r \to 1).$$

Since $G^*(e^{it}) \in L^p([-\pi, \pi])$ and $|B_r| \leqslant 1$, by the Lebesgue-dominated convergence theorem, we have

$$\|(B_r - B^*)G^*\|_p^p \to 0 \quad (r \to 1).$$

Therefore,

$$\|f_r - f^*\|_p^p \to 0 \quad (r \to 1).$$

This proves the conclusion required in (b). Since $m_p(f, r)$ is monotonically increasing, we know that the first equality in (c) holds. Using the result from (b), we can infer the conclusion in (c). □

Theorem 5.1.8. *If $f \in H^p(U)$ and $f^*(e^{it}) \in L^q([-\pi, \pi])$, $0 < p < q$, then $f \in H^q(U)$.*

Proof. Since an analytic function is harmonic, when $p > 1$, this can be deduced from Theorems 4.5.4 and 4.5.7 that $f(z)$ is the Poisson integral of $f^*(e^{it}) \in L^q([-\pi, \pi])$, hence $f \in H^q(U)$. Therefore, we only need to consider the case when $0 < p \leqslant 1$. Let $f(z) = G(z)B(z)$, where $B(z)$ is a Blaschke product formed by the zeros of $f(z)$, $G(z) \neq 0$, $G \in H^p$, and $\|G\|_{H^p} = \|f\|_{H^p}$. Let n be a positive integer such that $pn > 1$, since $G(z)$ has no zeros in the unit disk U, there exists $h(z) \in H(U)$ such that $h^n(z) = G(z)$. Consequently, $h \in H^{np}$ and $\|G\|_{H^p}^p = \|h\|_{H^{np}}^{np}$. Thus, h has almost everywhere non-tangential boundary values $h^*(e^{it})$, implying that G has non-tangential limits $G^*(e^{it}) = (h^*(e^{it}))^n \in L^q([-\pi, \pi])$ almost everywhere. This means that f has non-tangential limits $f^*(e^{it}) = B^*(e^{it})G^*(e^{it}) \in L^q([-\pi, \pi])$ almost everywhere. Hence, the non-tangential limit of h, $h^*(e^{it}) \in L^{nq}([-\pi, \pi])$. The conclusion we want to prove is derived from Theorems 4.5.4 and 4.5.7, which imply that $h(z)$ is a Poisson integral of $h^*(e^{it}) \in L^{qn}([-\pi, \pi])$ therefore $h \in H^{qn}(U)$. Consequently, it can be seen that $G = h^n \in H^q(U)$, and thus, $f \in H^q(U)$. This leads to the desired conclusion. □

Theorem 5.1.9. *If $f \in H^1(U)$, then $f(z)$ is the Poisson integral and Cauchy integral of its non-tangential limit $f^*(e^{it})$ given by*

$$f(z) = P(f^*)(re^{i\theta}) = \frac{1}{2\pi} \int_{-\pi}^{\pi} P(e^{it}, re^{i\theta}) f^*(e^{it}) dt \quad (z = re^{i\theta})$$

(5.1.10)

and

$$f(z) = \frac{1}{2\pi i} \int_{|\zeta|=1} \frac{f^*(\zeta)}{\zeta - z} d\zeta, \qquad (5.1.11)$$

where

$$P(\zeta, z) = \mathrm{Re}\left(\frac{\zeta + z}{\zeta - z}\right) = \frac{1 - |z|^2}{|\zeta - z|^2} = \frac{1 - r^2}{1 + r^2 - 2r\cos(\theta - t)}$$

$$(\zeta = e^{it}, \ z = re^{i\theta})$$

is the Poisson kernel on the unit disk U.

Proof. Let $0 < s < 1$, then

$$f(sz) = \frac{1}{2\pi} \int_{-\pi}^{\pi} P(e^{it}, re^{i\theta}) f(se^{it}) dt \quad (z = re^{i\theta}) \qquad (5.1.12)$$

and

$$f(sz) = \frac{1}{2\pi i} \int_{|\zeta|=1} \frac{f(s\zeta)}{\zeta - z} d\zeta. \qquad (5.1.13)$$

Since $\|f_s - f^*\|_1 \to 0$ as $s \to 1$, taking $s \to 1$ in (5.1.12) and (5.1.13), we obtain (5.1.10) and (5.1.11). □

Theorem 5.1.10. *If $f \in H^0(U)$, $f(z) \not\equiv 0$, $f^*(e^{it})$ is its non-tangential limit, then for almost all $t \in [-\pi, \pi]$, $f^*(e^{it}) \neq 0$.*

Proof. From (5.1.4) and Fatou's lemma, we have

$$\int_{-\pi}^{\pi} |\log|f^*(e^{it})|| dt < \infty,$$

thus for almost all $t \in [-\pi, \pi]$ we have $f^*(e^{it}) \neq 0$. □

Theorem 5.1.11. *Suppose $f_j \in H^0(U)$, $f_j(z) \not\equiv 0$ $(j = 1, 2)$, $f_1^*(e^{it})$ and $f_2^*(e^{it})$ are their respective non-tangential limits. If there exists a positive measurable set $E \subset [-\pi, \pi]$ such that $f_1^*(e^{it}) = f_2^*(e^{it})$ for $t \in E$, then $f_1 \equiv f_2$.*

Proof. It suffices to apply Theorem 5.1.10 to $f_1 - f_2$. □

H^p Spaces

Theorem 5.1.12 (F. Riesz–M. Riesz theorem). *If μ is a finite Borel measure on $[-\pi, \pi]$, and*

$$\int_{[-\pi,\pi]} e^{ikt} d\mu(t) = 0, \quad k = 1, 2, \ldots,$$

then μ is absolutely continuous with respect to the Lebesgue measure on $[-\pi, \pi]$, i.e., there exists $f \in L^1([-\pi, \pi])$ such that $d\mu(t) = f(t)dt$.

Proof. Let $F(z) = P(\mu)(z)$ be the Poisson integral of μ, defined as

$$F(z) = P(\mu)(re^{i\theta}) = \frac{1}{2\pi} \int_{-\pi}^{\pi} P(e^{it}, re^{i\theta}) d\mu(t), \quad (5.1.14)$$

then

$$F(re^{i\theta}) = \frac{1}{2\pi} \int_{-\pi}^{\pi} \sum_{k=-\infty}^{\infty} r^{|k|} e^{ik(\theta-t)} d\mu(t) = \sum_{k=0}^{\infty} a_k r^k e^{ik\theta},$$

where

$$a_k = \frac{1}{2\pi} \int_{[-\pi,\pi]} e^{-ikt} d\mu(t)$$

are the non-negative integer Fourier coefficients of μ, satisfying

$$|a_k| \leq \frac{1}{2\pi} \int_{[-\pi,\pi]} d|\mu|(t), \quad k = 0, 1, 2, \ldots.$$

Therefore, $F \in H(U)$ and

$$\frac{1}{2\pi} \int_{[-\pi,\pi]} |F(re^{it})| dt \leq \frac{1}{2\pi} \int_{[-\pi,\pi]} d|\mu|(t).$$

Thus, $F \in H^1(U)$. Let $F^*(e^{it})$ be the non-tangential limit of F, and define $f(t) = F^*(e^{it})$, then $f \in L^1([-\pi, \pi])$ and $F(z) = P(\mu)(z) = P(f)(z)$, therefore, by Theorem 4.5.6(b), we have $d\mu(t) = f(t)dt$, implying that μ is absolutely continuous with respect to the Lebesgue measure on $[-\pi, \pi]$. □

Theorem 5.1.13. *If $0 < p < \infty$, the space $H^p(U)$ is separable and complete.*

Proof. Let $\{f_n\}$ be a Cauchy sequence in $H^p(U)$, i.e., $\|f_n - f_m\|_{H^p} \to 0$ as $n, m \to \infty$. If $|z| \leqslant r < R < 1$, since $|f_n(z) - f_m(z)|^p$ is subharmonic and continuous in the unit disk U, we have

$$|f_n(z) - f_m(z)|^p \leqslant \frac{1}{2\pi} \int_{-\pi}^{\pi} \operatorname{Re}\left(\frac{Re^{it} + z}{Re^{it} - z}\right) |f_n(Re^{it}) - f_m(Re^{it})|^p dt, \tag{5.1.15}$$

hence

$$|f_n(z) - f_m(z)|^p \leqslant \frac{1}{2\pi} \int_{-\pi}^{\pi} \frac{R+r}{R-r} |f_n(Re^{it}) - f_m(Re^{it})|^p dt, \tag{5.1.16}$$

therefore

$$|f_n(z) - f_m(z)|^p \leqslant \frac{R+r}{R-r} \|f_n - f_m\|_{H^p}^p \to 0 \quad (n, m \to \infty).$$

Consequently, $\{f_n\}$ converges uniformly on compact subsets of U to $f \in H(U)$. For any $\varepsilon > 0$, there exists an m such that for any $n > m$, we have $\|f_n - f_m\|_{H^p}^p < \varepsilon$. Consequently, by Fatou's lemma, we have

$$\frac{1}{2\pi} \int_{-\pi}^{\pi} |f(Re^{it}) - f_m(Re^{it})|^p dt \leqslant \liminf_{n \to \infty} \frac{1}{2\pi}$$

$$\int_{-\pi}^{\pi} |f_n(Re^{it}) - f_m(Re^{it})|^p dt \leqslant \varepsilon,$$

which implies $\|f_n - f\|_{H^p} \to 0$ as $n \to \infty$. This completes the proof of completeness.

To prove the separability of the space $H^p(U)$, it suffices to show that polynomials with rational coefficients are dense in $H^p(U)$. Indeed, for any $f \in H^p(U)$, let $f(z) = \sum_{k=0}^{\infty} a_k z^k$, we define $f_r(z) = f(rz)$, then we have

$$\lim_{r \to 1} \int_{-\pi}^{\pi} |f(re^{it}) - f^*(e^{it})|^p dt = 0,$$

which implies $\|f_r - f\|_{H^p} \to 0$ as $r \to 1$. For any $\varepsilon > 0$, there exists $r_0 \in (0, 1)$ such that $\|f_{r_0} - f\|_{H^p} < \frac{\varepsilon}{4}$. Then there exists N such that

$$\left|\sum_{k=0}^{N} a_k r_0^k z^k - f_{r_0}(z)\right| \leqslant \frac{\varepsilon}{4}, \quad |z| \leqslant 1.$$

Furthermore, we can find a polynomial $p(z) = \sum_{k=0}^{N} c_k z^k$ with rational coefficients such that

$$\left| \sum_{k=0}^{N} a_k r_0^k z^k - \sum_{k=0}^{N} c_k z^k \right| \leq \frac{\varepsilon}{4}, \quad |z| \leq 1.$$

Therefore, we have $\|f - P\|_{H^p} < \varepsilon$. □

Theorem 5.1.14. *For any $f \in H^0(U)$ with $f \not\equiv 0$, we can write $f(z)$ as the quotient of two functions in $H^\infty(U)$, $b_1(z)$ and $b_2(z)$, $f(z) = b_1(z)/b_2(z)$,*

$$b_1(z) = cB(z) \exp\left\{ -\frac{1}{2\pi} \int_{-\pi}^{\pi} \frac{e^{it}+z}{e^{it}-z} d\mu^-(t) \right\}, \quad (5.1.17)$$

$$b_2(z) = \exp\left\{ -\frac{1}{2\pi} \int_{-\pi}^{\pi} \frac{e^{it}+z}{e^{it}-z} d\mu^+(t) \right\}, \quad (5.1.18)$$

where c is a constant with absolute value 1, $B(z)$ is the Blaschke product of zeros of $f(z)$, μ^+ and μ^- are two positive measures on $[-\pi, \pi]$, the non-tangential limit $f^*(e^{i\theta})$ of f exists for almost every θ and $\log|f^*(e^{i\theta})| \in L^1([-\pi, \pi])$. Furthermore, f has the following factorization:

$$f(z) = cB(z)G(z)S(z), \quad (5.1.19)$$

where

$$G(z) = \exp\left\{ \frac{1}{2\pi} \int_{-\pi}^{\pi} \frac{e^{it}+z}{e^{it}-z} \log|f^*(e^{it})| dt \right\}, \quad (5.1.20)$$

$$S(z) = \exp\left\{ \frac{1}{2\pi} \int_{-\pi}^{\pi} \frac{e^{it}+z}{e^{it}-z} d\mu_s(t) \right\}, \quad (5.1.21)$$

and μ_s is the unique singular sign measure determined by f.

Proof. Due to the harmonicity of $\log|f/B|$ in the unit disk U and $f/B \in H^0(U)$, we have

$$\sup\left\{ \frac{1}{2\pi} \int_{-\pi}^{\pi} |\log|f(re^{i\theta})/B(re^{i\theta})|| d\theta : 0 < r < 1 \right\} < \infty.$$

Thus, there exists a finite real signed measure μ such that

$$\log\left|\frac{f(z)}{B(z)}\right| = \mathrm{Re}\left(\frac{1}{2\pi}\int_{-\pi}^{\pi}\frac{e^{it}+z}{e^{it}-z}d\mu(t)\right).$$

Consequently, there exists a constant c, $|c|=1$ such that

$$f(z) = cB(z)\exp\left\{\frac{1}{2\pi}\int_{-\pi}^{\pi}\frac{e^{it}+z}{e^{it}-z}d\mu(t)\right\}.$$

Since the real signed measure μ can be decomposed into the difference of two non-negative measures μ^+ and μ^-, $\mu = \mu^+ - \mu^-$, then $f(z)$ can be written as the quotient of two functions $b_1(z)$ and $b_2(z)$ in $H^\infty(U)$, $f(z) = b_1(z)/b_2(z)$, where $b_1(z)$ and $b_2(z)$ are defined by (5.1.17) and (5.1.18). By the Lebesgue decomposition theorem, $d\mu^+(t) = \phi^+(t)dt + d\mu_s^+(t)$, $d\mu^-(t) = \phi^-(t)dt + d\mu_s^-(t)$, where $\phi^+, \phi^- \in L^1([-\pi,\pi])$, μ_s^+ and μ_s^- are singular positive measures. Therefore, $|b_1(z)| \leq 1, 0 < |b_2(z)| \leq 1$, and the non-tangential limits $b_1^*(e^{i\theta})$ and $b_2^*(e^{i\theta})$ exist for almost all θ. By Fatou's theorem, $\log|b_1^*(e^{i\theta})| = -\phi^-(\theta)$ and $\log|b_2^*(e^{i\theta})| = -\phi^+(\theta)$. Hence, the non-tangential limit $f^*(e^{i\theta})$ exists for almost all θ, and $\log|f^*(e^{i\theta})| = \phi^+(\theta) - \phi^-(\theta) \in L^1([-\pi,\pi])$. Therefore, there exists a constant c, $|c|=1$ such that

$$f(z) = cB(z)\exp\left\{\frac{1}{2\pi}\int_{-\pi}^{\pi}\frac{e^{it}+z}{e^{it}-z}\log|f^*(e^{it})|dt\right\}$$

$$\exp\left\{\frac{1}{2\pi}\int_{-\pi}^{\pi}\frac{e^{it}+z}{e^{it}-z}d\mu_s(t)\right\},$$

where $d\mu_s(t) = d\mu_s^+(t) - d\mu_s^-(t)$ is the singular signed measure uniquely determined by μ. This proves Theorem 5.1.14. $\square$

Definition 5.1.15. Let $M \in H^\infty(U)$, and let $M^*(e^{i\theta})$ denote the non-tangential limit of M, if $|M^*(e^{i\theta})| = 1$ holds for almost all $\theta \in [-\pi,\pi]$, then M is called an inner function. If $\varphi(t) > 0$ holds for almost all $t \in [-\pi,\pi]$ and $\log\varphi(t) \in L^1([-\pi,\pi])$, and let c be constant with $|c|=1$, then the function $Q(z)$ in the unit disk U defined by

$$Q(z) = c\exp\left\{\frac{1}{2\pi}\int_{-\pi}^{\pi}\frac{e^{it}+z}{e^{it}-z}\log\varphi(t)dt\right\} \qquad (5.1.22)$$

is called an outer function.

Theorem 5.1.16. *Let c be a constant with $|c| = 1$, $B(z)$ be a Blaschke product in U, and μ_s be a singular positive measure. Then, the function*

$$M(z) = cB(z) \exp\left\{-\frac{1}{2\pi} \int_{-\pi}^{\pi} \frac{e^{it} + z}{e^{it} - z} d\mu_s(t)\right\} \quad (5.1.23)$$

is an inner function. Conversely, every inner function has the form given in (5.1.23) and $\|M\|_{H^\infty} = 1$.

Proof. If (5.1.23) holds, then $\log|g|$ is the Poisson integral of $-\mu_s$, where $g = M/B$. Thus, $\log|g| \leq 0$, implying $g \in H^\infty(U)$. Since μ_s is a singular positive measure, we have $\mu'_s(t) = 0$ almost everywhere, and consequently, the non-tangential limit of $\log|g|$ is almost everywhere zero. Furthermore, since $|B^*(e^{i\theta})| = 1$ holds for almost all $\theta \in [-\pi, \pi]$, it follows that M is an inner function.

Conversely, let $A = \log\|M\|_{H^\infty}$, since M is an inner function, we have $A \geq 0$. Let $B(z)$ be the Blaschke product formed by the zeros $\{z_n\}$ of M, and define $g = M/B$. According to Theorem 5.1.6, $\log\|g\|_{H^\infty} = A < \infty$. Hence, $u(z) = A - \log|g(z)|$ is a non-negative harmonic function in the unit disk U. Let $\rho \in (0, 1)$, and consider $u_\rho(z) = u(\rho z)$, then $u_\rho(z) = P(u_\rho)(z)$ for $z = re^{i\theta} \in U$, in other words, we have

$$u(\rho r e^{i\theta}) = \frac{1}{2\pi} \int_{-\pi}^{\pi} \frac{1 - r^2}{1 - 2r\cos(\theta - t) + r^2} u_\rho(e^{it}) dt.$$

Let $\rho \to 1$, since the non-tangential limit $g^*(e^{i\theta})$ of g exists and satisfies $|g^*(e^{i\theta})| = 1$ for almost every $\theta \in [-\pi, \pi]$, by the Fatou lemma, we have

$$A - \log|g(re^{i\theta})| \geq \frac{1}{2\pi} \int_{-\pi}^{\pi} \frac{1 - r^2}{1 - 2r\cos(\theta - t) + r^2} \liminf_{\rho \to 1} u_\rho(e^{it}) dt = A,$$

which implies $|g| \leq 1$. Therefore, we have $A = 0$. Hence, there exists a positive singular measure μ such that $\log|g|$ is the Poisson integral of $-\mu$. Since $|M^*(e^{i\theta})| = |B^*(e^{i\theta})| = 1$ that holds for almost every $\theta \in [-\pi, \pi]$, we have $\mu'(t) = 0$ that holds for almost every $\theta \in [-\pi, \pi]$, which means that μ is a singular positive measure. Since $\log|g|$ is the real part of

$$h(z) = -\frac{1}{2\pi} \int_{-\pi}^{\pi} \frac{e^{it} + z}{e^{it} - z} d\mu(t),$$

there exists a constant c with $|c| = 1$ such that $g = ce^h$. Thus, M has the form given in (5.1.23). □

Theorem 5.1.17. *If $\varphi(t) > 0$ holds for almost all $t \in [-\pi, \pi]$ and $\log \varphi(t) \in L^1([-\pi, \pi])$, c with $|c| = 1$ is a constant, then*

$$Q(z) = c \exp\left\{ \frac{1}{2\pi} \int_{-\pi}^{\pi} \frac{e^{it} + z}{e^{it} - z} \log \varphi(t) dt \right\}$$

is analytic function in the unit disk U and satisfies the following:

(a) $\log |Q|$ *is the Poisson integral of* $\log \varphi(t)$: $\log |Q(z)| = P(\log \varphi)(z)$,
(b) $\lim_{r \to 1} |Q(re^{i\theta})| = \varphi(\theta)$ *holds for almost all $\theta \in [-\pi, \pi]$,*
(c) $Q \in H^p(U)$ *if and only if $\varphi \in L^p([-\pi, \pi])$. If this condition is satisfied, then*

$$2\pi \|Q\|_{H^p}^p = \int_{-\pi}^{\pi} |\varphi(t)|^p dt = \|\varphi\|_p^p.$$

Proof. By Theorem 4.5.4, (a) and (b) are obvious. If $Q \in H^p(U)$, then by Fatou's lemma, we have $\|Q^*\|_p^p \leqslant 2\pi \|Q\|_{H^p}^p$, which implies $2\pi \|Q\|_{H^p}^p \geqslant \|\varphi\|_p^p$. Conversely, if $\varphi \in L^p([-\pi, \pi])$, then

$$|Q(re^{i\theta})|^p = \exp\left\{ \frac{1}{2\pi} \int_{-\pi}^{\pi} P(e^{i(t-\theta)}, r) \log \varphi^p(t) dt \right\}$$

$$\leqslant \frac{1}{2\pi} \int_{-\pi}^{\pi} P(e^{i(t-\theta)}, r) \varphi^p(t) dt.$$

Therefore, when $0 < p < \infty$, $2\pi \|Q\|_{H^p}^p \leqslant \|\varphi\|_p^p$. When $p = \infty$, $\|Q\|_{H^p} \leqslant \|\varphi\|_p$ is obviously true. □

Theorem 5.1.18. *If $0 < p \leqslant \infty$, $f \in H^p(U)$, $f \not\equiv 0$, then $\log |f^*(e^{it})| \in L^1([-\pi, \pi])$, the function $Q_f(z)$ defined by*

$$Q_f(z) = \exp\left\{ \frac{1}{2\pi} \int_{-\pi}^{\pi} \frac{e^{it} + z}{e^{it} - z} \log |f^*(e^{it})| dt \right\} \in H^p(U) \quad (5.1.24)$$

H^p Spaces 263

is an outer function and there exists an inner function M_f such that

$$f = M_f Q_f. \tag{5.1.25}$$

Furthermore, we have

$$\log|f(0)| \leqslant \frac{1}{2\pi} \int_{-\pi}^{\pi} \log|f^*(e^{it})|dt. \tag{5.1.26}$$

The equality in (5.1.26) holds if and only if M_f is a constant.

Proof. First, assume $p = 1$, $f \in H^1(U)$, and $f(z)$ has no zeros in U. Since $H^1(U) \subset H^0(U)$, we have $\log|f^*(e^{it})| \in L^1([-\pi, \pi])$. By Theorem 5.1.17, it follows that $Q_f \in H^1(U)$. Let $0 < R < 1$, and define $f_R(z) = f(Rz)$. Then, for $z \in U$, we have

$$\log|f_R(z)| = P(\log^+|f_R|)(z) - P(\log^-|f_R|)(z). \tag{5.1.27}$$

Since for any real numbers u and v, $|\log^+ u - \log^+ v| \leqslant |u - v|$ and $\|f_R - f^*\|_1 \to 0$ as $R \to 1$, it follows that $P(\log^+|f_R|)(z) \to P(\log^+|f^*|)(z)$ as $R \to 1$. Therefore, by Fatou's lemma, we have

$$P(\log^-|f^*|)(z) \leqslant \liminf_{R \to 1} P(\log^-|f_R|)(z)$$
$$= P(\log^+|f_R|)(z) - \log|f(z)|. \tag{5.1.28}$$

This leads to

$$\log|f(z)| \leqslant P(\log|f^*|)(z), \tag{5.1.29}$$

which implies that $|f(z)| \leqslant |Q_f(z)|$. Since $Q_f \in H^1(U)$ and $|Q_f^*| = |f^*| \neq 0$ almost everywhere, we know that f/Q_f is an inner function, the decomposition in (5.1.26) is established. Setting $z = 0$ in (5.1.29), we know that (5.1.26) holds. The equality in (5.1.26) holds if and only if $|f(0)| = |Q_f(0)|$, which is equivalent to $|M_f(0)| = 1$. Since $\|M_f\|_\infty = 1$, the equality in (5.1.26) holds if and only if M_f is constant. If $1 \leqslant p \leqslant \infty$, then $H^p(U) \subset H^1(U)$. If $f \in H^p(U)$, and $B(z)$ is the Blaschke product of zeros of f, let $g = f/B$. By Theorem 5.1.10, we have $g \in H^p(U)$ and $|g^*(e^{i\theta})| = |f^*(e^{i\theta})|$ for almost all $\theta \in [-\pi, \pi]$. By Theorem 5.1.12, we know that $|f^*(e^{i\theta})| \in L^p([-\pi, \pi])$. Therefore, by Fatou's lemma, it follows that $Q_f \in H^p(U)$. This shows that the theorem holds for $1 \leqslant p \leqslant \infty$.

If $0 < p < 1$, let $f(z) = G(z)B(z) \in H^p(U)$, where $B(z)$ is the Blaschke product of zeros of $f(z)$, $G(z) \neq 0$, and $G \in H^p(U)$

with $\|G\|_{H^p} = \|f\|_{H^p}$. Let n be a positive integer such that $pn > 1$, since $G(z)$ has no zeros in the unit disk U, there exists $h(z) \in H(U)$ such that $h^n(z) = G(z)$, therefore $h \in H^{np}(U)$ and $\|G\|_{H^p}^p = \|h\|_{H^{np}}^{np}$. Hence, h has a non-tangential limit $h^*(e^{it})$ almost everywhere, which implies that the non-tangential limit of G is $G^*(e^{it}) = (h^*(e^{it}))^n \in L^p([-\pi, \pi])$. Thus, the non-tangential limit of f is $f^*(e^{it}) = B^*(e^{it})G^*(e^{it}) \in L^p([-\pi, \pi])$. Consequently, the non-tangential limit of h is $h^*(e^{it}) \in L^{np}([-\pi, \pi])$. By Theorems 4.5.4 and 4.5.7, it follows that $h(z)$ is the Poisson integral of $h^*(e^{it}) \in L^{pn}([-\pi, \pi])$, which implies $h \in H^{pn}(U)$. Therefore, by Fatou's lemma, $Q_f \in H^p(U)$. This shows that the theorem holds for $0 < p < 1$. $\square$

Theorem 5.1.19. *If $1 < p < \infty$, let u be a real harmonic function in the unit disk U satisfying (4.5.19), and let u^* be the non-tangential limit of u, then define $G(u)$ as follows:*

$$G(u)(z) = \frac{1}{2\pi} \int_{-\pi}^{\pi} \frac{e^{it} + z}{e^{it} - z} u^*(e^{it}) dt. \tag{5.1.30}$$

Let $v(z) = \operatorname{Im} G(u)(z)$ be the harmonic conjugate of u in U, where $v(0) = 0$. Then we have the inequality

$$\int_{-\pi}^{\pi} |v(re^{it})|^p dt \leq A_p \int_{-\pi}^{\pi} |u^*(e^{it})|^p dt, \tag{5.1.31}$$

where $A_p = 2^p \max\{\frac{p}{p-1}, p^{p-1}\}$.

Proof. First, suppose $1 < p \leq 2$, $u \geq 0, u \not\equiv 0$ and $f = u + iv \in H(U)$, where $v(0) = 0$. Then, by the minimum modulus principle, we have $u > 0$, and hence, $u^p, |f|^p \in C^2(U)$. Since $2\frac{\partial u}{\partial u} = f'$, we can calculate

$$\Delta u^p = p(p-1)|f'|^2 u^{p-2} \quad \text{and} \quad \Delta |f|^p = p^2 |f'|^2 |f|^{p-2}.$$

Let $h(z) = \frac{p}{p-1}(u(z))^p - |f(z)|^p$, then $h \in C^2(\mathbb{C}_+)$, and we have

$$\Delta h(z) = p^2 |f'|^2 (u^{p-2} - |f|^{p-2}) \geq 0. \tag{5.1.32}$$

Thus, $h(z)$ is subharmonic in U. By Theorem 4.3.8, the function

$$\int_{-\pi}^{\pi} h(re^{i\theta}) d\theta$$

is monotonically increasing in $(0,1)$, therefore

$$\frac{1}{2\pi}\int_{-\pi}^{\pi} h(re^{i\theta})d\theta \geqslant \lim_{r\to 0}\frac{1}{2\pi}\int_{-\pi}^{\pi} h(re^{i\theta})d\theta = \frac{(u(0))^p}{p-1} > 0.$$

As a result,

$$\|f_r\|_p^p = \int_{-\pi}^{\pi}|f(re^{it})|^p dt \leqslant \frac{p}{p-1}\int_{-\pi}^{\pi}|u(re^{it})|^p dt \leqslant \frac{p}{p-1}\|u^*\|_p^p.$$

If u is a real harmonic function in the unit disk U and satisfies (4.5.19), and u^* is the non-tangential limit of u, then $u^* \in L^p([-\pi,\pi])$. Therefore, both the positive part $u^{*+} \in L^p([-\pi,\pi])$ of u^* and the negative part u^{*-} of u^* are in $L^p([-\pi,\pi])$. We have $u = P(u^{*+}) - P(u^{*-})$, where $P(u^{*+})$ and $P(u^{*-})$ are non-negative harmonic functions in U that satisfy (4.5.19), and $G(u) = G(P(u^{*+})) - G(P(u^{*-}))$. Thus,

$$\|(G(u))_r\|_p \leqslant \left(\frac{p}{p-1}\right)^{\frac{1}{p}} \left(\|(P(u^{*+}))_r\|_p + \|(P(u^{*-}))_r\|_p\right)$$

$$\leqslant 2\left(\frac{p}{p-1}\right)^{\frac{1}{p}}\|u^*\|_p.$$

Therefore, Theorem 5.1.19 holds for $1 < p \leqslant 2$.

When $2 \leqslant p < \infty$, let $\frac{1}{q}+\frac{1}{p}=1$, suppose $g(z)=\sum_{n=0}^{m}a_n z^n$ is a polynomial, where a_0 is a real number, and $a_n = \alpha_n + \beta_n$, Define $\alpha(z) = \operatorname{Re} g(z)$ and $\beta(z) = \operatorname{Im} g(z)$, then $\beta(0)=0$. $f=u+iv \in H(U)$ with $v(0)=0$, then $\beta u_r+\alpha v_r = \operatorname{Im} gf_r$ is harmonic in U and vanishes at 0. Therefore,

$$\int_{-\pi}^{\pi} u(re^{it})\beta(e^{it})dt = -\int_{-\pi}^{\pi} v(re^{it})\alpha(e^{it})dt.$$

Using the Hölder inequality, we have

$$\left|\int_{-\pi}^{\pi} v_r(e^{it})\alpha(e^{it})dt\right| = \left|\int_{-\pi}^{\pi} u_r(e^{it})\beta(e^{it})dt\right|$$

$$\leqslant \|u_r\|_p\|\beta\|_q \leqslant 2\left(\frac{q}{q-1}\right)^{\frac{1}{q}}\|u_r\|_p\|\alpha\|_q.$$

Since real trigonometric polynomials $\alpha(e^{i\theta}) = a_0 + \sum_{n=1}^{m}(\alpha_n \cos n\theta - \beta_n \sin n\theta)$ is dense in real $L^q([-\pi, \pi])$, we have

$$\|v_r\|_p^p = \int_{-\pi}^{\pi} |v_r(e^{it})|^p dt \leqslant 2^p \left(\frac{q}{q-1}\right)^{\frac{p}{q}} \|u_r\|_p^p \leqslant 2^p p^{p-1} \|u^*\|_p^p.$$

This proves the theorem. □

Suppose $R > 0$, $z_0 \in D_+(0, R) = \{z : |z| < R, \operatorname{Re} z > 0\}$. The function $s = \phi_1(z) = -\left(\frac{z+iR}{z-iR}\right)^2$ is a conformal mapping from $D_+(0, R)$ to the upper half-plane $\mathbb{C}_+$, the function $w = \phi_2(s) = \frac{s - \phi_1(z_0)}{s - \overline{\phi_1(z_0)}}$ is a conformal mapping from $\mathbb{C}_+$ to U. Therefore, the function

$$w = \frac{\phi_1(z) - \phi_1(z_0)}{\phi_1(z) - \overline{\phi_1(z_0)}} = \frac{(z - z_0)(R^2 + z z_0)(\overline{z}_0 + iR)^2}{(z + \overline{z}_0)(R^2 - \overline{z}_0 z)(z_0 - iR)^2} \quad (5.1.33)$$

maps the upper half-disk $D_+(0, R)$ conformally onto the unit disk U. Let $w = \phi_{z_0}(z) = \phi_2(\phi_1(z))$, then $\phi_{z_0} \in H(\overline{D_+(0, R)})$, $\phi_{z_0}(z_0) = 0$ and is a homeomorphism from $\overline{D_+(0, R)}$ onto $\overline{U}$. Let $e^{i\varphi_R'} = \phi_{z_0}(-iR)$, $e^{i\varphi_R''} = \phi_{z_0}(iR)$, $\varphi_R' < \varphi_R'' < \varphi_R' + 2\pi$, and let $\varphi_1(\theta)$ be a strictly increasing continuous function from $[-\frac{\pi}{2}, \frac{\pi}{2}]$ onto $[\varphi_R', \varphi_R'']$ such that $e^{i\varphi_1(\theta)} = \phi_{z_0}(Re^{i\theta})$. Similarly, there exists a strictly decreasing continuous function $\varphi_2(t)$ from $[-R, R]$ onto $[\varphi_R'', \varphi_R' + 2\pi]$ such that $e^{i\varphi_2(t)} = \phi_{z_0}(it)$. Suppose $w = \phi_{z_0}(z)$ is the inverse function of $z = \phi_{z_0}^{-1}(w)$. If $f \in H(D_+(0, R))$ and $\log^+ |f|$ has a harmonic majorant function in $D_+(0, R)$, then the function $F(w) = f(\phi_{z_0}^{-1}(w)) \in H(U)$, and $\log^+ |F|$ has a harmonic majorant function in U. Hence, $F \in H^0(U)$, and thus, for almost every $\varphi \in (-\pi, \pi)$, the non-tangential limit $F^*(e^{i\varphi})$ exists. Therefore, for almost every $t \in (-R, R)$, the non-tangential limit $f^*(it)$ exists, and for almost every $\theta \in (-\frac{\pi}{2}, \frac{\pi}{2})$, the non-tangential limit $f^*(Re^{i\theta})$ exists.

Theorem 5.1.20. *If $R > 0$, $f \in H(D_+(0, R))$, and Λ_R is the zero set of f in $D_+(0, R)$ (counting multiplicities). If $\log^+ |f|$ has a harmonic majorant function in $D_+(0, R)$, then for almost all $t \in (-R, R)$, the non-tangential limit $f^*(it)$ exists, and for almost all $\theta \in (-\frac{\pi}{2}, \frac{\pi}{2})$, the non-tangential limit $f^*(Re^{i\theta})$ exists, there exist* singular sign measures $\mu_{1,s}$ on $[-R, R]$ and $\mu_{2,s}$ on $[-\frac{\pi}{2}, \frac{\pi}{2}]$ such that

$$\int_{-\frac{\pi}{2}}^{\frac{\pi}{2}} \cos\theta |\log|f^*(Re^{i\theta})||d\theta + \int_{-R}^{R} (R^2 - t^2)|\log|f^*(it)||dt$$

$$+ \int_{-\frac{\pi}{2}}^{\frac{\pi}{2}} \cos\theta d|\mu_{2,s}|(\theta) + \int_{-R}^{R} (R^2 - t^2)d|\mu_{1,s}|(t) < \infty \quad (5.1.34)$$

and

$$\sum_{\lambda \in \Lambda_R} \operatorname{Re}\lambda(R^2 - |\lambda|^2) < \infty \tag{5.1.35}$$

hold and there exists a decomposition $f(z) = cG(z)B(z)S(z)$ for all $z \in D_+(0, R)$, where c is a constant with modulus 1:

$$G(z) = \exp\left\{\frac{1}{2\pi}\int_{-\frac{\pi}{2}}^{\frac{\pi}{2}}\left(\frac{Re^{i\theta}+z}{Re^{i\theta}-z} - \frac{Re^{-i\theta}-z}{Re^{-i\theta}+z}\right)\log|f^*(Re^{i\theta})|d\theta\right.$$

$$\left. + \frac{1}{\pi}\int_{-R}^{R}\left(\frac{1}{z-it} - \frac{z}{R^2+itz}\right)\log|f^*(it)|dt\right\}, \quad (5.1.36)$$

$$B(z) = \prod_{\lambda \in \Lambda_R} \left(\frac{(z-\lambda)(R^2+\bar\lambda z)}{(z+\bar\lambda)(R^2-\bar\lambda z)}\right) e^{i\theta(\lambda, z_0)}, \tag{5.1.37}$$

$$e^{i\theta(\lambda, z_0)} = -\frac{(\bar\lambda - \bar z_0)(\bar\lambda + z_0)(R^2 - \bar\lambda z_0)(R^2 + \bar\lambda \bar z_0)}{|\bar\lambda - \bar z_0||\bar\lambda + z_0||R^2 - \bar\lambda z_0||R^2 + \bar\lambda \bar z_0|},$$

$$S(z) = \exp\left\{-\frac{1}{2\pi}\int_{-\frac{\pi}{2}}^{\frac{\pi}{2}}\left(\frac{Re^{i\theta}+z}{Re^{i\theta}-z} - \frac{Re^{-i\theta}-z}{Re^{-i\theta}+z}\right)d\mu_{2,s}(\theta)\right.$$

$$\left. - \frac{1}{\pi}\int_{-R}^{R}\left(\frac{1}{z-it} - \frac{z}{R^2+itz}\right)d\mu_{1,s}(t)\right\}. \quad (5.1.38)$$

Moreover, if there exists a positive number p such that $|f(z)|^p$ has a harmonic majorant function in $D_+(0, R)$, then the singular sign measures $\mu_{1,s}$ on $[-R, R]$ and $\mu_{2,s}$ on $[-\frac{\pi}{2}, \frac{\pi}{2}]$ can both be taken as measures.

Proof. If $f \in H(D_+(0, R))$, and $\log^+|f|$ has a harmonic majorant function in $D_+(0, R)$, let $z_0 \in D_+(0, R)$, and let $w = \phi_{z_0}(z)$

be the function defined by (5.1.33). Then the function $F(w) = f(\phi_{z_0}^{-1}(w)) \in H(U)$, and $\log^+ |F|$ has a harmonic majorant function in U. Therefore, $F \in H^0(U)$, without loss of generality, let $F(0) \neq 0$, by Theorem 5.1.14, F has a decomposition of the following form: $F(w) = c_1 G_1(w) B_1(w) S_1(w)$, where c_1 is a constant with modulus 1, $G_1(w)$ is an outer function on the unit disk U, given by

$$G_1(w) = \exp\left\{\frac{1}{2\pi} \int_{\varphi_R'}^{\varphi_R'+2\pi} \frac{e^{i\varphi} + w}{e^{i\varphi} - w} \log|F^*(e^{i\varphi})| d\varphi\right\},$$

with $\log|F^*(e^{i\varphi})| \in L^1([\varphi_R', \varphi_R' + 2\pi))$; $S_1(w)$ is the ratio of two singular inner functions on the unit disk U, given by

$$S_1(w) = \exp\left\{-\frac{1}{2\pi} \int_{\varphi_R'}^{\varphi_R'+2\pi} \frac{e^{i\varphi} + w}{e^{i\varphi} - w} d\nu_s(\varphi)\right\},$$

where ν_s s a singular sign measure of finite total variation on $[\varphi_R', \varphi_R' + 2\pi)$; (if there exists a positive number p such that $|f(z)|^p$ has a harmonic majorant function in $D_+(0, R)$, then $F \in H^p(U)$, and ν_s can be taken as a finite singular measure on $[\varphi_R', \varphi_R' + 2\pi)$.

$$B_1(w) = \prod_{\lambda \in \Lambda_R} \left(\frac{\phi_{z_0}(\lambda) - w}{1 - \overline{\phi_{z_0}(\lambda)}w}\right)\left(\frac{\overline{\phi_{z_0}(\lambda)}}{|\phi_{z_0}(\lambda)|}\right)$$

is a Blaschke product on the unit circle U and satisfies

$$\sum_{\lambda \in \Lambda_R} (1 - |\phi_{z_0}(\lambda)|^2) \leqslant 2 \sum_{\lambda \in \Lambda_R} (1 - |\phi_{z_0}(\lambda)|) < \infty.$$

Therefore,

$$\log|G_1(\phi_{z_0}(z))| = \frac{R}{2\pi} \int_{-\frac{\pi}{2}}^{\frac{\pi}{2}} \frac{(1 - |\phi_{z_0}(z)|^2)|\phi_{z_0}'(Re^{i\theta})|}{|\phi_{z_0}(Re^{i\theta}) - \phi_{z_0}(z)|^2}$$

$$\times \log|F^*(\phi_{z_0}(Re^{i\theta}))| d\theta$$

$$+ \frac{1}{2\pi} \int_{-R}^{R} \frac{1 - |\phi_{z_0}(z)|^2}{|\phi_{z_0}(it) - \phi_{z_0}(z)|^2} |\phi_{z_0}'(it)|$$

$$\times \log|F^*(\phi_{z_0}(it))| dt,$$

$$\log |S_1(\phi_{z_0}(z))| = \frac{1}{2\pi} \int_{-\frac{\pi}{2}}^{\frac{\pi}{2}} \frac{1 - |\phi_{z_0}(z)|^2}{|\phi_{z_0}(Re^{i\theta}) - \phi_{z_0}(z)|^2} d\nu_s(\varphi_1(\theta))$$

$$+ \frac{1}{2\pi} \int_{-R}^{R} \frac{1 - |\phi_{z_0}(z)|^2}{|\phi_{z_0}(it) - \phi_{z_0}(z)|^2} d\nu_s(\varphi_2(t)),$$

Using (5.1.33), for any $\zeta \in \overline{D_+(0,R)}$ and $z \in D_+(0,R)$, where $\zeta \neq z$, the calculation yields

$$\phi_{z_0}(\zeta) - \phi_{z_0}(z) = \frac{(\zeta - z)(z_0 + \bar{z}_0)(R^2 - |z_0|^2)(R^2 + \zeta z)(\bar{z}_0 + iR)^2}{(\zeta + \bar{z}_0)(z + \bar{z}_0)(R^2 - \bar{z}_0 z)(R^2 - \bar{z}_0 \zeta)(z_0 - iR)^2}$$

and

$$1 - \overline{\phi_{z_0}(\zeta)}\phi_{z_0}(z) = \frac{(z + \bar{\zeta})(z_0 + \bar{z}_0)(R^2 - |z_0|^2)(R^2 - \bar{\zeta} z)}{(\bar{\zeta} + z_0)(z + \bar{z}_0)(R^2 - \bar{\zeta} z_0)(R^2 - \bar{z}_0 z)}.$$

Therefore,

$$\phi'_{z_0}(z) = \frac{(z_0 + \bar{z}_0)(R^2 - |z_0|^2)(R^2 + z^2)(\bar{z}_0 + iR)^2}{(z + \bar{z}_0)^2(R^2 - \bar{z}_0 z)^2(z_0 - iR)^2}$$

and

$$1 - |\phi_{z_0}(z)|^2 = \frac{(z + \bar{z})(z_0 + \bar{z}_0)(R^2 - |z_0|^2)(R^2 - |z|^2)}{|z + \bar{z}_0|^2 |R^2 - \bar{z}_0 z|^2}.$$

Thus,

$$1 - |\phi_{z_0}(z)|^2 \geq \frac{\mathrm{Re} z \, \mathrm{Re} z_0 (R^2 - |z|^2)(R^2 - |z_0|^2)}{4R^6},$$

$$\frac{1 - |\phi_{z_0}(z)|^2}{|\phi_{z_0}(\zeta) - \phi_{z_0}(z)|^2} |\phi'_{z_0}(\zeta)| = 2\mathrm{Re} z \frac{(R^2 - |z|^2)|R^2 + \zeta^2|}{|\zeta - z|^2 |R^2 + z\zeta|^2},$$

$$\left(\frac{\phi_{z_0}(\lambda) - \phi_{z_0}(z)}{1 - \overline{\phi_{z_0}(\lambda)}\phi_{z_0}(z)}\right) \left(\frac{\overline{\phi_{z_0}(\lambda)}}{|\phi_{z_0}(\lambda)|}\right) = \left(\frac{(z - \lambda)(R^2 + \lambda z)}{(z + \bar{\lambda})(R^2 - \bar{\lambda} z)}\right) e^{i\theta(\lambda, z_0)}.$$

so that

$$\frac{1 - |\phi_{z_0}(z)|^2}{|\phi_{z_0}(Re^{i\theta}) - \phi_{z_0}(z)|^2} |\phi'_{z_0}(Re^{i\theta})|R = \frac{R^2 - |z|^2}{|Re^{i\theta} - z|^2} - \frac{R^2 - |z|^2}{|Re^{-i\theta} + z|^2}$$

$$= \mathrm{Re}\left(\frac{Re^{i\theta} + z}{Re^{i\theta} - z} - \frac{Re^{-i\theta} - z}{Re^{-i\theta} + z}\right)$$

and

$$\frac{1-|\phi_{z_0}(z)|^2}{|\phi_{z_0}(it)-\phi_{z_0}(z)|^2}|\phi'_{z_0}(it)| = \frac{2\operatorname{Re} z}{|it-z|^2} - \frac{2R^2\operatorname{Re} z}{|R^2+itz|^2}$$
$$= 2\operatorname{Re}\left(\frac{1}{z-it} - \frac{z}{R^2+itz}\right).$$

We define the singular sign measures on $[-R, R]$ and on $[-\frac{\pi}{2}, \frac{\pi}{2}]$ as follows: $d\mu_{1,s}(t) = |\phi'_{z_0}(it)|^{-1}d\nu_s(\varphi_1(t))$ and $d\mu_{2,s}(\theta) = R^{-1}|\phi'_{z_0}(Re^{i\theta})|^{-1}d\nu_s(\varphi_2(\theta))$. Then, there exist constants c_2 and c_3 with modulus 1 such that $G_1(\phi_{z_0}(z)) = c_2 G(z), S_1(\phi_{z_0}(z)) = c_3 S(z)$. Letting $c = c_1 c_2 c_3$ and $B(z) = B_1(\phi_{z_0}(z))$, we can conclude that $f(z) = cG(z)B(z)S(z)$, and (5.1.34) and (5.1.35) hold. $\square$

For the half-disk, we have the following generalized Nevanlinna formula.

Theorem 5.1.21 (Generalized Nevanlinna Formula). *If $R > 0$, $f \in H(D_+(0,R))$, and Λ_R is the zero set of f in $D_+(0,R)$ (counting multiplicities). If $\log^+ |f|$ has a harmonic majorant function in $D_+(0,R)$, then for almost all $t \in (-R, R)$, the non-tangential limit $f^*(it)$ exists, and for almost all $\theta \in (-\frac{\pi}{2}, \frac{\pi}{2})$, the non-tangential limit $f^*(Re^{i\theta})$ exists, there exist singular sign measures $\mu_{1,s}$ on $[-R, R]$ and $\mu_{2,s}$ on $[-\frac{\pi}{2}, \frac{\pi}{2}]$ such that (5.1.34) and (5.1.35) hold and for $z \in D_+(0,R)$,*

$$\log|f(z)| = \frac{1}{2\pi}\int_{-\frac{\pi}{2}}^{\frac{\pi}{2}} \left(\frac{R^2-|z|^2}{|Re^{i\theta}-z|^2} - \frac{R^2-|z|^2}{|Re^{-i\theta}+z|^2}\right) \log|f^*(Re^{i\theta})| d\theta$$
$$+ \frac{1}{\pi}\int_{-R}^{R} \left(\frac{\operatorname{Re} z}{|it-z|^2} - \frac{R^2\operatorname{Re} z}{|R^2+itz|^2}\right) \log|f^*(it)| dt$$
$$+ \sum_{\lambda \in \Lambda_R} \log\left|\frac{z-\lambda}{R^2-\bar{\lambda}z} \frac{R^2+\lambda z}{z+\bar{\lambda}}\right|$$
$$- \frac{1}{2\pi}\int_{-\frac{\pi}{2}}^{\frac{\pi}{2}} \left(\frac{R^2-|z|^2}{|Re^{i\theta}-z|^2} - \frac{R^2-|z|^2}{|Re^{-i\theta}+z|^2}\right) d\mu_{2,s}(\theta)$$
$$- \frac{1}{\pi}\int_{-R}^{R} \left(\frac{\operatorname{Re} z}{|it-z|^2} - \frac{R^2\operatorname{Re} z}{|R^2+itz|^2}\right) d\mu_{1,s}(t). \quad (5.1.39)$$

Moreover, if there exists a positive number p such that $|f(z)|^p$ has a harmonic majorant function in $D_+(0, R)$, then the singular sign measures $\mu_{1,s}$ on $[-R, R]$ and $\mu_{2,s}$ on $[-\frac{\pi}{2}, \frac{\pi}{2}]$ can both be taken as measures.

Proof. This follows from Theorem 5.1.20. □

Theorem 5.1.22 (Generalized Carleman Formula). *If $R > \rho > 0$, $f \in H(D_+(0, R))$, and Λ_r is the zero set of f in $D_+(0, r)$ ($0 < r \leqslant R$) (counting multiplicities). If $\log^+ |f|$ has a harmonic majorant function in $D_+(0, R)$, then for almost all $t \in (-R, R)$, the non-tangential limit $f^*(it)$ exists, and for almost all $\theta \in (-\frac{\pi}{2}, \frac{\pi}{2})$, the non-tangential limit $f^*(Re^{i\theta})$ exists, There exist singular sign measures $\mu_{1,s}$ on $[-R, R]$ and $\mu_{2,s}$ on $[-\frac{\pi}{2}, \frac{\pi}{2}]$ such that (5.1.34) and (5.1.35) hold and for any positive integer q,*

$$\frac{1}{q} \sum_{\lambda_k \in \Lambda_R \setminus \Lambda_\rho} \left(\frac{1}{r_k^q} - \frac{r_k^q}{R^{2q}} \right) \sin q \left(\theta_k + \frac{\pi}{2} \right)$$

$$= \frac{1}{\pi R^q} \int_{-\frac{\pi}{2}}^{\frac{\pi}{2}} \sin q \left(\theta + \frac{\pi}{2} \right) (\log |f^*(Re^{i\theta})| d\theta - d\mu_{2,s}(\theta))$$

$$+ \frac{(-1)^{q+1}}{2\pi} \int_{\rho < |t| < R} \left(\frac{1}{t^{q+1}} - \frac{t^{q-1}}{R^{2q}} \right) (\log |f^*(it)| dt - d\mu_{1,s}(t))$$

$$+ c(\rho, f, q) + R^{-2q} d(\rho, f, q), \tag{5.1.40}$$

where $\lambda_k = r_k e^{i\theta_k}$ is the zero of $f(z)$,

$$d(\rho, f, q) = \frac{1}{q} \sum_{\lambda_k \in \Lambda_\rho} r_k^q \sin q \left(\theta_k + \frac{\pi}{2} \right)$$

$$+ \frac{(-1)^q}{2\pi} \int_{|t| < \rho} t^{q-1} (\log |f^*(it)| dt - d\mu_{1,s}(t)), \tag{5.1.41}$$

and $c(\rho, f, q)$ is independent of R. All series and integrals converge absolutely. In particular, when $q = 1$, we have the following *generalized Carleman formula:*

$$\sum_{\lambda_k \in \Lambda_R \setminus \Lambda_\rho} \left(\frac{1}{r_k} - \frac{r_k}{R^2}\right) \cos \theta_k$$

$$= \frac{1}{\pi R} \int_{-\frac{\pi}{2}}^{\frac{\pi}{2}} \cos\theta (\log|f^*(Re^{i\theta})|d\theta - d\mu_{2,s}(\theta))$$

$$+ \frac{1}{2\pi} \int_{\rho<|t|<R} \left(\frac{1}{t^2} - \frac{1}{R^2}\right) (\log|f^*(it)|dt - d\mu_{1,s}(t))$$

$$+ c(\rho, f, 1) + R^{-2} d(\rho, f, 1).$$

Moreover, if there exists a positive number p such that $|f(z)|^p$ has a harmonic majorant function in $D_+(0, R)$, then the singular sign measures $\mu_{1,s}$ on $[-R, R]$ and $\mu_{2,s}$ on $[-\frac{\pi}{2}, \frac{\pi}{2}]$ can both be taken as measures.

Proof. According to (5.1.35), $\sum_{\lambda_k \in \Lambda_\rho} \text{Re} \lambda_k < \infty$, where $\lambda_k = r_k e^{i\theta_k}$ is the zero of $f(z)$, thus

$$B_\rho(z) = \prod_{\lambda_k \in \Lambda_\rho} \frac{z - \lambda_k}{z + \overline{\lambda_k}}$$

is analytic in $\mathbb{C} \setminus \overline{\{-\overline{\lambda} : \lambda \in \Lambda_\rho\}}$. The function

$$G_\rho(z) = \exp\left\{-\frac{1}{\pi} \int_{-\rho}^{\rho} \frac{1}{it - z} (\log|f(it)|dt - d\mu_{1,s}(t))\right\}$$

is analytic in $\mathbb{C} \setminus [-i\rho, i\rho]$. The function B_ρ and G_ρ is independent of R, only dependent on f and ρ. Therefore, using (5.1.36)–(5.1.38), the function $f_\rho(z) = \frac{f(z)}{B_\rho(z) G_\rho(z)}$ can be extended to an analytic function in $D(0, \rho)$ without zeros. Thus, the function $\log|f_\rho(z)|$ can be extended to a harmonic function in $D(0, \rho)$ with $\log|f_\rho(iy)| = 0$ ($|y| < \rho$). Hence, $\log|f_\rho(re^{i\psi})|$ can be expanded as a Fourier series in $[-\pi, \pi]$:

$$\log|f_\rho(re^{i\psi})| = \sum_{q=1}^{\infty} c_q r^q \sin q\left(\psi + \frac{\pi}{2}\right),$$

where
$$c_q = \frac{2}{\pi r^q} \int_{-\frac{\pi}{2}}^{\frac{\pi}{2}} (\log |f_\rho(re^{i\psi})|) \sin q\left(\psi + \frac{\pi}{2}\right) d\psi,$$

which is independent of $r \in (0, \rho)$. On the other hand, from (5.1.39),

$$\log |f_\rho(z)| = \frac{1}{2\pi} \int_{-\frac{\pi}{2}}^{\frac{\pi}{2}} \left(\frac{R^2 - |z|^2}{|Re^{i\theta} - z|^2} - \frac{R^2 - |z|^2}{|Re^{-i\theta} + z|^2} \right)$$
$$\times (\log |f^*(Re^{i\theta})| d\theta - d\mu_{2,s}(\theta))$$
$$- \frac{1}{\pi} \int_{-R}^{R} \left(\frac{R^2 \text{Re} z}{|R^2 + itz|^2} \right) (\log |f^*(it)| dt - d\mu_{1,s}(t))$$
$$+ \frac{1}{\pi} \int_{\rho<|t|<R} \left(\frac{\text{Re} z}{|it - z|^2} \right) (\log |f^*(it)| dt - d\mu_{1,s}(t))$$
$$+ \sum_{\lambda_k \in \Lambda_R \setminus \Lambda_\rho} \log \left| \frac{z - \lambda_k}{z + \overline{\lambda_k}} \right| + \sum_{\lambda_k \in \Lambda_R} \log \left| \frac{R^2 + \lambda_k z}{R^2 - \overline{\lambda_k} z} \right|.$$

$z = re^{i\psi}, 0 < r < \rho < R$, by computation, we have

$$\frac{R^2 - |z|^2}{|Re^{i\theta} - z|^2} - \frac{R^2 - |z|^2}{|Re^{-i\theta} + z|^2} = 4 \sum_{q=1}^{\infty} \frac{r^q}{R^q} \sin q\left(\theta + \frac{\pi}{2}\right) \sin q\left(\psi + \frac{\pi}{2}\right);$$

$$\frac{\text{Re} z}{|it - z|^2} = \sum_{q=1}^{\infty} (-1)^{q+1} \frac{r^q}{t^{q+1}} \sin q\left(\psi + \frac{\pi}{2}\right);$$

$$\frac{R^2 \text{Re} z}{|R^2 + itz|^2} = \sum_{q=1}^{\infty} (-1)^{q+1} \frac{t^{q-1} r^q}{R^{2q}} \sin q\left(\psi + \frac{\pi}{2}\right);$$

$$\log \left| \frac{z - \lambda_k}{z + \overline{\lambda_k}} \right| = - \sum_{q=1}^{\infty} \frac{2 r^q}{q r_k^q} \sin q\left(\theta_k + \frac{\pi}{2}\right) \sin q\left(\psi + \frac{\pi}{2}\right);$$

$$\log \left| \frac{R^2 + \lambda_k z}{R^2 - \overline{\lambda_k} z} \right| = \sum_{q=1}^{\infty} \frac{2(r_k r)^q}{q R^{2q}} \sin q\left(\theta_k + \frac{\pi}{2}\right) \sin q$$
$$\times \left(\psi + \frac{\pi}{2}\right).$$

By the uniqueness of Fourier coefficients, we obtain

$$c_q = \frac{2}{\pi R^q} \int_{-\frac{\pi}{2}}^{\frac{\pi}{2}} \sin q\left(\theta + \frac{\pi}{2}\right) (\log|f^*(Re^{i\theta})|d\theta - d\mu_{2,s}(\theta))$$

$$- \frac{(-1)^{q+1}}{\pi R^{2q}} \int_{-R}^{R} t^{q-1}(\log|f^*(it)|dt - d\mu_{1,s}(t))$$

$$+ \frac{(-1)^{q+1}}{\pi} \int_{\rho<|t|<R} \frac{1}{t^{q+1}}(\log|f^*(it)|dt - d\mu_{1,s}(t))$$

$$- \frac{2}{q} \sum_{\lambda_k \in \Lambda_R \setminus \Lambda_\rho} \frac{1}{r_k^q} \sin q\left(\theta_k + \frac{\pi}{2}\right) + \frac{2}{q} \sum_{\lambda_k \in \Lambda_R} \frac{r_k^q}{R^{2q}} \sin q\left(\theta_k + \frac{\pi}{2}\right).$$

Putting $c(\rho, f, q) = -\frac{c_q}{2}$, we see that (5.1.40) and (5.1.41) hold. This proves the theorem. □

5.2 H^p Space in the Upper Half-Plane

Definition 5.2.1. Let $0 \leq p \leq \infty$. The space $H^p(\mathbb{C}_+) = \{f \in H(\mathbb{C}_+) : \|f\|_{\widetilde{H}^p} < \infty\}$ is called H^p space in the upper half-plane $\mathbb{C}_+ = \{z = x + iy : y > 0\}$, where

$$\|f\|_{\widetilde{H}^p} = \sup\{M_p(f, y) : 0 < y < \infty\},$$

$$M_p(f, y) = \|f_y\|_p = \left(\int_{-\infty}^{\infty} |f(x+iy)|^p dx\right)^{\frac{1}{p}} \quad (f_y(x) = f(x+iy)),$$

$$M_\infty(f, y) = \|f_y\|_\infty = \sup\{|f(x+iy)| : x \in \mathbb{R}\}.$$

The Nevanlinna class in the upper half-plane $\mathbb{C}_+$ is defined as the set $H^0(\mathbb{C}_+) = \{f \in H(\mathbb{C}_+) : \|f\|_{\widetilde{H}^0} < \infty\}$, where

$$\|f\|_{\widetilde{H}^0} = \sup\{M_0(f, y) : 0 < y < \infty\},$$

$$M_0(f, y) = \exp\left\{\frac{1}{2\pi} \int_{-\infty}^{\infty} \log^+ |f(x+iy)| dx\right\}.$$

Due to the fact that $\log^+|f|, |f|^p$ are subharmonic, by Theorem 4.6.13, $M_p(f, y)$ is decreasing in $(0, \infty)$, which implies that

$$\|f\|_{\widetilde{H}^p} = \lim_{y \to 0} M_p(f, y) \quad (0 \leq p \leq \infty). \qquad (5.2.1)$$

$H^p(\mathbb{C}_+)$ possesses the following properties:

When $p \geqslant 1$, $\|\cdot\|_{\widetilde{H}^p}$ is a norm, satisfying the triangle inequality

$$\|f+g\|_{\widetilde{H}^p} \leqslant \|f\|_{\widetilde{H}^p} + \|g\|_{\widetilde{H}^p},$$

thus making $H^p(\mathbb{C}_+)$ a linear normed space. When $0 < p < 1$, $\rho(f,g) = \|f-g\|_{\widetilde{H}^p}^p$ is a metric on $H^p(\mathbb{C}_+)$, hence $H^p(\mathbb{C}_+)$ is a linear metric space.

Theorem 5.2.2. *If $f \in H^0(\mathbb{C}_+)$, $f(z) \not\equiv 0$, then f has a non-tangential limit $f^*(x)$ almost everywhere on $\mathbb{R}$, and*

$$\int_{-\infty}^{\infty} \frac{|\log|f^*(t)||}{1+t^2} dt < \infty. \tag{5.2.2}$$

Thus, for almost all $t \in \mathbb{R}$, we have $f^*(t) \neq 0$.

Proof. In fact, since $\log^+|f|$ is subharmonic and satisfies (4.6.17), by Theorem 4.6.13, there exists a positive measure $\mu \in \mathscr{M}(\mathbb{R})$ such that u is the Poisson–Stieltjes integral of μ, given by $u(x+iy) = P_y * \mu$ and for any $z \in \mathbb{C}_+$, we have $\log^+|f|(z) \leqslant u(z)$. The function

$$\alpha(w) = i\frac{1-w}{1+w} \tag{5.2.3}$$

maps the unit disk U conformally onto the upper half-plane $\mathbb{C}_+$, and its inverse mapping is

$$\beta(z) = \frac{i-z}{z+i}. \tag{5.2.4}$$

The function $\log^+|f|(\alpha(w))$ has harmonic majorant $u(\alpha(w))$ in U, and by Theorem 4.3.20, $\log^+|f|(\alpha(w)) \in H^0(U)$. By Theorem 5.1.14, $F(w) = f(\alpha(w))$ has a non-tangential limit $F^*(e^{i\theta})$ almost everywhere on ∂U and $\log|F^*(e^{i\theta})| \in L^1([-\pi,\pi])$. Since α and β are conformal mappings, f has a non-tangential limit $f^*(t) = F^*(\beta(t))$ almost everywhere on $\mathbb{R}$. Using the fact that $t = \alpha(e^{i\theta}) = \tan\frac{\theta}{2}$ and $2dt = (1+t^2)d\theta$, we obtain

$$\int_{-\infty}^{\infty} \frac{2|\log|f^*(t)||}{1+t^2} dt = \int_{-\pi}^{\pi} |\log|F^*(e^{i\theta})||d\theta < \infty.$$

This proves (5.2.2). □

Theorem 5.2.3. *Let $f \in H^p(\mathbb{C}_+)$ with $p > 0$, $f(z) \not\equiv 0$, then f has a non-tangential limit $f^*(x)$ almost everywhere on $\mathbb{R}$, and it satisfies (5.2.2) and*

$$\int_{-\infty}^{\infty} |f^*(x)|^p dx \leqslant \|f\|_{H^p}^p < \infty. \tag{5.2.5}$$

Proof. In fact, since $|f|^p$ is subharmonic and $p\log^+ |f| \leqslant |f|^p$, so $f \in H^0(\mathbb{C}_+)$. By Theorem 5.2.2, f has a non-tangential limit $f^*(x)$ almost everywhere on $\mathbb{R}$, and (5.2.2) holds. According to Theorem 4.6.13, there exists a positive measure $\mu \in \mathcal{M}(\mathbb{R})$ such that $\|\mu\| = \|f\|_{H^p}^p$ and $|f(z)|^p$ has a harmonic majorant function $u(z)$, where u is Poisson–Stieltjes integral of μ defined by $u(x+iy) = P_y * \mu$. For any $z \in \mathbb{C}_+$, we have $|f(z)|^p \leqslant u(z)$. This implies

$$\int_{-\infty}^{\infty} |f(x+iy)|^p dx \leqslant \int_{-\infty}^{\infty} (P_y * \mu)(x) dx = \|\mu\| = \|f\|_{H^p}^p.$$

By Fatou's Lemma,

$$\int_{-\infty}^{\infty} |f^*(x)|^p dx \leqslant \liminf_{y \to 0} \int_{-\infty}^{\infty} |f(x+iy)|^p dx \leqslant \|\mu\| = \|f\|_{H^p}^p.$$

This proves (5.2.5). □

Theorem 5.2.4. *Let $z_1, z_2, \ldots, z_n = r_n e^{i\theta_n}, \ldots$ be an infinite sequence of non-zero points in the upper half-plane $\mathbb{C}_+$ satisfying*

$$\sum_{n=1}^{\infty} \frac{r_n \sin \theta_n}{1 + |z_n|^2} < \infty. \tag{5.2.6}$$

Let

$$\varphi_n = \begin{cases} 0, & |z_n| < 1, \\ -2\theta_n, & |z_n| \geqslant 1, \end{cases} \tag{5.2.7}$$

$$b_n(z) = e^{i\varphi_n} \left(\frac{z - z_n}{z - \bar{z}_n} \right). \tag{5.2.8}$$

Then the Blaschke product

$$B(z) = \prod_{n=1}^{\infty} b_n(z) \tag{5.2.9}$$

converges uniformly on compact subsets of $\mathbb{C}_+$, $B \in H(\mathbb{C}_+)$ and $|B(z)| \leqslant 1$ for $z \in \mathbb{C}_+$, the zeros of $B(z)$ in $\mathbb{C}_+$ are $z_1, z_2, \ldots, z_n, \ldots,$

B has a non-tangential limit $B^*(x)$, satisfying $|B^*(x)| = 1$ almost everywhere on $\mathbb{R}$.

Proof. For $r_n = |z_n| < 1$, we have

$$|1 - b_n(z)| = \frac{2r_n \sin \theta_n}{|z - \bar{z}_n|} \leqslant \frac{4r_n \sin \theta_n}{y(1 + r_n^2)},$$

and when $r_n = |z_n| \geqslant 1$, we have

$$|1 - b_n(z)| = \frac{2|z| \sin \theta_n}{|z - \bar{z}_n|} \leqslant \frac{(12|z|^2 + 2|z|)r_n \sin \theta_n}{y(1 + r_n^2)}.$$

Note that for any compact set $K \subset \mathbb{C}_+$, there exists $\delta > 0$ and $R > \delta$ such that for any $z \in K$, we have $\delta \leqslant y \leqslant |z| \leqslant R$, therefore $B(z)$ converges uniformly on K, which implies that $B \in H(\mathbb{C}_+)$ and $|B(z)| \leqslant 1$ for $y > 0$. The zeros of $B(z)$ in $\mathbb{C}_+$ (counted with multiplicity) are $z_1, z_2, \ldots, z_n, \ldots$. Since $B \in H^0(\mathbb{C}_+)$, by Theorem 5.2.2, B has non-tangential limits $B^*(x)$ almost everywhere on $\mathbb{R}$. According to Theorem 5.1.5, $|B^*(x)| = 1$ almost everywhere. □

Theorem 5.2.5. *Suppose $f \in H^0(U)$, $f(z) \not\equiv 0$, and the zeros of $f(z)$ in $\mathbb{C}_+$ are $z_1, z_2, \ldots, z_n, \ldots$ (counted with multiplicity). Then (5.2.6) holds.*

Proof. Without loss of generality, let $\{z_j\}$ be an infinite set. The function $z = \alpha(w)$ defined by (5.2.3) maps the unit disk U conformally onto the upper half-plane $\mathbb{C}_+$. Its inverse mapping is given by the function $w = \beta(z)$ defined by (5.2.4). Since $\log^+ |f|$ is subharmonic and satisfies (4.6.17), by Theorem 4.6.13, there exist a harmonic function $u(z)$ and a positive measure $\mu \in \mathscr{M}(\mathbb{R})$ such that u is the Poisson–Stieltjes integral of μ defined by $u(x+iy) = P_y * \mu$ and for any $z \in \mathbb{C}_+$, we have $\log^+ |f(z)| \leqslant u(z)$. The function $\log^+ |f(\alpha(w))|$ has a harmonic majorant function $u(\alpha(w))$ in the unit disk U, and by Theorem 4.3.23, $f(\alpha(w)) \in H^0(U)$. By Theorem 5.1.4 and $|\beta(z)| < 1$, we have

$$\sum_{n=1}^{\infty} \frac{4r_n \sin \theta_n}{|z_n + i|^2} = \sum_{n=1}^{\infty} (1 - |\beta(z_n)|^2) < \infty.$$

Thus, (5.2.6) holds. □

Theorem 5.2.6 ($H^p(\mathbb{C}_+)$ Decomposition Theorem). *For any $f \in H^p(\mathbb{C}_+)$ with $0 \leqslant p \leqslant \infty$, the following decomposition holds:*

$$f(z) = G(z)B(z), \qquad (5.2.10)$$

where $B(z)$ is a Blaschke product formed by the zeros of $f(z)$, $G(z) \neq 0$, $G \in H^p(\mathbb{C}_+)$, and $\|G\|_{\widetilde{H}^p} = \|f\|_{\widetilde{H}^p}$.

Proof. Let

$$B_n(z) = \prod_{j=1}^n b_j(z), \quad B(z) = \prod_{j=1}^\infty b_j(z),$$

$G(z) = f(z)/B(z)$, where $b_j(z)$ is defined by (5.2.8). Since $\log|B_n(x+iy)|$ converges to 0 uniformly for $x \in \mathbb{R}$ as $y \to 0$, so

$$\lim_{y \to 0} M_p(f, y) = \lim_{y \to 0} M_p(\frac{f}{B_n}, y) \geqslant M_p(\frac{f}{B_n}, y).$$

As $n \to \infty$, $\frac{|f(z)|}{|B_n(z)|}$ monotonically increases toward $|G(z)|$. When $p > 0$, by using the Lebesgue monotone convergence theorem, we know that

$$\|f\|_{\widetilde{H}^p} \geqslant M_p(G, y).$$

Thus, $\|f\|_{\widetilde{H}^p} \geqslant \|G\|_{\widetilde{H}^p}$. On the other hand, it is obvious that $|f(z)| \leqslant |G(z)|$, so $\|f\|_{\widetilde{H}^p} \leqslant \|G\|_{\widetilde{H}^p}$. This proves that $\|f\|_{\widetilde{H}^p} = \|G\|_{\widetilde{H}^p}$. $\square$

Theorem 5.2.7. *For $f \in H^p(\mathbb{C}_+)$ with $0 < p < \infty$, the following conclusion holds:*

(a) *For any $\alpha > 0$, for almost every $t \in \mathbb{R}$, the non-tangential limit*

$$\lim_{z \in \Gamma_\alpha(t), z \to t} f(z) = f^*(t)$$

exists and $f^* \in L^p(\mathbb{R})$;

(b)

$$\|f_y - f^*\|_p^p = \int_{-\infty}^\infty |f(x+iy) - f^*(x)|^p dx \to 0 \quad (y \to 0),$$

where $f_y(x) = f(x+iy)$;

(c)
$$\|f\|_{\widetilde{H}^p} = \lim_{y \to 0} \left(\int_{-\infty}^{\infty} |f(x+iy)|^p dx \right)^{\frac{1}{p}} = \left(\int_{-\infty}^{\infty} |f^*(x)|^p dx \right)^{\frac{1}{p}}.$$

Proof. Since analytic function is harmonic function, when $p > 1$, the conclusions of Theorem 5.2.7 can be deduced from Theorems 4.5.4 and 4.5.7. Therefore, it suffices to consider the case $0 < p \leq 1$. Let $f(z) = G(z)B(z)$, where $B(z)$ is the Blaschke product whose zeros are the zeros of $f(z)$, $G(z) \neq 0$, $G \in H^p(\mathbb{C}_+)$, and $\|G\|_{\widetilde{H}^p} = \|f\|_{\widetilde{H}^p}$. Let n be a positive integer such that $pn > 1$. Since $G(z)$ has no zeros in the upper half-plane $\mathbb{C}_+$, there exists $h(z) \in H(\mathbb{C}_+)$ such that $h^n(z) = G(z)$, then $h \in H^{np}(\mathbb{C}_+)$ and $\|G\|_{\widetilde{H}^p}^p = \|h\|_{\widetilde{H}^{np}}^{np}$. Thus, h has a non-tangential limit $h^*(t)$ almost everywhere, which implies that G has a non-tangential limit $G^*(t) = (h^*(t))^n \in L^p(\mathbb{R})$ almost everywhere. Therefore, f has a non-tangential limit $f^*(t) = B^*(t)G^*(t) \in L^p(\mathbb{R})$ almost everywhere. This proves the conclusion (a). According to Theorems 4.6.5 and 4.6.10, we know that as $y \to 0$, $\|h_y - h^*\|_{np} \to 0$, where $h_y(t) = h(t+iy)$. Hence,

$$\|f_y - f^*\|_p^p = \|B_y(G_y - G^*) + (B_y - B^*)G^*\|_p^p$$
$$\leq \|B_y(G_y - G^*)\|_p^p + \|(B_y - B^*)G^*\|_p^p.$$

Since $|B_y| \leq 1$ and

$$|G_y - G^*| \leq |h_y - h^*| \sum_{k=0}^{n-1} |h_y|^{n-1-k} |h^*|^k \leq n|h_y - h^*|(|h_y| + |h^*|)^{n-1},$$

using the Hölder inequality $(n^{-1} + (\frac{n}{n-1})^{-1} = 1)$, we have

$$\|G_y - G^*\|_p^p \leq n^p \|h_y - h^*\|_{np}^p \||h_y| + |h^*|\|_{pn}^{p(n-1)}$$
$$\leq 2^{p(n-1)} n^p \|h_y - h^*\|_{np}^p \|h^*\|_{pn}^{p(n-1)}.$$

Since $\|h^*\|_{pn}^{pn} = \|G^*\|_p^p$, we have

$$\|B_y(G_y - G^*)\|_p^p \leq \|G_y - G^*\|_p^p$$
$$\leq 2^{p(n-1)} n^p \|h_y - h^*\|_{np}^p \|G^*\|_p^{p(1-n^{-1})} \to 0 \quad (y \to 0).$$

Since $G^* \in L^p(\mathbb{R}), |B_y| \leq 1$, by the Lebesgue-dominated convergence theorem, we have

$$\|(B_y - B^*)G^*\|_p^p \to 0 \quad (y \to 0).$$

Therefore,

$$\|f_y - f^*\|_p^p \to 0 \quad (y \to 0).$$

This proves the desired conclusion (b). Since $M_p(f, y)$ is decreasing on $(0, \infty)$, we know that the first equality in (c) holds. By using (b), we can deduce (c). □

Theorem 5.2.8. *If $f \in H^p(\mathbb{C}_+)$ and $f^*(t) \in L^q(\mathbb{R})(0 < p < q)$, then $f \in H^q(\mathbb{C}_+)$.*

Proof. It can be deduced from Theorems 4.6.10 and 4.6.5 that when $p > 1$, since analytic functions are harmonic functions, $f(z)$ is the Poisson integral of $f^*(t) \in L^q(\mathbb{R})$, implying $f \in H^q(\mathbb{C}_+)$. Therefore, we only need to consider the case of $0 < p \leq 1$. Let $f(z) = G(z)B(z)$, where $B(z)$ is a Blaschke product composed of the zeros of $f(z)$, $G(z) \neq 0$, and $G \in H^p(\mathbb{C}_+)$, with $\|G\|_{\widetilde{H}^p} = \|f\|_{\widetilde{H}^p}$. Let n be a positive integer such that $pn > 1$, since $G(z)$ has no zeros in the upper half-plane $\mathbb{C}_+$, there exists $h(z) \in H(\mathbb{C}_+)$ such that $h^n(z) = G(z)$, thus $h \in H^{np}(\mathbb{C}_+)$ and $\|G\|_{\widetilde{H}^p}^p = \|h\|_{\widetilde{H}^{np}}^{np}$. Consequently, h has a non-tangential limit $h^*(t)$ almost everywhere, and thus, G has a non-tangential limit $G^*(t) = (h^*(t))^n \in L^p(\mathbb{R})$, therefore f has a non-tangential limit $f^*(t) = B^*(t)G^*(t) \in L^q(\mathbb{R})$. Therefore, it follows that the non-tangential limit $h^*(t)$ belongs to $L^{nq}(\mathbb{R})$. According to Theorems 4.6.5 and 4.6.10, $h(z)$ is the Poisson integral of $h^*(t) \in L^{qn}(\mathbb{R})$, implying $h \in H^{qn}(\mathbb{C}_+)$. Therefore, $G = h^n \in H^q(\mathbb{C}_+)$, and thus, $f \in H^q(\mathbb{C}_+)$. This proves the desired result. □

Theorem 5.2.9. *If $f \in H^p(\mathbb{C}_+)$ $(1 \leq p \leq \infty)$, then $f(z)$ is the Poisson integral and the Cauchy integral of its boundary value $f^*(t)$, i.e.,*

$$f(x+iy) = (P_y * f^*)(x) = \int_{-\infty}^{\infty} P_y(x-t)f^*(t)dt, \quad (5.2.11)$$

$$f(z) = \frac{1}{2\pi i} \int_{\mathbb{R}} \frac{f^*(\zeta)}{\zeta - z} d\zeta, \quad y = \mathrm{Im}\, z > 0, \quad (5.2.12)$$

and
$$\frac{1}{2\pi i} \int_{\mathbb{R}} \frac{f^*(\zeta)}{\zeta - z} d\zeta = 0, \quad y = Imz < 0, \qquad (5.2.13)$$

where
$$P_y(x) = \operatorname{Re}\left(\frac{i}{\pi z}\right) = \frac{y}{\pi(x^2 + y^2)}$$

is the Poisson kernel on the upper half-plane $\mathbb{C}_+$.

Proof. According to Theorem 5.2.7, for almost all $t \in \mathbb{R}$, the non-tangential limit $f^*(t)$ exists and $f^* \in L^p(\mathbb{R})$. The function $z = \alpha(w)$ defined by (5.2.3) maps the unit disk U conformally onto the upper half-plane $\mathbb{C}_+$, and its inverse mapping is given by $w = \beta(z)$, as defined in (5.2.4). Since $|f|^p$ is subharmonic and satisfies inequality (4.6.17), by Theorem 4.6.13, there exist a harmonic function u and a positive measure $\mu \in \mathscr{M}(\mathbb{R})$ such that u is the Poisson–Stieltjes integral of μ, i.e., $u(x+iy) = P_y * \mu$, and for any $z \in \mathbb{C}_+$, we have $|f(z)|^p \leqslant u(z)$. The function $F(w) = f(\alpha(w))$ has a harmonic majorant function $u(\alpha(w))$ in the unit disk U, and by Theorem 4.3.23, $F(w) = f(\alpha(w)) \in H^p(U)$. For almost all $t \in [-\pi, \pi]$, the non-tangential limit $F^*(e^{it}) = f^*(\alpha(e^{it}))$ exists and $F^* \in L^p([-\pi, \pi])$. By Theorem 5.1.9, we have

$$F(w) = \frac{1}{2\pi} \int_{-\pi}^{\pi} P(e^{it}, w) F^*(e^{it}) dt.$$

Since $t = \alpha(e^{i\theta}) = \tan\frac{\theta}{2}$ and $2dt = (1+t^2)d\theta$, we obtain (5.2.11). Since $f^* \in L^p(\mathbb{R})$,

$$\widetilde{f}(z) = \frac{1}{2\pi i} \int_{\mathbb{R}} \frac{f^*(\zeta)}{\zeta - z} d\zeta \in H(\mathbb{C} \setminus \mathbb{R}), \qquad (5.2.14)$$

and for $z \in \mathbb{C}$, we have $\widetilde{f}(z) - \widetilde{f}(\bar{z}) = (P_y * f^*)(x) = f(z)$, which implies that $\widetilde{f}(z), \widetilde{f}(\bar{z}) = \widetilde{f}(z) - f(z) \in H(\mathbb{C}_+)$, therefore $\widetilde{f}(\bar{z})$ is a constant. However, $\widetilde{f}(-iy) \to 0$ as $y \to \infty$, so $\widetilde{f}(\bar{z}) \equiv 0$. This proves (5.2.12) and (5.2.13). □

Theorem 5.2.10. *If $f_1, f_2 \in H^0(\mathbb{C}_+)$, where $f(z) \not\equiv 0$, $f_1^*(t)$ and $f_2^*(t)$ are their respective non-tangential limits. If there exists a positive measurable set $E \subset \mathbb{R}$ such that $f_1^*(t) = f_2^*(t)$ for all $t \in E$, then $f_1 \equiv f_2$.*

Proof. By applying Theorem 5.2.2 to the function $f_1^*(t) - f_2^*(t)$, we have that if $f_1^*(t) - f_2^*(t)$ for almost every $t \in E$, then $f_1(z) - f_2(z) = 0$ for all $z \in \mathbb{C}_+$. $\square$

Theorem 5.2.11. *For any $f \in H^0(\mathbb{C}_+)$ with $f \not\equiv 0$, we can write $f(z)$ as a quotient $f(z) = b_1(z)/b_2(z)$ of two functions $b_1(z)$ and $b_2(z)$ in $H^\infty(\mathbb{C}_+)$, where*

$$b_1(z) = cB(z) \exp\left\{ ic_1 z - \frac{1}{\pi i} \int_{-\infty}^{\infty} \left(\frac{1}{t-z} - \frac{t}{1+t^2} \right) d\mu^-(t) \right\} \tag{5.2.15}$$

and

$$b_2(z) = \exp\left\{ ic_2 z - \frac{1}{\pi i} \int_{-\infty}^{\infty} \left(\frac{1}{t-z} - \frac{t}{1+t^2} \right) d\mu^+(t) \right\}, \tag{5.2.16}$$

with $c_1 \geq 0, c_2 \geq 0$, μ^+ and μ^- being two non-negative measures on $\mathbb{R}$ and f having a non-tangential limit $f^*(t)$ almost everywhere such that $(1+t^2)^{-1} \log |f^*(t)| \in L^1(\mathbb{R})$. Moreover, f has the following factorization:

$$f(z) = c e^{i\tau z} B(z) G(z) S(z), \tag{5.2.17}$$

where $|c| = 1$, $\tau = c_1 - c_2$ is a real number, $B(z)$ is the Blaschke product of zeros of $f(z)$, and

$$G(z) = \exp\left\{ \frac{1}{\pi i} \int_{-\infty}^{\infty} \left(\frac{1}{t-z} - \frac{t}{1+t^2} \right) \log |f^*(t)| dt \right\}, \tag{5.2.18}$$

$$S(z) = \exp\left\{ \frac{1}{\pi i} \int_{-\infty}^{\infty} \left(\frac{1}{t-z} - \frac{t}{1+t^2} \right) d\mu_s(t) \right\}, \tag{5.2.19}$$

where μ_s is the unique singular sign measure determined by f.

Proof. The function $z = \alpha(w)$ defined by (5.2.3) maps the unit disk U conformally onto the upper half-plane $\mathbb{C}_+$. Its inverse mapping is given by the function $w = \beta(z)$ defined in (5.2.4). Since $\log^+ |f|$ is subharmonic and satisfies (4.6.17), by Theorem 4.6.13, there exists a positive harmonic function u in $\mathbb{C}_+$ such that for any $z \in \mathbb{C}_+$, we have $\log^+ |f|(z) \leq u(z)$. The function $\log^+ |f|(\alpha(w))$ has a harmonic majorant function $u(\alpha(w))$ in the unit disk U. By Theorem 4.3.20, $f(\alpha(w)) \in H^0(U)$. According to Theorem 5.1.14, $f(z)$ can be written as the quotient of two functions $b_1(z)$ and $b_2(z)$ in $H^\infty(\mathbb{C}_+)$, $f(z) = b_1(z)/b_2(z)$, where

$$b_1(z) = cB(z)\exp\left\{-\frac{1}{2\pi}\int_{-\pi}^{\pi}\frac{e^{i\theta}+\beta(z)}{e^{i\theta}-\beta(z)}d\nu^-(\theta)\right\}$$

and

$$b_2(z) = \exp\left\{-\frac{1}{2\pi}\int_{-\pi}^{\pi}\frac{e^{i\theta}+\beta(z)}{e^{i\theta}-\beta(z)}d\nu^+(\theta)\right\},$$

where ν^+ and ν^- are two non-negative measures on $[-\pi,\pi]$. Let f^* be the non-tangential limit of f, then we have

$$\log|f^*(\alpha(e^{i\theta}))| \in L^1([-\pi,\pi]).$$

Since

$$\frac{\beta(t)+\beta(z)}{\beta(t)-\beta(z)} = \frac{i(1+tz)}{z-t} = it - \frac{i(1+t^2)}{t-z}, \quad t = \alpha(e^{i\theta}) = \tan(\frac{\theta}{2})$$

and $2dt = (1+t^2)d\theta$, we get

$$b_1(z) = cB(z)\exp\left\{ic_1 z - \frac{1}{\pi i}\int_{-\infty}^{\infty}\left(\frac{1}{t-z}-\frac{t}{1+t^2}\right)d\mu^-(t)\right\}$$

and

$$b_2(z) = \exp\left\{ic_2 z - \frac{1}{\pi i}\int_{-\infty}^{\infty}\left(\frac{1}{t-z}-\frac{t}{1+t^2}\right)d\mu^+(t)\right\},$$

where

$$c_1 = \frac{1}{2\pi}\nu^-(\{-\pi,\pi\}) \geq 0,] \quad c_2 = \frac{1}{2\pi}\nu^+(\{-\pi,\pi\}) \geq 0$$

and

$$2d\mu^-(t) = (1+t^2)d\nu^-(2\arctan t),$$
$$2d\mu^+(t) = (1+t^2)d\nu^+(2\arctan t)$$

are two non-negative measures on $\mathbb{R}$. By the Lebesgue decomposition theorem, $d\mu^+(t) = \phi^+(t)dt + d\mu_s^+(t)$ and $d\mu^-(t) = \phi^-(t)dt + d\mu_s^-(t)$, where $\phi^+, \phi^- \in L^1(\mathbb{R})$, and μ_s^+ and μ_s^- are singular positive measures. Therefore, we have $|b_1(z)| \leq 1$ and $0 < |b_2(z)| \leq 1$.

The non-tangential limits $b_1^*(e^{i\theta})$ and $b_2^*(e^{i\theta})$ exist for almost all θ, and $-\log|b_1^*(t)| = \phi^+(t)$ and $-\log|b_2^*(t)| = \phi^-(t)$. Thus, the non-tangential limit $f^*(t)$ exists for almost all t, and $\log|f^*(t)| = \phi^+(t) - \phi^-(t) \in L^1(\mathbb{R})$. Therefore, there exists a constant c with $|c| = 1$ such that (5.2.18) and (5.2.19) hold, where $d\mu_s(t) = d\mu_s^+(t) - d\mu_s^-(t)$ is a singular signed measure uniquely determined by μ. This completes the proof of Theorem 5.2.11. $\square$

Definition 5.2.12. Let $M \in H^\infty(\mathbb{C}_+)$, and let $M^*(t)$ denote the non-tangential limit of M. If $|M^*(t)| = 1$ holds for almost every $t \in \mathbb{R}$, this is called an inner function. Moreover, if $\varphi(t) > 0$ holds for almost every $t \in \mathbb{R}$ and $\log\varphi(t)/(1+t^2) \in L^1(\mathbb{R})$, where c is a constant with $|c| = 1$, then the function

$$Q(z) = c\exp\left\{\frac{1}{\pi i}\int_{-\infty}^{\infty}\left(\frac{1}{t-z} - \frac{t}{1+t^2}\right)\log\varphi(t)dt\right\} \quad (5.2.20)$$

is called an outer function in the upper half-plane $\mathbb{C}_+$.

Theorem 5.2.13. Let c be a constant with $|c| = 1$, let $z_1, z_2, \ldots$, $z_n = r_n e^{i\theta_n}, \ldots$ be a sequence of in the upper half-plane $\mathbb{C}_+$ satisfying (5.2.6), let $B(z)$ be a Blaschke product formed by $\{z_n\}$, $c_1 \geq 0$, and μ_s be a singular positive measure. Then the function

$$M(z) = cB(z)\exp\left\{ic_1 z - \frac{1}{\pi i}\int_{-\infty}^{\infty}\left(\frac{1}{t-z} - \frac{t}{1+t^2}\right)d\mu_s(t)\right\} \quad (5.2.21)$$

is an inner function. Conversely, every inner function can be represented in the form given by (5.2.21).

Proof. If (5.2.21) holds, then $\log|g|$ is the Poisson integral of $-\mu_s$, where $g = M/B$. Therefore, $\log|g| \leq 0$, which implies that $g \in H^\infty(\mathbb{C}_+)$, since μ_s is a singular positive measure, it follows that for almost every t, $\mu_s'(t) = 0$, thus the non-tangential limit of

$$\log|g(z)| = c_1 y - \frac{y}{\pi}\int_{-\infty}^{\infty}\frac{d\mu_s(t)}{|t-z|^2}$$

is zero almost everywhere. Since $|B^*(t)| = 1$ holds for almost every $t \in \mathbb{R}$, we can conclude that M is an inner function.

Conversely, let $B(z)$ be the zeros of M given by $\{z_n\}$ as a Blaschke product. Define $G(w) = g(\alpha(w)) = M(\alpha(w))/B(\alpha(w))$, then $\log|G(w)|$ is harmonic in the unit disk U and $|G| \leq 1$ holds. Moreover, the non-tangential limit $G^*(e^{i\theta})$ satisfies $|G^*(e^{i\theta})| = 1$ for almost every $\theta \in [-\pi, \pi]$. Thus, there exists a positive measure μ such that $\log|G|$ is the Poisson integral of $-\mu$. Since $|B^*(\alpha(e^{i\theta}))| = 1$ holds for almost every $\theta \in [-\pi, \pi]$, it follows that $\mu'(t) = 0$ holds for almost every $\theta \in [-\pi, \pi]$. Therefore, μ is a singular positive measure. Since $\log|g(z)| = \log|G(\beta(z))|$ is a real part of

$$h(z) = -\frac{1}{2\pi}\int_{-\pi}^{\pi}\frac{e^{i\theta}+\beta(z)}{e^{i\theta}-\beta(z)}d\mu(\theta)$$

and

$$h(z) = ic_1 z - \frac{1}{2\pi i}\int_{-\infty}^{\infty}\left(\frac{1}{t-z} - \frac{t}{1+t^2}\right)d\mu(2\arctan t),$$

where $c_1 = \frac{1}{2\pi}\mu(\{-\pi, \pi\}) \geq 0$. There exists a constant c with $|c| = 1$ such that $g(z) = ce^{h(z)}$. Thus, M has the form of (5.2.21). □

Theorem 5.2.14. *If $\varphi(t) > 0$ holds for almost every $t \in \mathbb{R}$ and $\log\varphi(t)/(1+t^2) \in L^1(\mathbb{R})$, c is a constant with $|c| = 1$, then the function*

$$Q(z) = c\exp\left\{\frac{1}{\pi i}\int_{-\infty}^{\infty}\left(\frac{1}{t-z} - \frac{t}{1+t^2}\right)\log\varphi(t)dt\right\} \quad (5.2.22)$$

defined in the upper half-plane $\mathbb{C}_+$ has the following properties:

(a) *$\log|Q|$ is the Poisson integral of $\log\varphi(t)$, i.e., $\log|Q(z)| = (P_y * \log\varphi)(x)$;*
(b) *$\lim_{y\to 0}|Q(x+iy)| = \varphi(x)$ holds for almost every $x \in \mathbb{R}$;*
(c) *$Q \in H^p(\mathbb{C}_+)$ if and only if $\varphi \in L^p(\mathbb{R})$. Moreover, under this condition, we have*

$$\|Q\|_{\widetilde{H}^p}^p = \int_{-\infty}^{\infty}\varphi^p(t)dt = \|\varphi\|_p^p. \quad (5.2.23)$$

Proof. By Theorem 4.6.5, (a) and (b) are obvious. If $Q \in H^p(\mathbb{C}_+)$, Fatou's lemma implies that $\|Q^*\|_p^p \leq \|Q\|_{\widetilde{H}^p}^p$. Therefore,

$\|Q\|_{\widetilde{H}^p}^p \geq \|\varphi\|_p^p$. Conversely, if $\varphi \in L^p(\mathbb{R})$, then

$$|Q(x+iy)|^p = \exp\left\{\int_{-\infty}^{\infty} P_y(x-t)\log\varphi^p(t)dt\right\}$$
$$\leq \int_{-\infty}^{\infty} P_y(x-t)\varphi^p(t)dt.$$

For $0 < p < \infty$, we have $\|Q\|_{\widetilde{H}^p}^p \leq \|\varphi\|_p^p$. If $p = \infty$, it is obvious that $\|Q\|_{\widetilde{H}^p} \leq \|\varphi\|_p$. □

Theorem 5.2.15. *If $0 < p \leq \infty$, $f \in H^p(\mathbb{C}_+)$ and $f \not\equiv 0$. Then $(1+t^2)^{-1}\log|f^*(t)| \in L^1(\mathbb{R})$, the outer function of f is*

$$Q_f(z) = \exp\left\{\frac{1}{\pi i}\int_{-\infty}^{\infty}\left(\frac{1}{t-z} - \frac{t}{1+t^2}\right)\log|f^*(e^{it})|dt\right\} \in H^p(\mathbb{C}_+), \tag{5.2.24}$$

and there exists an inner function M_f such that

$$f = M_f Q_f. \tag{5.2.25}$$

Furthermore, we have

$$\log|f(i)| \leq \frac{1}{\pi}\int_{-\infty}^{\infty}\frac{\log|f^*(t)|dt}{1+t^2}, \tag{5.2.26}$$

where the equality in (5.2.26) holds if and only if M_f is a constant.

Proof. The function $z = \alpha(w)$ defined by (5.2.3) maps the unit disk U conformally onto the upper half-plane $\mathbb{C}_+$, and its inverse mapping is given by the function $w = \beta(z)$ defined by (5.2.4). Since $|f|^p$ is subharmonic and satisfies (4.6.17), by Theorem 4.6.13, there exists a positive harmonic function u in $\mathbb{C}_+$ such that $|f(z)|^p \leq u(z)$ for $z \in \mathbb{C}_+$. The function $|f(\alpha(w))|^p$ has a harmonic majorant function $u(\alpha(w))$ in the unit disk U, and hence, by Theorem 4.3.23, $F(w) = f(\alpha(w)) \in H^p(U)$. By Theorem 5.1.18, $\log|F^*(e^{it})| \in L^1([-\pi,\pi])$, and $F(w)$ can be written as the product of two functions $M(w)$ and $Q(w)$: $F(w) = M(w)Q(w)$, where $M \in H^\infty(U)$, $M^*(e^{i\theta})$ is the non-tangential limit of M with $|M^*(e^{i\theta})| = 1$ almost everywhere on $\theta \in [-\pi, \pi]$ and

$$Q(w) = \exp\left\{\frac{1}{2\pi}\int_{-\pi}^{\pi}\frac{e^{it}+w}{e^{it}-w}\log|F^*(e^{it})|dt\right\} \in H^p(U), \quad (5.2.27)$$

with

$$\log|F(0)| \leqslant \frac{1}{2\pi}\int_{-\pi}^{\pi}\log|F^*(e^{it})|dt, \quad (5.2.28)$$

where the equality in (5.2.28) holds if and only if M is a constant. Hence, $Q_f(z) = Q(\beta(z))$ has the expression given by (5.2.24). By Theorem 5.2.14, $Q_f \in H^p(\mathbb{C}_+)$, and $M_f(z) = M(\beta(z))$ is an inner function in $\mathbb{C}_+$, we have $F(0) = f(i)$, and

$$\frac{1}{2\pi}\int_{-\pi}^{\pi}\log|F^*(e^{it})|dt = \frac{1}{\pi}\int_{-\infty}^{\infty}\frac{\log|f^*(t)|dt}{1+t^2}. \quad (5.2.29)$$

Thus, f has the decomposition given by (5.2.25) and the inequality (5.2.26) holds, where the equality in (5.2.26) holds if and only if M_f is a constant. □

Theorem 5.2.16. *Suppose that* $1 < p < \infty$, $u^* \in L^p(\mathbb{R})$,

$$G(u^*)(z) = \frac{1}{\pi i}\int_{-\infty}^{\infty}\frac{u^*(t)dt}{t-z}, \quad (5.2.30)$$

then $G(u^*) \in H^p(\mathbb{C}_+)$ *and*

$$\int_{-\infty}^{\infty}|G(u^*)(t+iy)|^p dt \leqslant A_p \int_{-\infty}^{\infty}|u^*(t)|^p dt, \quad (5.2.31)$$

where $A_p = \max\{\frac{p2^p}{p-1}, 2^p p^{p-1}\}$.

Proof. First, we assume that $1 < p \leqslant 2$, $u^* \geqslant 0$ is a real function with compact support, $\{t : u^*(t) \neq 0\} \subset [-a, a]$. Let $f = G(u^*) = u+iv \in H(\mathbb{C}_+)$. By the minimum modulus principle, we have $u > 0$, and thus, u^p and $|f|^p \in C^2(\mathbb{C}_+)$ are twice continuously differentiable on $\mathbb{C}_+$. Since $2\frac{\partial u}{\partial \bar{z}} = f'$, we can compute that

$$\Delta u^p = p(p-1)|f'|^2 u^{p-2} \quad \text{and} \quad \Delta |f|^p = p^2|f'|^2|f|^{p-2}.$$

Let $h(z) = \frac{p}{p-1}(u(z))^p - |f(z)|^p$, then $h \in C^2(\mathbb{C}_+)$, and we have

$$\Delta h(z) = p^2|f'|^2(u^{p-2} - |f|^{p-2}) \geqslant 0. \qquad (5.2.32)$$

Thus, $h(z)$ is subharmonic in $\mathbb{C}_+$ and by Minkowski's inequality, we obtain

$$\int_{-\infty}^{\infty} |h(x+iy)|dx \leqslant \frac{2p}{p-1}\|f_y\|_p^p$$

$$\leqslant \frac{2p(2a)^{\frac{p}{q}}}{\pi^p(p-1)y^{p-1}} \frac{2p}{p-1}\|u^*\|_p^p \int_{-\infty}^{\infty} \frac{dx}{|x+i|^p}.$$

By Theorem 4.6.13, the function

$$\int_{-\infty}^{\infty} h(x+iy)dx$$

is non-negative and decreasing on $(0, \infty)$. Therefore,

$$\frac{p}{p-1} \int_{-\infty}^{\infty} (u(x+iy))^p dx \geqslant \int_{-\infty}^{\infty} |f(x+iy)|^p dx = \|f_y\|_p^p.$$

Hence,

$$\frac{p}{p-1}\|u^*\|_p^p \geqslant \|G(u^*)\|_{\widetilde{H}^p}^p.$$

If $1 < p \leqslant 2$, u^* has compact support with $\{t : u^*(t) \neq 0\} \subset [-a, a]$. Let u^* be decomposed into its positive part u^{*+} and its negative part u^{*-}. Then we have $u(z) = P_y * u^{*+}(x) - P_y * u^{*-}(x)$, where $P_y * u^{*+}$ and $P_y * u^{*-}$ are both positive harmonic functions in the upper half-plane $\mathbb{C}_+$, where $G(u^*) = G(u^{*+}) - G(u^{*-})$. Therefore,

$$\|G(u^*)\|_{\widetilde{H}^p} \leqslant \|G(u^{*+})\|_{\widetilde{H}^p} + \|G(u^{*-})\|_{\widetilde{H}^p}$$

$$\leqslant \left(\frac{p}{p-1}\right)^{\frac{1}{p}} (\|u^{*+}\|_p + \|u^{*-}\|_p) \leqslant 2\left(\frac{p}{p-1}\right)^{\frac{1}{p}} \|u^*\|_p.$$

If $1 < p \leq 2$, and $u^* \in L^p(\mathbb{R})$, define $u_n^*(t)$ as follows: for $|t| \leq n$, $u_n^*(t) = u^*(t)$; for $|t| > n$, let $u_n^*(t) = 0$. Then $\|u_n^* - u^*\|_p \to 0$ as $n \to \infty$ and

$$|G(u_n^*)(x+iy) - G(u^*)(x+iy)| \to 0 \ (n \to \infty).$$

Therefore, by Fatou's lemma, we have

$$\int_{-\infty}^{\infty} |G(u^*)(x+iy)|^p dx \leq \liminf_{n \to \infty} \int_{-\infty}^{\infty} |G(u_n^*)(x+iy)|^p dx$$

$$\leq \liminf_{n \to \infty} \frac{p 2^p}{p-1} \|u_n^*\|_p^p = \frac{p 2^p}{p-1} \|u^*\|_p^p.$$

Hence, Theorem 5.2.16 holds for $1 < p \leq 2$.

When $2 \leq p < \infty$, let q satisfy $\frac{1}{q} + \frac{1}{p} = 1$, since $u^* \in L^p(\mathbb{R})$, we have $G(u^*)(z) \in H(\mathbb{C}_+)$. For any $y > 0$, let

$$L_y(g) = \int_{-\infty}^{\infty} g(x) \overline{G(u^*)(x+iy)} dx \quad (g \in L^q(\mathbb{R})),$$

then

$$L_y(g) = \int_{-\infty}^{\infty} \overline{u^*(t)} G(g)(t+iy) dt \quad (g \in L^q(\mathbb{R})).$$

Thus,

$$|L_y(g)| \leq \|G(g)\|_q \|u^*\|_p \leq 2 \left(\frac{q}{q-1}\right)^{\frac{1}{q}} \|u^*\|_p \|g\|_q \quad (g \in L^q(\mathbb{R})),$$

and therefore,

$$\|L_y\|^p = \int_{-\infty}^{\infty} |G(u^*)(x+iy)|^p dx$$

$$\leq 2^p \left(\frac{q}{q-1}\right)^{\frac{p}{q}} \|u^*\|_p^p = 2^p p^{p-1} \|u^*\|_p^p.$$

Hence, (5.2.31) holds. □

Theorem 5.2.17. *If $0 < p < \infty$, then $H^p(\mathbb{C}_+)$ is complete and separable.*

Proof. Suppose $\{f_n\}$ is a Cauchy sequence in $H^p(\mathbb{C}_+)$, i.e., $\|f_n - f_m\|_{\widetilde{H}^p} \to 0$ $(n, m \to \infty)$. If $y > y_1 > 0$, then $|f_n(z) - f_m(z)|^p$ is subharmonic and continuous in the upper half-plane $\mathbb{C}_+$, then by Theorem 4.6.12,

$$|f_n(z) - f_m(z)|^p \leqslant \frac{2\|f_n - f_m\|_{\widetilde{H}^p}^p}{\pi y_1} \to 0 \quad (n, m \to \infty).$$

As a result, $\{f_n\}$ uniformly converges on each compact of $\mathbb{C}_+$ to some $f \in H(\mathbb{C}_+)$. For any $\varepsilon > 0$, there exists m such that for $n > m$, $\|f_n - f_m\|_{\widetilde{H}^p}^p < \varepsilon$. Therefore, by Fatou's lemma, we have

$$\int_{-\infty}^{\infty} |f(t+iy) - f_m(t+iy)|^p dt \leqslant \liminf_{n \to \infty}$$

$$\int_{-\infty}^{\infty} |f_n(t+iy) - f_m(t+iy)|^p dt \leqslant \varepsilon,$$

which implies that $\|f_n - f\|_{\widetilde{H}^p} \to 0$ $(n \to \infty)$. This proves completeness.

We will first prove that $H^p(\mathbb{C}_+)$ is separable when $1 < p < \infty$. In fact, since $L^p(\mathbb{R})$ is separable, there exists a countable set $B(p) \subset L^p(\mathbb{R}) \cap L^\infty(\mathbb{R})$ such that $B(p)$ is dense in $L^p(\mathbb{R})$. Let $T(g) = \frac{G(g)}{2}$ for $g \in L^p(\mathbb{R})$, where $G(g)$ is defined by (5.2.30). Then, the set $T(B(p)) = \{T(g) : g \in B(p)\} \subset H^p(\mathbb{C}_+)$ is countable. For any $f \in H^p(\mathbb{C}_+)$ and for any $\varepsilon > 0$, there exists $g \in B(p)$ such that $\|f^* - g\|_p < \frac{\varepsilon}{2A_p}$, where $A_p = \max\{\frac{2p}{p-1}, 2^p p^{p-1}\}$. Then, by (5.2.12) and (5.2.31), we have $\|f - T(g)\|_p = \|T(f^* - g)\|_p < \frac{\varepsilon}{2}$. This implies that the set $T(B(p))$ is dense in $H^p(\mathbb{C}_+)$. Therefore, when $1 < p < \infty$, $H^p(\mathbb{C}_+)$ is separable.

When $\frac{1}{2} < p \leqslant 1$, there exists a countable set $B(2p) \subset L^{2p}(\mathbb{R}) \cap L^\infty(\mathbb{R})$ such that $B(2p)$ is dense in $L^{2p}(\mathbb{R})$. Let $T(g) = \frac{G(g)}{2}$, $g \in L^{2p}(\mathbb{R})$, where $G(g)$ is defined by (5.2.30), then the set $T(B(2p)) = \{T(g) : g \in B(2p)\} \subset H^{2p}(\mathbb{C}_+)$ is countable, dense in $H^{2p}(\mathbb{C}_+)$. For any $f \in H^p(\mathbb{C}_+)$, according to Theorem 5.2.7, we have $\|f_\delta - f\|_{\widetilde{H}^p} \to 0$ as $\delta \to 0$, where $f_\delta(z) = f(\delta + z)$. The function $|f(z)|^p$ is subharmonic and continuous in the upper half-plane $\mathbb{C}_+$. Therefore, by Theorem 4.6.12, we have $|f_\delta(z)|^p \leqslant \frac{2\|f\|_{\widetilde{H}^p}^p}{\pi \delta}$. For any $\varepsilon > 0$, there exists $\delta > 0$ such that $\|f_\delta - f\|_{\widetilde{H}^p} < \frac{\varepsilon}{4}$. Let $f_\delta(z) = G(z)B(z)$, where

$B(z)$ is the Blaschke product of the zeros of $f_\delta(z)$, $G(z) \neq 0$ with $G \in H^p(\mathbb{C}_+) \cap H^\infty(\mathbb{C}_+)$, and $\|G\|_{\widetilde{H}^p} = \|f\|_{\widetilde{H}^p}$. Since $G(z)$ has no zeros in the upper half-plane $\mathbb{C}_+$, there exists $h(z) \in H(\mathbb{C}_+)$ such that $h^2(z) = G(z)$, thus $h \in H^{2p}(\mathbb{C}_+) \cap H^\infty(\mathbb{C}_+)$ and $\|G\|_{\widetilde{H}^p}^p = \|h\|_{\widetilde{H}^{2p}}^{2p}$. Therefore, h has an almost everywhere non-tangential limit $h^*(t)$, and G also has an almost everywhere non-tangential limit $G^*(t) = (h^*(t))^2 \in L^p(\mathbb{R})$. Let $h_1(z) = B(z)h(z)$ and $h_2(z) = h(z)$, then $h_j \in H^{2p}(\mathbb{C}_+) \cap H^\infty(\mathbb{C}_+)$ and h_j has an almost everywhere non-tangential limit $h_j^*(t)$, there exists $g_j \in B(2p)$ such that $\|h_j^* - g_j\|_{2p} < \frac{\varepsilon}{8A_{2p}}$ for $j = 1, 2$. Using (5.2.12) and (5.2.31), we have $\|h_j - T(g_j)\|_{\widetilde{H}^{2p}} = \|T(h_j^* - g_j)\|_{\widetilde{H}^{2p}} < \frac{\varepsilon}{8}$, therefore

$$\|f_\delta - T(g_1)T(g_2)\|_{\widetilde{H}^p}^p \leq \|(h_1 - T(g_1))h_2\|_{\widetilde{H}^p}^p$$
$$+ \|(h_2 - T(g_2))T(g_1)\|_{\widetilde{H}^p}^p$$
$$\leq \|h_1 - T(g_1)\|_{\widetilde{H}^{2p}}^p \|h_2\|_{\widetilde{H}^{2p}}^p + \|h_2 - T(g_2)\|_{\widetilde{H}^{2p}}^p \|T(g_1)\|_{\widetilde{H}^{2p}}^p$$
$$\leq \varepsilon^p (1 + (A_{2p})^p) \|G\|_{\widetilde{H}^p}^{\frac{p}{2}}.$$

The set $\{T(g_1)T(g_2) : g_1, g_2 \in B(2p)\} \subset H^p(\mathbb{C}_+)$ is countable and dense in $H^p(\mathbb{C}_+)$. Therefore, when $\frac{1}{2} < p < \infty$, $H^p(\mathbb{C}_+)$ is separable. By induction, when $\frac{1}{2^n} < p < \infty$ for $n = 1, 2, 3, \ldots$, $H^p(\mathbb{C}_+)$ is separable. This proves Theorem 5.2.17. $\square$

5.3 Holomorphic Fourier Transform

Definition 5.3.1. Let $f \in L^1(\mathbb{R})$. For $x \in \mathbb{R}$, the expression

$$\hat{f}(x) = \int_\mathbb{R} f(t) e^{-ixt} dt$$

is called the Fourier transform of f.

Theorem 5.3.2. *Fourier transform has the following fundamental properties:*

(a) *If $f \in L^1(\mathbb{R})$, then $\hat{f}$ is uniformly continuous on $\mathbb{R}$ and $|\hat{f}(x)| \leq \|f\|_1$;*
(b) **Riemann–Lebesgue Lemma:** *if $f \in L^1(\mathbb{R})$, then $\hat{f} \in C_0(\mathbb{R})$.*

Proof. (a) For any $x \in \mathbb{R}$ and $h \in \mathbb{R}$, applying the Lebesgue-dominated convergence theorem, we have

$$|\hat{f}(x+h) - \hat{f}(x)| \leqslant \int_{\mathbb{R}} |f(t)||e^{-iht} - 1|dt \to 0 \quad (|h| \to 0).$$

Therefore, $|\hat{f}(x)| \leqslant \|f\|_1$ is obvious. Now, let's consider (b). From result (a), it is sufficient to prove that $\lim_{|x| \to \infty} \hat{f}(x) = 0$. Let $I = [a,b]$ be an interval in $\mathbb{R}$, then

$$\widehat{\chi_I}(x) = \int_a^b e^{-ixt} dt = \frac{i}{x}(e^{-ibx} - e^{-iax}) \to 0, \quad |x| \to \infty.$$

This implies that the conclusion (b) holds for indicator functions of intervals in $\mathbb{R}$. Now, consider a general function $f \in L^1(\mathbb{R})$. Since the set of all indicator functions is dense in $L^1(\mathbb{R})$, for any $\varepsilon > 0$, there exists a simple function g such that $\|f - g\|_1 < \varepsilon/2$, and for sufficiently large $|x|$, $|\hat{g}(x)| < \varepsilon/2$. Therefore, and for sufficiently large $|x|$, using result (a), we have

$$|\hat{f}(x)| \leqslant |\hat{f}(x) - \hat{g}(x)| + |\hat{g}(x)|$$
$$= |\widehat{(f-g)}(x)| + |\hat{g}(x)| \leqslant \|f - g\|_1 + |\hat{g}(x)| < \varepsilon. \qquad \square$$

Theorem 5.3.3 (Multiplication Formula). *If $f, g \in L^1(\mathbb{R})$, then $\widehat{(f*g)}(x) = \hat{f}(x) \cdot \hat{g}(x)$ and*

$$\int_{\mathbb{R}} f(t)\hat{g}(t)dt = \int_{\mathbb{R}} \hat{f}(x)g(x)dx. \qquad (5.3.1)$$

Theorem 5.3.4. *If $y > 0$, then*

$$\frac{1}{2\pi}\int_{\mathbb{R}} e^{-y|t|}e^{-ixt}dt = \frac{y}{\pi(y^2 + x^2)} = P_y(x), \qquad (5.3.2)$$

$$\int_{\mathbb{R}} P_y(x)e^{-ixt}dx = e^{-y|t|}. \qquad (5.3.3)$$

Proof.

$$\int_{\mathbb{R}} e^{-y|t|}e^{-ixt}dt = \frac{2y}{y^2 + x^2} = 2\pi P_y(x).$$

Let
$$I(t,y) = \int_{\mathbb{R}} P_y(x) e^{-ixt} dx,$$
then $I(-t,y) = I(t,y)$. When $t > 0$ and $y > 0$, we have $I(t,y) = I(yt, 1)$, and
$$I(|t|,1) = \int_{-\infty}^{\infty} \frac{e^{ix|t|} dx}{1+x^2} = 2\pi i \operatorname{Res}\left(\frac{e^{iz|t|}}{1+z^2}, i\right) = \pi e^{-|t|},$$
this implies (5.3.3). $\square$

Theorem 5.3.5. *If $f \in L^1(\mathbb{R})$, then for any $y > 0$,*
$$\frac{1}{2\pi} \int_{\mathbb{R}} \hat{f}(t) e^{ixt} e^{-y|t|} dt = \int_{\mathbb{R}} f(s) P_y(x-s) ds. \tag{5.3.4}$$

Proof. Applying Fubini's theorem and (5.3.2), we obtain
$$\frac{1}{2\pi} \int_{\mathbb{R}} \hat{f}(t) e^{ixt} e^{-y|t|} dt = \frac{1}{2\pi} \int_{\mathbb{R}} \int_{\mathbb{R}} f(s) e^{-ist} e^{-y|t|} e^{ixt} dt ds$$
$$= \int_{\mathbb{R}} f(s) P_y(x-s) ds.$$
$\square$

Next, we discuss under what conditions the Fourier integral $\hat{f}$ of a function f in $L^1(\mathbb{R})$ shows a way of returning from the transforms to the function which is almost everywhere equal to f, which is known as the inversion problem of the Fourier transform.

Theorem 5.3.6. *If f and $\hat{f} \in L^1(\mathbb{R})$, then at a Lebesgue point x of f, we have*
$$f(x) = \frac{1}{2\pi} \int_{\mathbb{R}} \hat{f}(t) e^{ixt} dt, \tag{5.3.5}$$

In particular, (5.3.5) holds for almost every $x \in \mathbb{R}$.

Proof. From Theorem 4.6.5 (d), we know that at a Lebesgue point x of f, we have $\lim_{y\to 0}(f * P_y)(x) = f(x)$. On the other hand, from (5.3.4), we have
$$\lim_{y\to 0}(f * P_y)(x) = \lim_{y\to 0} \frac{1}{2\pi} \int_{\mathbb{R}} \hat{f}(t) e^{ixt} e^{-y|t|} dt.$$

Since $\hat{f} \in L^1(\mathbb{R})$, applying the Lebesgue-dominated convergence theorem, we obtain

$$f(x) = \lim_{y\to 0}(f*P_y)(x) = \frac{1}{2\pi}\int_{\mathbb{R}} \hat{f}(t)e^{ixt}\lim_{y\to 0}e^{-y|t|}dt\mathbb{N}$$

$$= \frac{1}{2\pi}\int_{\mathbb{R}} \hat{f}(t)e^{ixt}dt.$$
□

Theorem 5.3.7. *Suppose $f \in L^1(\mathbb{R})$. If $\hat{f}(x) \geq 0$ and f is continuous at $x = 0$, then $\hat{f} \in L^1(\mathbb{R})$. Hence, for almost every $x \in \mathbb{R}$, we have*

$$f(x) = \frac{1}{2\pi}\int_{\mathbb{R}} \hat{f}(t)e^{ixt}dt. \tag{5.3.6}$$

In particular, $f(0) = \frac{1}{2\pi}\int_{\mathbb{R}} \hat{f}(x)dx$.

Proof. First, using (5.3.1) and (5.3.2), and noting that f is continuous at $x = 0$, we have

$$\lim_{y\to 0}\frac{1}{2\pi}\int_{\mathbb{R}} \hat{f}(t)e^{-y|t|}dt = \lim_{y\to 0}\int_{\mathbb{R}} f(t)P_y(t)dt = f(0).$$

Since $\hat{f}(x) \geq 0$, by Fatou's lemma, we have

$$\frac{1}{2\pi}\int_{\mathbb{R}} \hat{f}(t)dt = \frac{1}{2\pi}\int_{\mathbb{R}}\lim_{y\to 0}\hat{f}(t)e^{-y|t|}dt$$

$$\leq \liminf_{y\to 0}\frac{1}{2\pi}\int_{\mathbb{R}} \hat{f}(t)e^{-y|t|}dt = f(0).$$

Thus, $\hat{f} \in L^1(\mathbb{R})$. From Theorem 5.3.6, we know that (5.3.6) holds. □

Theorem 5.3.8 (Uniqueness of the Fourier Transform). *If $f_1, f_2 \in L^1(\mathbb{R})$, and for every $x \in \mathbb{R}, \hat{f}_1(x) = \hat{f}_2(x)$, then for almost every $x \in \mathbb{R}$, $f_1(x) = f_2(x)$.*

Proof. Let $f = f_1 - f_2$. Then $\hat{f}(x) = \hat{f}_1(x) - \hat{f}_2(x) \equiv 0$. Therefore, by Theorem 5.3.6, for almost every $x \in \mathbb{R}$,

$$f(x) = \frac{1}{2\pi}\int_{\mathbb{R}} \hat{f}(t)e^{ixt}dt = 0.$$

Then, for almost every $x \in \mathbb{R}$, $f_1(x) = f_2(x)$. □

We present a very important property of the Fourier transform.

H^p Spaces

Theorem 5.3.9. *There does not exist a function $f \in L^1(\mathbb{R})$ such that $\|f\|_1 > 0$ and both $\mathrm{supp}\, f$ and $\mathrm{supp}\, \hat{f}$ are compact sets.*

Proof. Let $f \in L^1(\mathbb{R})$ satisfy $\|f\|_1 > 0$ if $\mathrm{supp} f$ is a compact set in $\mathbb{C}$. Then, for any point $z = x + iy$ in the complex plane $\mathbb{C}$, define

$$F(z) = F(x+iy) = \int_{-\infty}^{\infty} e^{-i(x+iy)t} f(t) dt.$$

Since f) has compact support, it follows that F is an entire function in $\mathbb{C}$ and $F(x) = \hat{f}(x)$. On the other hand, if $\mathrm{supp}\,\hat{f}$ is compact, then $F(z) \equiv 0$, which implies that $\hat{f}(x) \equiv 0$. Therefore, by the uniqueness property of the Fourier transform for L^1 function, we conclude that for almost all $t \in \mathbb{R}$, $f(t) = 0$. However, this contradicts the assumption that $\|f\|_1 > 0$. □

Theorem 5.3.10. *If $f \in L^1(\mathbb{R}) \cap L^2(\mathbb{R})$, then $\hat{f} \in L^2(\mathbb{R})$, and $\|\hat{f}\|_2 = 2\pi \|f\|_2$.*

Proof. For $f \in L^1(\mathbb{R}) \cap L^2(\mathbb{R})$, let $g(x) = \overline{f(-x)}$, then $g \in L^1(\mathbb{R}) \cap L^2(\mathbb{R})$, and $\hat{g} = \overline{\hat{f}}$. On the other hand, if we define $h = f * g$, then by Fubini's theorem, $h \in L^1(\mathbb{R})$, and by Theorem 5.3.3,

$$\hat{h}(x) = \hat{f}(x) \cdot \hat{g}(x) = \hat{f}(x) \cdot \overline{\hat{f}(x)} = |\hat{f}(x)|^2 \geq 0. \quad (5.3.7)$$

Now, we will show that h is uniformly continuous on $\mathbb{R}$. In fact, for any $x \in \mathbb{R}$ and $\delta > 0$,

$$|h(x+\delta) - h(x)| = \left| \int_{\mathbb{R}} f(x+\delta-t) g(t) dt - \int_{\mathbb{R}} f(x-t) g(t) dt \right|$$

$$\leq \int_{\mathbb{R}} |f(x+\delta-t) - f(x-t)| |g(t)| dt$$

$$\leq \omega_2(f)(\delta) \cdot \|g\|_2 \to 0 (\delta \to 0),$$

where $\omega_2(f)(\delta) = \sup_{|u| < \delta} \left(\int_{\mathbb{R}} |f(x+u) - f(x)|^2 dx \right)^{1/2}$ is the modulus of integral continuity for f. From (5.3.7) and Theorem 5.3.6, we can conclude that

$$\|\hat{f}\|_2^2 = \int_{\mathbb{R}} |\hat{f}(x)|^2 dx = \int_{\mathbb{R}} \hat{h}(x)dx = 2\pi h(0)$$

$$= 2\pi \int_{\mathbb{R}} f(t)g(-t)dt = 2\pi \int_{\mathbb{R}} |f(t)|^2 dt = 2\pi \|f\|_2^2.$$

$\square$

Now, we can define the Fourier transform in $L^2(\mathbb{R})$. Note $L^1(\mathbb{R}) \cap L^2(\mathbb{R})$ is dense in $L^2(\mathbb{R})$. For any $g \in L^2(\mathbb{R})$, we can take a sequence $\{g_k\}$ in $L^1(\mathbb{R}) \cap L^2(\mathbb{R})$ such that $\{g_k\}$ converges to g in L^2 norm. By Theorem 5.3.10, we have

$$\|\hat{g}_k - \hat{g}_j\|_2 = \|(g_k - g_j)\hat{\;}\|_2 = 2\pi \|g_k - g_j\|_2.$$

Thus, $\{\hat{g}_k\}$ forms a Cauchy sequence in $L^2(\mathbb{R})$, and there exists a limit $\tilde{g}$ in $L^2(\mathbb{R})$ such that $\|\hat{g}_k - \tilde{g}\|_2 \to 0$ as $k \to \infty$.

Now, we show that $g \in L^2(\mathbb{R})$, $\tilde{g}$ is unique almost everywhere. Assume there exist sequences $\{g_k\}$ and $\{h_k\}$ in $L^1(\mathbb{R}) \cap L^2(\mathbb{R})$ such that $\|g_k - g\|_2 \to 0$ and $\|h_k - g\|_2 \to 0$ as $k \to \infty$, let $\{\hat{g}_k\}$ and $\{\hat{h}_k\}$ in $L^2(\mathbb{R})$ be the limits of $\tilde{g}$ and $\tilde{h}$, respectively. Then,

$$\|\hat{g}_k - \hat{h}_k\|_2 = \|(g_k - h_k)\hat{\;}\|_2 = \|g_k - h_k\|_2$$
$$\leqslant \|g_k - g\|_2 + \|h_k - g\|_2 \to 0 \quad (k \to \infty).$$

Therefore, we have

$$\|\tilde{g} - \tilde{h}\|_2 \leqslant \|\tilde{g} - \hat{g}_k\|_2 + \|\tilde{h} - \hat{h}_k\|_2 + \|\hat{g}_k - \hat{h}_k\|_2 \to 0 \quad \text{as} \quad k \to \infty.$$

This implies that for any $g \in L^2(\mathbb{R})$, we can define $\hat{g}(x) = \tilde{g}(x)$. Using Theorem 5.3.10, we obtain

$$\|\hat{g}\|_2 = \lim_{k \to \infty} \|\hat{g}_k\|_2 = \lim_{k \to \infty} 2\pi \|g_k\|_2 = 2\pi \|g\|_2. \qquad (5.3.8)$$

Thus, we have defined the Fourier transform $\tilde{g}$ of a function g in $L^2(\mathbb{R})$, denoted by $\mathscr{F}(g) = \tilde{g}$. We also use the notation $\hat{g} = \mathscr{F}(g)$. The following theorem holds.

Theorem 5.3.11 (Parseval's Equality for Fourier Transform). If $f \in L^2(\mathbb{R})$, then $\hat{f} \in L^2(\mathbb{R})$, and $\|\hat{f}\|_2^2 = 2\pi \|f\|_2^2$.

From the previous discussion, we know that the Fourier transform $\hat{f} = \mathscr{F}(f)$ of f in $L^2(\mathbb{R})$ is independent of the choice of the fundamental sequence. Therefore, for $f \in L^2(\mathbb{R})$, for any positive integer k, let

$$f_k(x) = \begin{cases} f(x), & |x| \leqslant k, \\ 0, & |x| > k, \end{cases}$$

then $\|\hat{f}_k - \hat{f}\|_2 = 2\pi\|f_k - f\|_2 \to 0$ as $k \to \infty$.

Theorem 5.3.12. For any $f \in L^2(\mathbb{R})$ and $g \in L^2(\mathbb{R})$,

(a) **Multiplication Formula:**

$$\int_{\mathbb{R}} \hat{f}(x)g(x)dx = \int_{\mathbb{R}} f(x)\hat{g}(x)dx; \qquad (5.3.9)$$

(b) **Parseval's Equality for Fourier Transform:**

$$2\pi \int_{\mathbb{R}} f(x)\overline{g(x)}dx = \int_{\mathbb{R}} \hat{f}(x)\overline{\hat{g}(x)}dx. \qquad (5.3.10)$$

Proof. Theorem 5.3.3 implies that (5.3.9) holds for $f, g \in L^1(\mathbb{R}) \cap L^2(\mathbb{R})$. Now, let's consider $f \in L^2(\mathbb{R})$ and $g \in L^1(\mathbb{R}) \cap L^2(\mathbb{R})$, then there exist a sequence $\{f_k\} \subset L^1(\mathbb{R}) \cap L^2(\mathbb{R})$ such that $\lim_{k\to\infty} \|f_k - f\|_2 = 0$. Since $\hat{g} \in L^2(\mathbb{R})$, we have

$$\int_{\mathbb{R}} f(x)\hat{g}(x)dx = \lim_{k\to\infty} \int_{\mathbb{R}} f_k(x)\hat{g}(x)dx.$$

On the other hand, using $\lim_{k\to\infty} \|\hat{f}_k - \hat{f}\|_2 = 0$ and the multiplication formula (5.3.1), we get

$$\lim_{k\to\infty} \int_{\mathbb{R}} f_k(x)\hat{g}(x)dx = \lim_{k\to\infty} \int_{\mathbb{R}} \hat{f}_k(x)g(x)dx = \int_{\mathbb{R}} \hat{f}(x)g(x)dx.$$

Combining these results, we have

$$\int_{\mathbb{R}} f(x)\hat{g}(x)dx = \int_{\mathbb{R}} \hat{f}(x)g(x)dx,$$

which shows that (5.3.9) holds for $g \in L^1(\mathbb{R}) \cap L^2(\mathbb{R})$ and $f \in L^2(\mathbb{R})$.

Now, suppose $f, g \in L^2(\mathbb{R})$ and take $\{g_k\} \subset L^1(\mathbb{R}) \cap L^2(\mathbb{R})$ such that $\lim_{k \to \infty} \|g_k - g\|_2 = 0$. Using a similar proof strategy, we can show that the multiplication formula still holds for $f, g \in L^2(\mathbb{R})$,

$$\int_{\mathbb{R}} \hat{f}(x)g(x)dx = \lim_{k\to\infty} \int_{\mathbb{R}} \hat{f}(x)g_k(x)dx = \lim_{k\to\infty} \int_{\mathbb{R}} f(x)\hat{g}_k(x)dx$$
$$= \int_{\mathbb{R}} f(x)\hat{g}(x)dx,$$

For $f, g \in L^2(\mathbb{R})$, by the identity,

$$4f\bar{g} = |f + g|^2 - |f - g|^2 + i|f + ig|^2 - i|f - ig|^2,$$

integrating from $-\infty$ to ∞, we obtain the polarization identity,

$$8\pi \int_{\mathbb{R}} f(x)\overline{g(x)}dx = 2\pi(\|f + g\|_2^2 - \|f - g\|_2^2 + i\|f$$
$$+ ig\|_2^2 - i\|f - ig\|_2^2)$$
$$= \|\hat{f} + \hat{g}\|_2^2 - \|\hat{f} - \hat{g}\|_2^2 + i\|\hat{f} + i\hat{g}\|_2^2 - i\|\hat{f}N$$
$$- i\hat{g}\|_2^2 = \int_{\mathbb{R}} \hat{f}(x)\overline{\hat{g}(x)}dx.$$

This proves that (5.3.10) holds. □

Theorem 5.3.13. *If $f \in L^2(\mathbb{R})$, then for almost every $x \in \mathbb{R}$, $\mathscr{F}^2 f(x) = \mathscr{F}(\mathscr{F}f)(x) = 2\pi f(-x)$.*

Proof. For $f \in L^2(\mathbb{R})$ and $g \in L^1(\mathbb{R}) \cap L^2(\mathbb{R})$, let $g_1(x) = g(-x)$, then $(\mathscr{F}g_1)(-t) = \overline{(\mathscr{F}g)(t)}$. By (5.3.9) and (5.3.10), we have

$$\int_{\mathbb{R}} \mathscr{F}^2 f(x)\overline{g(-x)}dx = \int_{\mathbb{R}} \mathscr{F}f(t)(\mathscr{F}g_1)(t)dt$$
$$= \int_{\mathbb{R}} (\mathscr{F}f)(t)\overline{(\mathscr{F}g)(t)}dt = 2\pi \int_{\mathbb{R}} f(-x)\overline{g(-x)}dx.$$

Now, for $g \in L^2(\mathbb{R})$, let $\{g_k\} \subset L^1(\mathbb{R}) \cap L^2(\mathbb{R})$ be a sequence such that $\lim_{k \to \infty} \|f_k - f\|_2 = 0$. Then, we have

$$\int_\mathbb{R} \mathscr{F}^2 f(x)\overline{g(-x)}dx = \lim_{k \to \infty} \int_\mathbb{R} \mathscr{F}^2 f(x)\overline{g_k(-x)}dx$$

$$= 2\pi \lim_{k \to \infty} \int_\mathbb{R} f(-x)\overline{g_k(-x)}dx$$

$$= 2\pi \int_\mathbb{R} f(-x)\overline{g(-x)}dx.$$

This implies that for any $f \in L^2(\mathbb{R})$, $\mathscr{F}^2 f(x) = 2\pi f(-x)$ for almost every $x \in \mathbb{R}$. Therefore, Theorem 5.3.13 holds. $\square$

Theorem 5.3.14 (The Uniqueness of the Fourier Transform of an L^2 function). *Let $f \in L^2(\mathbb{R})$. If $\hat{f}(x) \equiv 0$, then $f(x) = 0$ for almost every $x \in \mathbb{R}$.*

Theorem 5.3.15 (Plancherel Theorem and the Inversion Formula). *The Fourier transform $\mathscr{F}$ is a one-to-one, bounded linear operator from $L^2(\mathbb{R})$ onto $L^2(\mathbb{R})$. It satisfies the property that for any $f \in L^2(\mathbb{R})$, $\|\mathscr{F}f\|_2^2 = 2\pi\|f\|_2^2$. The Fourier inverse transform $\mathscr{F}^{-1}$ satisfies the property that for any $g \in L^2(\mathbb{R})$, $2\pi\mathscr{F}^{-1}(g)(x) = \mathscr{F}(g)(-x)$. Additionally, $\frac{1}{(2\pi)^2}\mathscr{F}^4$ is the identity operator on $L^2(\mathbb{R})$, i.e., for any $f \in L^2(\mathbb{R})$, $\mathscr{F}^4 f = (2\pi)^2 f$.*

Proof. According to Theorems 5.3.11 and 5.3.14, the Fourier transform $\mathscr{F}$ is a one-to-one, bounded linear operator from $L^2(\mathbb{R})$ onto $L^2(\mathbb{R})$. For any $f \in L^2(\mathbb{R})$, we have $\|\mathscr{F}f\|_2^2 = 2\pi\|f\|_2^2$. According to Theorem 5.3.14, the Fourier inverse transform $\mathscr{F}^{-1}$ on $L^2(\mathbb{R})$ satisfies the property that for any $g \in L^2(\mathbb{R})$, $2\pi\mathscr{F}^{-1}(g)(x) = \mathscr{F}(g)(-x)$. Moreover, for any $f \in L^2(\mathbb{R})$, we have $\mathscr{F}^4 f = (2\pi)^2 f$, which implies that the transformation $\frac{1}{(2\pi)^2}\mathscr{F}$ is the identity operator on $L^2(\mathbb{R})$. $\square$

Theorem 5.3.16. *If $f \in L^2(\mathbb{R})$, $g \in L^1(\mathbb{R})$, then $\widehat{(f * g)}(x) = \hat{f}(x) \cdot \hat{g}(x)$ for almost every $x \in \mathbb{R}$.*

Proof. Let $\{f_k\} \subset L^1(\mathbb{R}) \cap L^2(\mathbb{R})$ be such that $\lim_{k \to \infty} \|f_k - f\|_2 = 0$. It is easy to see that in the sense of L^2, $\widehat{(f_k * g)}$ converges to $\widehat{(f * g)}$.

On the other hand, for $g \in L^1(\mathbb{R})$, we have $\widehat{(f_k * g)}(x) = \widehat{f_k}(x) \cdot \hat{g}(x)$ and $\|\hat{g}\|_\infty \leqslant \|g\|_1$. Therefore,

$$\|\widehat{f_k}\hat{g} - \hat{f}\hat{g}\|_2 = \left(\int_{\mathbb{R}^n} |\widehat{f_k}(x) - \hat{f}(x)|^2 |\hat{g}(x)|^2 dx\right)^{1/2}$$

$$\leqslant \|g\|_1 \|\widehat{f_k} - \hat{f}\|_2 \to 0 \quad (k \to \infty).$$

Thus, $\widehat{(f_k * g)}(x) = \widehat{f_k}(x)\hat{g}(x)$ converges to $\hat{f}(x)\hat{g}(x)$ in the sense of L^2. By the uniqueness of the L^2 limit, we conclude that $\widehat{(f * g)}(x) = \hat{f}(x) \cdot \hat{g}(x)$ holds for almost every $x \in \mathbb{R}$. $\square$

Let $F \in L^2(-\infty, \infty)$ be such that $F(t) = 0$ for $t \in (-\infty, 0)$, i.e., $F \in L^2(0, \infty)$. Define

$$f(z) = \int_0^\infty F(t)e^{itz} dt \quad (z \in \mathbb{C}_+), \tag{5.3.11}$$

where $\mathbb{C}_+ = \{z = x + iy : y > 0\}$ is the upper half-plane. If $z \in \mathbb{C}_+$, then $|e^{itz}| = e^{-ty}$ is Lebesgue-integrable on $t \in (0, \infty)$ for $y > 0$. For any $\delta > 0$, if Im $z \geqslant \delta$, then Schwarz's inequality implies that $|F(t)e^{itz}| \leqslant |F(t)|e^{-\delta t} \in L^1(0, \infty)$. By the Lebesgue-dominated convergence theorem, f is continuous in the half-plane Im $z \geqslant \delta$. Since $\delta > 0$ is arbitrary, f is continuous in $\mathbb{C}_+$. By Fubini's theorem and Cauchy's theorem, for any piecewise smooth closed Jordan curve γ in $\mathbb{C}_+$, we have $\int_\gamma f(z)dz = 0$. Applying Morera's theorem, we conclude that $f \in H(\mathbb{C}_+)$.

The following is the Paley–Wiener theorem that we want to prove.

Theorem 5.3.17 (Paley–Wiener Theorem). *(a) The necessary and sufficient condition for $f \in H^2(\mathbb{C}_+)$ is the existence of $F \in L^2(0, \infty)$ such that (5.3.11) holds. Assuming (5.3.11) holds, then*

$$2\pi \int_0^\infty |F(t)|^2 dt = \|f\|_{\widetilde{H}^2}^2. \tag{5.3.12}$$

(b) If $0 < p < 2$, and $f \in H^p(\mathbb{C}_+)$, then there exist positive constants A_p and a measurable function F on $\mathbb{R}$ such that for any $\delta > 0$, we have

$$2\pi \int_{-\infty}^\infty |F(t)|^2 e^{-2|t|\delta} dt = \int_{-\infty}^\infty |f(t+i\delta)|^2 dt \leqslant A_p^p \|f\|_{\widetilde{H}^p}^p \delta^{1-\frac{2}{p}}, \tag{5.3.13}$$

and (5.3.11) holds. If $0 < p \leq 1$, it is also possible to choose F such that F is continuous on $\mathbb{R}$ and satisfies

$$|F(t)| \leq A_p |t|^{\frac{1}{p}-1}. \tag{5.3.14}$$

Proof. (a) If there exists $F \in L^2(0, \infty)$ such that (5.3.11) holds, fix $y > 0$. By the Plancherel theorem, we have

$$\frac{1}{2\pi} \int_{-\infty}^{\infty} |f(x+iy)|^2 dx = \int_0^{\infty} |F(t)|^2 e^{-2ty} dt \leq \int_0^{\infty} |F(t)|^2 dt.$$

Therefore, $f \in H^2(\mathbb{C}_+)$. Conversely, if $f \in H^2(\mathbb{C}_+)$, then by Theorem 5.2.3, the non-tangential limit $f^*(t)$ exists for almost every $t \in \mathbb{R}$ and $f^* \in L^2(\mathbb{R})$. By the Plancherel theorem, $2\pi \|f\|_{\tilde{H}^2}^2 = 2\pi \|f^*\|_2^2 = \|\hat{f}^*\|_2^2$. For any $y = \operatorname{Im} z > 0$, let the function $\sigma(s)$ be defined as follows: For $s \geq 0$, $\sigma(s) = \frac{e^{izs}}{2\pi}$ and for $s < 0$, $\sigma(s) = 0$, then

$$\hat{\sigma}(t) = \frac{1}{2\pi} \int_0^{\infty} e^{izs} e^{-its} ds = \frac{1}{2\pi i (t-z)}.$$

Using (5.2.12), we have

$$f(z) = \frac{1}{2\pi i} \int_{-\infty}^{\infty} \frac{f^*(t) dt}{t-z} = \int_{-\infty}^{\infty} f^*(t) \hat{\sigma}(t) dt.$$

Thus, by the multiplication formula (5.3.9), we obtain

$$f(z) = \int_{-\infty}^{\infty} f^*(t) \hat{\sigma}(t) dt = \int_0^{\infty} \hat{f}^*(s) \sigma(s) ds = \int_0^{\infty} F(t) e^{itz} dt,$$

where $F(t) = \frac{\hat{f}^*(t)}{2\pi} \in L^2([0, \infty))$.

(b) If $0 < p < 2$, and $f \in H^p(\mathbb{C}_+)$, for any $\delta > 0$, let $f_\delta(z) = f(z + i\delta)$, Then, by (4.6.9), for any $y > 0$,

$$\int_{-\infty}^{\infty} |f_\delta(x+iy)|^2 dx \leq \int_{-\infty}^{\infty} |f_\delta(x+iy)|^p |f_\delta(x+iy)|^{2-p} dx$$

$$\leq C_p^p \|f\|_{\tilde{H}^p}^p \delta^{1-\frac{2}{p}},$$

where C_p is a constant defined by Theorem 4.6.8, which only depends on p. Thus, by the proof of part (a), we have

$$f_\delta(z) = \frac{1}{2\pi} \int_0^\infty \hat{f}_\delta^*(s) e^{itz} dt,$$

where $f_\delta^*(x) = f_\delta(x)$. On the other hand, by Theorem 5.2.9, when $y > 0$, $f_\delta(x+iy) = (P_y * f_\delta^*)(x)$. Since $f_\delta^* = f_\delta \in L^2(\mathbb{R})$, $P_y \in L^1(\mathbb{R})$ and $\hat{P}_y(s) = e^{-|s|y}$, we have by Theorem 5.3.16 that for almost every $s \in \mathbb{R}$, $\hat{f}_{\delta+y}^*(s) = \hat{f}_\delta^*(s) e^{-|s|y}$. Hence, for almost every $s \in \mathbb{R}$, $\hat{f}_{\delta+y}^*(s) e^{|s|(\delta+y)} = \hat{f}_\delta^*(s) e^{|s|\delta}$. In other words, the function $F(s) = \frac{1}{2\pi} \hat{f}_\delta^*(s) e^{|s|\delta}$ is independent of $\delta > 0$ and satisfies

$$(2\pi)^2 \int_{-\infty}^\infty |F(s)|^2 e^{-2|s|\delta} dt = \int_{-\infty}^\infty |\hat{f}_\delta^*(x)|^2 dx$$

$$= 2\pi \int_{-\infty}^\infty |f_\delta(x)|^2 dx \leqslant 2\pi C_p^p \|f\|_{\widetilde{H}^p}^p \delta^{1-\frac{2}{p}}.$$

This proves (5.3.13). If $0 < p \leqslant 1$, for any $\delta > 0$, let $f_\delta(z) = f(z+i\delta)$. Then, by (4.6.9), for any $y > 0$,

$$\int_{-\infty}^\infty |f_\delta(x+iy)| dx \leqslant \int_{-\infty}^\infty |f_\delta(x+iy)|^p |f_\delta(x+iy)|^{1-p} dx$$

$$\leqslant C_p^p \|f\|_{\widetilde{H}^p}^p \delta^{1-\frac{1}{p}},$$

where C_p is defined by Theorem 4.6.8 and depends only on p. On the other hand, by Theorem 5.2.9, when $y > 0$, $f_\delta(x+iy) = (P_y * f_\delta^*)(x)$, since $f_\delta^* = f_\delta \in L^1(\mathbb{R})$, $P_y \in L^1(\mathbb{R})$ and $\hat{P}_y(s) = e^{-|s|y}$, by Theorems 5.3.2 and 5.3.3, we have $\hat{f}_{\delta+y}^*(s) = \hat{f}_\delta^*(s) e^{-|s|y} \in C_0(\mathbb{R})$. Therefore, for all $s \in \mathbb{R}$, we have $\hat{f}_{\delta+y}^*(s) e^{|s|(\delta+y)} = \hat{f}_\delta^*(s) e^{|s|\delta}$. This means the function $F(s) = \frac{1}{2\pi} \hat{f}_\delta^*(s) e^{|s|\delta}$ is independent of $\delta > 0$, continuous on $\mathbb{R}$ and satisfies, for any $\delta > 0$,

$$2\pi |F(s)| = |\hat{f}_\delta^*(s)| e^{|s|\delta} \leqslant \|\hat{f}_\delta^*\|_1 \leqslant C_p^p \|f\|_{\widetilde{H}^p}^p e^{|s|\delta} \delta^{-B_p},$$

where $B_p = \frac{1}{p} - 1 \geqslant 0$. Therefore,

$$2\pi |F(s)| \leqslant C_p^p \|f\|_{\widetilde{H}^p}^p \inf\{e^{|s|\delta} \delta^{-B_p} : \delta > 0\} = C_p^p \|f\|_{\widetilde{H}^p}^p B_p^{-B_p} e^{B_p} |s|^{B_p}.$$

This proves (5.3.14). □

Theorem 5.3.18. *If $f \in H^2(\mathbb{C}_+)$ and $f^*(t)$ is the non-tangential limit of f, then the Fourier transform $\hat{f}^*$ of $f^*(t)$ satisfies $\hat{f}^*(t) = 0$ for almost all $t \in (-\infty, 0)$.*

Proof. If $f \in H^2(\mathbb{C}_+)$, then by Theorem 5.2.3, for almost all $t \in \mathbb{R}$, the non-tangential limit $f^*(t)$ exists and belongs to $L^2(\mathbb{R})$. Fix $y = \operatorname{Im} z < 0$, and define the function $\sigma(s)$ as follows: For $s \leqslant 0$, $\eta(s) = \frac{e^{izs}}{2\pi}$ and for $s > 0$, let $\eta(s) = 0$, then

$$\hat{\eta}(t) = \frac{1}{2\pi} \int_{-\infty}^{0} e^{izs} e^{-its} ds = \frac{-1}{2\pi i (t - z)}.$$

Then, by (5.2.13),

$$0 = \frac{1}{2\pi i} \int_{-\infty}^{\infty} \frac{-f^*(t) dt}{t - z} = \int_{-\infty}^{\infty} f^*(t) \hat{\eta}(t) dt.$$

Using the multiplication formula (5.3.9), we have

$$0 = \int_{-\infty}^{\infty} f^*(t) \hat{\eta}(t) dt = \int_{0}^{\infty} \hat{f}^*(s) \eta(s) ds$$

$$= \frac{1}{2\pi} \int_{-\infty}^{0} \hat{f}^*(s) e^{izs} ds = \frac{1}{2\pi} \int_{-\infty}^{\infty} \left(\hat{f}^*(-s) e^{ys} \right) e^{-ixs} ds \quad (y < 0).$$

Since $\hat{f}^*(-s) e^{ys} \in L^2([0, \infty))$ $(y < 0)$, by Theorem 5.3.15, we have

$$0 = \int_{0}^{\infty} |\hat{f}^*(-s)|^2 e^{2sy} ds, \quad y < 0.$$

Therefore, for almost all $s > 0$, we have $\hat{f}^*(-s) = 0$. This proves Theorem 5.3.18. $\square$

Theorem 5.3.19. *Suppose $f \in H^1(\mathbb{C}_+)$, and $f^*(t)$ is the non-tangential limit of f, Then the Fourier transform $\hat{f}^*$ of $f^*(t)$ satisfies $\hat{f}^*(t) = 0$ for all $t \leqslant 0$.*

Proof. If $f \in H^1(\mathbb{C}_+)$, then by Theorem 5.2.3, the non-tangential limit $f^*(t)$ exists for almost all $t \in \mathbb{R}$, and $f^* \in L^1(\mathbb{R})$. $\hat{f}^*$ is uniformly continuous on $\mathbb{R}$ and $|\hat{f}^*(x)| \leqslant \|f^*\|_1$. Fix $y_0 = \operatorname{Im} z_0 > 0$, using (5.2.12), we have

$$f_{y_0}(x) = f(x+iy_0) = \frac{1}{2\pi i}\int_{-\infty}^{\infty}\frac{f^*(t)dt}{t-x-iy_0}.$$

Applying the Minkowski inequality, we obtain

$$\left(\int_{-\infty}^{\infty}|f_{y_0}(x)|^2 dx\right)^{\frac{1}{2}} \leqslant \frac{1}{2\pi}\int_{-\infty}^{\infty}\left(\int_{-\infty}^{\infty}\frac{|f^*(t)|^2 dx}{(t-x)^2+y_0^2}\right)^{\frac{1}{2}} dt = \frac{\|f^*\|_1}{2\sqrt{y_0\pi}}.$$

Therefore, $f_{y_0}(z) \in H^2(\mathbb{C}_+)$, and $f_{y_0}(t)$ is the non-tangential limit of $f_{y_0}(z)$. By Theorem 5.3.18, the Fourier transform $\hat{f}_{y_0}$ of $f_{y_0}(t)$ satisfies $\hat{f}_{y_0}(t) = 0$ for all $t < 0$. Using Theorem 5.3.18 and the fact that $\|f_{y_0} - f\|_1 \to 0$ as $y_0 \to 0$, we conclude that for $t \leqslant 0$,

$$\hat{f}^*(t) = \lim_{y_0 \to 0}\hat{f}_{y_0}(t) = 0.$$

This proves Theorem 5.3.19. $\square$

Now, we give a definition of the Fourier–Stieltjes transform of a measure.

Definition 5.3.20. For $\mu \in \mathscr{M}(\mathbb{R})$, Fourier–Stieltjes transform of μ is defined as

$$\hat{\mu}(x) = \int_{\mathbb{R}} e^{-ixt} d\mu(t), \qquad x \in \mathbb{R}.$$

The Fourier–Stieltjes transform of a measure has the following fundamental properties, which can be easily proven.

Theorem 5.3.21. Let $\mu \in \mathscr{M}(\mathbb{R})$, then the Fourier–Stieltjes transform $\hat{\mu}$ of a measure μ satisfies

(a) $\|\hat{\mu}\|_\infty \leqslant \|\mu\|_{\mathscr{M}}$,
(b) $\hat{\mu}(x)$ is uniformly continuous on $\mathbb{R}$.

Theorem 5.3.22. Let $\mu, \nu \in \mathscr{M}(\mathbb{R})$ and $f \in L^p(\mathbb{R})(p \in \{1,2\})$, then

(a) **Multiplication Formula in** $\mathscr{M}(\mathbb{R})$: $\int_{\mathbb{R}} \hat{\mu}(x)d\nu(x) = \int_{\mathbb{R}} \hat{\nu}(x)d\mu(x)$.

(b) **(Fourier Transform of Measure Convolution):**
$\widehat{(f * \mu)}(x) = \hat{f}(x)\hat{\mu}(x)$.

(c) **Parseval's Identity for the Fourier–Stieltjes Transform:**
If f and $\hat{f}$ are both in $L^1(\mathbb{R})$ and f is continuous on $\mathbb{R}$, then

$$2\pi \int_{\mathbb{R}} f(x)d\mu(x) = \int_{\mathbb{R}} \hat{f}(\xi)\hat{\mu}(-\xi)d\xi \qquad (5.3.15)$$

and

$$\int_{\mathbb{R}} \hat{f}(x)d\mu(x) = \int_{\mathbb{R}} f(\xi)\hat{\mu}(\xi)d\xi. \qquad (5.3.16)$$

Proof. We prove that (5.3.15) holds. If $f, \hat{f} \in L^1(\mathbb{R})$ and f is continuous on $\mathbb{R}$, then from (5.3.5), we have

$$f(x) = \frac{1}{2\pi} \int_{\mathbb{R}} \hat{f}(\xi)e^{ix\xi}d\xi \qquad (x \in \mathbb{R}).$$

Therefore,

$$2\pi \int_{\mathbb{R}} f(x)d\mu(x) = \int_{\mathbb{R}}\int_{\mathbb{R}} \hat{f}(\xi)e^{ix\xi}d\mu(x)d\xi = \int_{\mathbb{R}} \hat{f}(\xi)\hat{\mu}(-\xi)d\xi.$$

This completes the proof of the theorem. □

Corollary 5.3.23. *A continuous function φ on $\mathbb{R}$ is the Fourier–Stieltjes transform of a measure μ in $\mathcal{M}(\mathbb{R})$ if and only if there exists a constant $C > 0$ such that for every $f \in C_0(\mathbb{R}) \cap L^1(\mathbb{R})$ with $\hat{f}$ having compact support, we have*

$$\left| \int_{\mathbb{R}} \hat{f}(\xi)\varphi(-\xi)d\xi \right| \leq C \sup_{x \in \mathbb{R}} |f(x)|. \qquad (5.3.17)$$

Proof. If $\varphi = \hat{\mu}, f \in C_0(\mathbb{R}) \cap L^1(\mathbb{R})$, by (5.3.15), we have

$$\left| \int_{\mathbb{R}} \hat{f}(\xi)\varphi(-\xi)d\xi \right| \leq 2\pi \|\mu\|_{\mathcal{M}} \sup_{x \in \mathbb{R}} |f(x)|.$$

Conversely, let $K = \{f \in C_0(\mathbb{R}) \cap L^1(\mathbb{R}) : \hat{f} \text{ has compact support}\}$, then K is dense in $C_0(\mathbb{R})$. If (5.3.17) holds, then the mapping

$$T : f \to \int_{\mathbb{R}} \hat{f}(\xi)\varphi(-\xi)d\xi$$

is a bounded linear functional defined on K, thus T has a unique bounded extension to $C_0(\mathbb{R})$. By the Riesz representation theorem, there exists $\mu \in \mathscr{M}(\mathbb{R})$ such that for all $f \in C_0(\mathbb{R})$, $T(f) = \int_{\mathbb{R}} f(x) d\mu(x)$ and $\|\mu\|_{\mathscr{M}} \leqslant \|T\|$. Applying (5.3.15) once more, we know that

$$\int_{\mathbb{R}} \hat{f}(\xi)[\hat{\mu}(-\xi) - \varphi(-\xi)] d\xi = 0 \quad (f \in K).$$

Therefore, $\varphi = \hat{\mu}$. □

Similar to the case of $L^1(\mathbb{R})$ functions, we can define the Fourier–Stieltjes integral for measures and discuss the inversion problem for the Fourier–Stieltjes integral of measures in the sense of * convergence. Let $\mu \in \mathscr{M}(\mathbb{R})$, then the Abel mean of the Fourier–Stieltjes integral of μ is defined as

$$A_y(\mu)(t) = \frac{1}{2\pi} \int_{\mathbb{R}} \hat{\mu}(x) e^{ixt} e^{-y|x|} dx, \qquad y > 0.$$

Theorem 5.3.24. *Suppose $\mu \in \mathscr{M}(\mathbb{R})$, $A_y(\mu)$ converges weakly * to μ, as $y \to 0$, i.e., $g \in C_0(\mathbb{R})$,*

$$\lim_{y \to 0} \int_{\mathbb{R}} A_y(\mu)(t) g(t) dt = \int_{\mathbb{R}} g(t) d\mu(t). \tag{5.3.18}$$

Proof. By Theorem 4.6.7(b), it suffices to show that for any $y > 0$,

$$A_y(\mu)(x) = (P_y * \mu)(x). \tag{5.3.19}$$

Let $d\nu(s) = e^{isx} e^{-y|s|} ds$. By Theorem 5.3.22 (a), we have

$$2\pi A_y(\mu)(x) = \int_{\mathbb{R}} \hat{\mu}(s) d\nu(x) = \int_{\mathbb{R}} \hat{\nu}(y) d\mu(y).$$

Now, we can calculate $\hat{\nu}(t)$ as follows:

$$\hat{\nu}(t) = \int_{\mathbb{R}} e^{-ist} e^{isx} e^{-y|s|} ds = \int_{\mathbb{R}} e^{-i(t-x)s} e^{-y|s|} ds = P_y(t-x).$$

Hence, (5.3.19) holds. □

Corollary 5.3.25 (Uniqueness of the Fourier–Stieltjes Transform for Measures). *Let $\mu_1, \mu_2 \in \mathcal{M}(\mathbb{R})$ and $\hat{\mu}_1 = \hat{\mu}_2$, then $\mu_1 = \mu_2$.*

Proof. Let $\mu = \mu_1 - \mu_2$. Since $\hat{\mu} \equiv 0$, we have $A_y(\mu) \equiv 0$. Applying (5.3.19), we obtain the result. □

Suppose $0 < A < \infty$, $F \in L^2(-A, A)$. Then the function

$$f(z) = \int_{-A}^{A} F(t)e^{itz} dt \tag{5.3.20}$$

is an entire function and satisfies the growth condition

$$|f(z)| \leqslant \int_{-A}^{A} |F(t)|e^{-ty} dt \leqslant e^{A|y|} \int_{-A}^{A} |F(t)| dt. \tag{5.3.21}$$

If we denote the last integral by C, then $C < \infty$, and from (5.3.21), we have

$$|f(z)| \leqslant Ce^{A|z|} \tag{5.3.22}$$

(an entire function satisfying (5.3.22) is called an exponential-type entire function). Therefore, for any function f of the form (5.3.20), f satisfies (5.3.22) and is an entire function, and when restricted to the real axis $\mathbb{R}$, it is an $L^2(\mathbb{R})$ function (by the Plancherel theorem).

Theorem 5.3.26. *Suppose A and C are positive constants, f is an entire function satisfying (5.3.22), and there exists $p \in (0, 2]$ such that*

$$\int_{-\infty}^{\infty} |f(x)|^p dx < \infty, \tag{5.3.23}$$

then there exists $F \in L^2(-A, A)$ such that (5.3.20) holds for any $z \in \mathbb{C}$.

Proof. Let $f_y(x) = f(x + iy)$, according to Theorem 4.3.19, we have $\|f_y\|_p \leqslant \|f_0\|_p e^{A|y|}$. Since $|f(z)|^p$ is subharmonic, for any real number t, we have

$$|f(t)|^p \leqslant \frac{1}{\pi} \int_{D(t,1)} |f(t+iz)|^p d\lambda(z) \leqslant \frac{1}{\pi} \int_{-1}^{1} \|f_y\|_p^p dy \leqslant \frac{2}{\pi} \|f_0\|_p^p e^{Ap}.$$

Therefore, $f(t)$ is bounded in $\mathbb{R}$, i.e., $|f(t)| \leq \frac{2}{\pi}\|f_0\|_p e^A$. Thus,

$$\int_{-\infty}^{\infty} |f(t)|^2 dt \leq \|f_0\|_p^{2-p} e^{A(2-p)} \|f_0\|_p^p < \infty.$$

According to Theorem 4.3.19, $\|f_y\|_2 \leq \|f_0\|_2 e^{A|y|}$. Therefore,

$$f(z)e^{Azi} \in H^2(\mathbb{C}_+), \quad f(-z)e^{Azi} \in H^2(\mathbb{C}_+).$$

By Theorem 5.3.18, the Fourier transform of $f_0(x)e^{iAx}$ is $\hat{f}_0(t-A)$, and for almost all $t < 0$, we have $\hat{f}_0(t-A) = 0$; the Fourier transform of $f_0(-x)e^{iAx}$ is $\hat{f}_0(t-A)$ and for almost all $t < 0$, we have $\hat{f}_0(-t+A) = 0$. Therefore, for almost all $|t| > A$, we have $\hat{f}_0(t) = 0$. By Theorem 5.3.13, for almost all real numbers $x \in \mathbb{R}$, $2\pi f_0(-x) = \mathscr{F}\hat{f}(x)$. Since $\hat{f}_0(t) = 0$ for almost all $|t| > A$, we have for almost all real numbers $x \in \mathbb{R}$,

$$f_0(x) = \frac{1}{2\pi} \int_{-A}^{A} \hat{f}_0(t) e^{itx} dt, \qquad (5.3.24)$$

and the functions

$$g(z) = \frac{1}{2\pi} \int_{-A}^{A} \hat{f}_0(t) e^{itz} dt$$

and $f(z)$ are both entire functions, and they almost everywhere coincide with $f_0(x)$ on the real axis $\mathbb{R}$, according to the uniqueness theorem for analytic functions, (5.3.24) holds in the entire complex plane $\mathbb{C}$, where $F(t) = \frac{\hat{f}_0(t)}{2\pi}$. This completes the proof of Theorem 5.3.26.

□

Theorem 5.3.27. *If μ is a finite Borel measure on $\mathbb{R}$ and its Poisson integral $f(z) = f(x+iy) = (P_y * \mu)(x) \in H(\mathbb{C}_+)$, then μ is absolutely continuous with respect to the Lebesgue measure on $\mathbb{R}$ and $d\mu(t) = f^*(t)dt$, where $f^*(t)$ is the non-tangential limit of $f(z)$, and $f^* \in L^1(\mathbb{R})$, $\|\mu\|_{\mathcal{M}} = \|f^*\|_1$.*

Proof. If μ is a finite Borel measure on $\mathbb{R}$ and $f(z) = f(x+iy) = (P_y * \mu)(x)$ is the Poisson integral of $\mu(t)$ such that $f \in H(\mathbb{C}_+)$, then by the Minkowski inequality, we know that $f \in H^1(\mathbb{C}_+)$. According to Theorem 5.2.9, $f(z)$ is the Poisson integral of its non-tangential limit $f^*(t)$: $f(z) = (P_y * f^*)(x)$. By Theorem 4.6.7, the Poisson–Stieltjes integral of $d\mu(t) - f^*(t)dt$ weak* converges to 0, so we have $d\mu(t) = f^*(t)dt$. This completes the proof of Theorem 5.3.27. □

Theorem 5.3.28. *If μ is a finite Borel measure on $\mathbb{R}$, and*

$$\int_{\mathbb{R}} \frac{d\mu(\zeta)}{\zeta - z} d\zeta = 0, \quad y = \operatorname{Im} z < 0, \tag{5.3.25}$$

then $f(z) \in H^1(\mathbb{C}_+)$ and μ is absolutely continuous with respect to the Lebesgue measure on $\mathbb{R}$ and can be expressed as $d\mu(t) = f^*(t)dt$, where $f^*(t)$ is the non-tangential limit of $f(z)$ with $f^* \in L^1(\mathbb{R})$, $\|\mu\|_{\mathcal{M}} = \|f^*\|_1$.

Proof. If μ is a finite Borel measure on $\mathbb{R}$ and satisfies (5.3.25), then due to the fact that the Poisson kernel satisfies

$$P_y(x) = \frac{1}{2\pi i}\left(\frac{1}{t-z} - \frac{1}{t-\bar{z}}\right),$$

the Poisson integral of $\mu(t)$ denoted as $f(z) = f(x+iy) = (P_y * \mu)(x)$ belongs to $H(\mathbb{C}_+)$. According to Theorem 5.3.27, μ is absolutely continuous with respect to the Lebesgue measure on $\mathbb{R}$ and can be expressed as $d\mu(t) = f^*(t)dt$ and $f(z) \in H^1(\mathbb{C}_+)$, where $f^*(t)$ is the non-tangential limit of $f(z)$ with $f^* \in L^1(\mathbb{R})$, $\|\mu\|_{\mathcal{M}} = \|f^*\|_1$. This proves Theorem 5.3.28. □

Theorem 5.3.29 (F. Riesz–M. Riesz). *If μ is a finite Borel measure on $\mathbb{R}$, and for any $s < 0$,*

$$\int_{\mathbb{R}} e^{-ist} d\mu(t) = 0, \tag{5.3.26}$$

then $f(z) \in H^1(\mathbb{C}_+)$ and μ is absolutely continuous with respect to the Lebesgue measure on $\mathbb{R}$ and can be expressed as $d\mu(t) = f^*(t)dt$, where $f^*(t)$ is the non-tangential limit of $f(z)$ with $f^* \in L^1(\mathbb{R})$, $\|\mu\|_{\mathcal{M}} = \|f^*\|_1$.

Proof. Suppose (5.3.25) holds. Fix $y = \operatorname{Im} z > 0$. According to (5.3.3), the Fourier transform of the Poisson kernel P_y is given by

$\hat{P}_y(s) = e^{-y|s|}$. Using (5.3.26), the Fourier transform of the Poisson integral $f_y(x) = (P_y * \mu)(x)$ of $\mu(t)$ is

$$\hat{f}_y(s) = \begin{cases} e^{-ys}\hat{\mu}(s), & s > 0, \\ 0, & s \leq 0. \end{cases}$$

Thus, f_y and $\hat{f}_y$ belong to $L^1(\mathbb{R})$. From (5.3.5), we have

$$f_y(x) = \frac{1}{2\pi}\int_{-\infty}^{\infty} \hat{f}_y(t)e^{ixt}dt = \frac{1}{2\pi}\int_0^{\infty} e^{i(x+iy)t}\hat{\mu}(t)dt.$$

Hence, $f(z) = f_y(x) = (P_y * \mu)(x) \in H(\mathbb{C}_+)$. According to Theorem 5.3.27, $f(z) \in H^1(\mathbb{C}_+)$ and μ is absolutely continuous with respect to the Lebesgue measure on $\mathbb{R}$ and can be expressed as $d\mu(t) = f^*(t)dt$, where $f^*(t)$ is the non-tangential limit of $f(z)$ with $f^* \in L^1(\mathbb{R})$, $\|\mu\|_{\mathcal{M}} = \|f^*\|_1$. This proves Theorem 5.3.29. □

Exercises V

1. Let $0 < p < s \leq \infty$. Show that $H^s(U)$ is a proper subspace of $H^p(U)$.
2. Suppose that $f \in H(U)$ and $f(U)$ is not dense in the complex plane. Then, for almost every point ζ on the unit circle ∂U, the limit $\lim_{r \to 1} f(r\zeta)$ exists.
3. Suppose that $0 < p \leq \infty$ and $f \in H(U)$. Show that the necessary and sufficient condition for $f \in H^p(U)$ is that there exists a harmonic function u in the unit disk U such that $|f(z)|^p \leq u(z)$ for any $U z \in U$. Furthermore, show that if there exists a harmonic majorant function u for $|f|^p$ in the unit disk U, then the least harmonic function u_f for $|f|^p$ in the unit disk U (i.e., $|f|^p \leq u_f$ and if $|f|^p \leq u$, where u is harmonic in the unit disk U, then $u_f \leq u$).
4. Suppose that $f \in H(U)$, $f(U) \subset \overline{U}$ with $|f'(0)| = \delta > 0$. Show that if $|z| < \eta < \delta$, then

$$|f(z)| \geq \left(\frac{\delta - \eta}{1 - \eta\delta}\right)|z|,$$

and for any $w \in D\left(0, \left(\frac{\delta-\eta}{1-\eta\delta}\right)\eta\right)$, $f(z) = w$ has exactly one root in the disk $D(0, \eta)$.

5. For any $\alpha > 0$, let $W_\alpha = \{z \in U : |1-z| < \alpha(1-|z|^2)\}$. Assume that $f \in H(U)$, $f(U) \subset \overline{U}$ and there exists a sequence $\{z_n\} \subset U$ such that $z_n \to 1$, $f(z_n) \to 1$, and
$$\frac{1-|f(z_n)|^2}{1-|z_n|^2} \to A < \infty.$$
Show that $f(W_\alpha) \subset f(W_{A\alpha})$.

6. Suppose $f(e^{it}) \in L^1([-\pi, \pi])$. Show that the necessary and sufficient conditions for the existence of $g \in H^1(U)$ such that $f(e^{it})$ is the non-tangential limit of $g(z)$, are one of the following:

   (a) Poisson integral of $f(e^{it})$ is $f(z) = f(x+iy) = P(f)(z) \in H(U)$,

   (b)
   $$\int_{|\zeta|=1} \frac{f(\zeta)}{\zeta-z} d\zeta = 0, \quad |z| > 1,$$

   (c)
   $$\int_{-\pi}^{\pi} f(e^{it}) h^*(e^{it}) dt = 0$$
   for any $h \in H^\infty(U)$ with $h(0) = 0$, where $h^*(e^{it})$ is the non-tangential limit of $h(z)$,

   (d)
   $$\hat{f}^*(-n) = \frac{1}{2\pi} \int_{-\pi}^{\pi} e^{int} f(e^{it}) dt = 0$$
   for any positive integer n.

7. Suppose that $f \in L^1(\mathbb{R})$. Show that the necessary and sufficient conditions for the existence of $g \in H^1(\mathbb{C}_+)$ such that $f(t)$ is the non-tangential limit of $g(z)$, are one of the following:

   (a) Poisson integral of $f(t)$ is $f(z) = f(x+iy) = (P_y * f)(x) \in H(\mathbb{C}_+)$,

   (b)
   $$\int_\mathbb{R} \frac{d\mu(\zeta)}{\zeta-z} d\zeta = 0, \quad y = \mathrm{Im}\, z < 0,$$

(c) for any $h \in H^\infty(\mathbb{C}_+)$,
$$\int_{-\infty}^{\infty} f(t)h^*(t)dt = 0,$$
where $h^*(t)$ is the non-tangential of $h(z)$,

(d) $\hat{f}(s) = 0$ for any $s < 0$.

8. Prove that $f \in H^p(\mathbb{C}_+)$, $1 < p < \infty$ if and only if $f(\alpha(w))\alpha'(w) \in H^p(U)$, where $\alpha(w) = i\frac{1-w}{1+w}$.

9. Suppose $f(z) = \sum_{n=0}^{\infty} a_n z^n \in H^1(U)$, $f(e^{it})$ is the non-tangential limit of $f(z)$. Prove that $f(e^{it})$ is almost everywhere equal to a bounded variation function on $[-\pi, \pi]$ if and only if $f'(z) \in H^1(U)$. In the case where $f'(z) \in H^1(U)$,
$$\sum_{n=0}^{\infty} |a_n| < \infty.$$

10. Suppose $f(z) \in H(U)$, $0 < p < 1$. If $\operatorname{Re} f(z) \geqslant 0$ for any $z \in U$, prove that $f \in L^p(U)$.

11. Prove the Hardy inequality: If $f(z) = \sum_{n=0}^{\infty} a_n z^n \in H^1(U)$, then
$$\sum_{n=0}^{\infty} \frac{|a_n|}{n+1} \leqslant \pi \|f\|_{H^1}.$$

12. Suppose $\{f_j\}$ is a Cauchy sequence in $L^1(\mathbb{R})$. Prove that there exists $f \in L^1(\mathbb{R})$ such that, as $j \to \infty$, the sequence $\{\hat{f}_j\}$ uniformly converges to $\hat{f}$ on $\mathbb{R}$.

13. Let $B(z)$ be the upper half-plane Blaschke product defined by (5.2.9). Prove that
$$\lim_{y \to 0} \int_{-\infty}^{\infty} \frac{\log |B(x+iy)|}{1+x^2} dx = 0.$$
Conversely, if $f \in H^\infty(\mathbb{C}_+)$, $|f(z)| < 1$, and
$$\lim_{y \to 0} \int_{-\infty}^{\infty} \frac{\log |f(x+iy)|}{1+x^2} dx = 0,$$
then $f(z) = e^{i(\gamma + \alpha z)} B(z)$, where γ is a real number, $\alpha \geqslant 0$, $B(z)$ is the upper half-plane Blaschke product defined by (5.2.9).

14. Suppose $f \in H^1(\mathbb{C}_+)$, $f^*(t)$ is the non-tangential limit of f. Prove that the Fourier transform $\hat{f}^*$ of $f^*(t)$ satisfies

$$\int_0^\infty \frac{|\hat{f}^*(t)|}{t} dt \leqslant \frac{1}{2} \int_{-\infty}^\infty |f^*(x)| dx.$$

15. Suppose that $f \in H^p(\mathbb{C}_+)$, $1 < p < \infty$, $f^*(t)$ is the non-tangential limit of f. Prove that

$$\int_0^\infty |f(x+iy)|^p dx \leqslant \frac{1}{2} \int_{-\infty}^\infty |f^*(x)|^p dx.$$

16. Let N be a positive integer, and let $\mathfrak{F}_N$ be the family of all bounded analytic functions $f(z)$ in $\mathbb{C}_+$, continuous in $\overline{\mathbb{C}}_+$ and $f(t)$ is indifferentiable on $\mathbb{R}$ satisfying

$$\lim_{z \in \overline{\mathbb{C}}_+, z \to \infty} |z|^N f(z) = 0.$$

Prove the following:

(a) For $0 < p < \infty$, and for $f \in H^p(\mathbb{C}_+)$, there exists a sequence $\{f_n\} \subset \mathfrak{F}_N$ such that

$$\lim_{n \to \infty} \|f_n - f\|_{\widetilde{H}^p} = 0.$$

(b) For $f \in H^p(\mathbb{C}_+)$, there exists a sequence $\{f_n\} \subset \mathfrak{F}_N$ such that $\|f_n\|_{\widetilde{H}^\infty} \leqslant \|f\|_{\widetilde{H}^\infty}$ and $\lim_{n \to \infty} f_n(t) = f(t)$ for almost all $t \in \mathbb{R}$.

Chapter 6

Uniform Approximation by Polynomials

In this chapter, we introduce the fundamental theory of uniform approximation of rational functions.

6.1 Uniform Approximation by Rational Function and Simply Connected Region

In this section, we discuss the polynomial approximation of analytic functions and, more generally, the rational function approximation. The main results of this section are Runge's theorem and its relationship with the simply connected region.

Theorem 6.1.1 (Runge's Theorem). *Let $K \subset \mathbb{C}$ be a compact. Let f be holomorphic on a neighborhood of K. Suppose that A is a subset of $\overline{\mathbb{C}} \setminus K$ containing one point from each connected component of $\overline{\mathbb{C}} \setminus K$. Then for any $\varepsilon > 0$, there is a rational function $R(z)$ with poles in A such that*

$$\sup\{|f(z) - R(z)| : z \in K\} < \varepsilon. \tag{6.1.1}$$

Note that $\overline{\mathbb{C}} \setminus K$ has at most countably many connected components. A special case is when A contains only ∞, where a rational function $R(z)$ with ∞ as its pole must be a polynomial. Thus, we have the following corollary.

Corollary 6.1.2. *Let $K \subset \mathbb{C}$ be a compact and K^c be connected. Let f be holomorphic on a neighborhood of $K \subset \mathbb{C}$ and K^c be connected. Then f can be uniformly approximated by a sequence of polynomials on K, i.e., there exists a sequence of polynomials $\{P_n\}$, such that*

$$\sup\{|f(z) - P_n(z)| : z \in K\} \to 0 \ (n \to \infty). \qquad (6.1.2)$$

Note that this corollary does not require K to be connected.

Example 6.1.3. There is a sequence of holomorphic polynomials $\{P_n\}$, such that for all $z \in \overline{U} = \{z : |z| \leqslant 1\}$, $\lim_{n\to\infty} P_n(z) = 1$, and for all $z \in \mathbb{C}\setminus\overline{U} = \{z : |z| > 1\}$, $\lim_{n\to\infty} P_n(z) = 0$. We can define $K_n = \overline{U} \cup \{z : 1 + \frac{1}{n} \leqslant |z| \leqslant n, 0 \leqslant \arg z \leqslant 2\pi - \frac{1}{n}\}$, then K_n^c is connected and $K_n \subset K_{n+1}$ $(n = 1, 2, \ldots)$. Let

$$f_n(z) = \begin{cases} 1, & |z| < 1 + \frac{1}{2n}, \\ 0, & |z| > 1 + \frac{1}{2n}. \end{cases}$$

Then $f_n \in H(K_n)$. Therefore, for any positive integer n, there exists a polynomial P_n such that

$$\sup\{|f(z) - P_n(z)| : z \in K\} < \frac{1}{n}.$$

These $\{P_n\}$ are the desired polynomials.

The condition in the theorem requires that the set A satisfies $A \cap V \neq \varnothing$ for any connected component V of $K^c = \overline{\mathbb{C}}\setminus K$, $A \cap V \neq \varnothing$. This condition is necessary. In fact, if a is a point in a connected component V of K^c, then the function $f(z) = (z-a)^{-1}$ is analytic in some neighborhood of K (if $a = \infty$, take $f(z) = z$). If $f(z)$ can be uniformly approximated on K by a rational function $R(z)$ without poles in V, then $(z-a)R(z) - 1$ converges uniformly to 0 on $\partial V \subset K$. By the maximum modulus principle, $(z-a)R(z) - 1 \equiv 0$ in V. This would lead to $1 = 0$ at $z = a$, which is a contradiction.

Proof of Theorem 6.1.1. We consider the Banach space $C(K)$ whose members are the continuous complex functions on K, with the supremum norm, The dual space $(C(K))^*$ is the space of (complex) measures on K. Let M be the subspace of $C(K)$ which consists of the restrictions to K of those rational functions which have all their poles in A. The theorem asserts that f is the closure of M. By the Hahn–Banach Theorem, this is equivalent to saying that every bounded linear functional on $C(K)$ which vanishes at f, and hence, the Riesz representation theorem shows that we must prove the following assertion:

If μ is a complex Borel measure on K such that

$$\int_K R(z) d\mu(z) = 0$$

for every rational function $R(z) \in M$ and if $f \in H(\Omega)$ (where Ω is a neighborhood of K)

$$\int_K f(z) d\mu(z) = 0. \tag{6.1.3}$$

Let

$$h(\zeta) = \int_K \frac{d\mu(z)}{z - \zeta} \quad (\zeta \in \overline{\mathbb{C}} \setminus K),$$

and let V be the component of $\overline{\mathbb{C}} \setminus K$ which contains $a \in V \cap A$. If $a \neq \infty$, and if $\zeta \in D(a, r) \subset\subset V$ is fixed in $D(a, r)$, then

$$\frac{1}{z - \zeta} = \sum_{n=0}^{\infty} \frac{(\zeta - a)^n}{(z - a)^{n+1}} \tag{6.1.4}$$

uniformly for $z \in K$. Each of the functions on the right-hand side of (6.1.4) is one to which (6.1.3) applies. Hence, $h(\zeta) = 0$ for all $\zeta \in D(a, r)$. This implies that $h(\zeta)$ for all $\zeta \in V$ by the uniqueness of analytic function. $a = \infty$ (6.1.4) is replaced by

$$\frac{1}{z - \zeta} = -\sum_{n=0}^{\infty} \frac{z^n}{\zeta^{n+1}} \quad (z \in K, |\zeta| > r), \tag{6.1.5}$$

which implies that $h(\zeta) = 0$ in $\{\zeta : |\zeta| > r\}$ and hence in V. We have to prove from (6.1.3) that $h \equiv 0$ in $\overline{\mathbb{C}} \setminus K$ for any $\phi \in C_0^\infty(\mathbb{C})$ such that it is constantly 1 in some neighborhood V_1 of K with $\overline{V_1} \subset \overline{\mathbb{C}} \backslash K$. For $z \in K$, we have

$$f(z) = f(z)\phi(z) = -\frac{1}{\pi} \int \frac{f(\zeta)\partial\phi/\partial\overline{\zeta}}{\zeta - z} d\lambda(\zeta).$$

Then, according to the Fubini theorem, and because $\partial\phi/\partial\overline{\zeta} = 0$ on K, and $h = 0$ outside K,

$$\int_K f(\zeta)d\mu(\zeta) = \frac{1}{\pi}\int_K f(\zeta)h(\zeta)(\partial\phi(\zeta)/\partial\overline{\zeta})d\lambda(\zeta) = 0.$$

Thus, (6.1.3) holds, and the proof is complete. □

Using the following simple lemma, we can derive several important corollaries of Runge's theorem.

Lemma 6.1.4. *Suppose $\Omega \subset \mathbb{C}$, then there are a sequence $\{K_n\}$ of compact sets such that*

$$K_1 \subset \text{int}(K_2) \subset \cdots \subset K_n \subset \text{int}(K_{n+1}) \subset \cdots \subset \Omega, \quad \Omega = \bigcup_{n=1}^{\infty} K_n$$

and such that each component V_n of $\overline{\mathbb{C}} \setminus K_n = K_n^c$ contains a component V of Ω^c.

Proof. For $n = 1, 2, \ldots$, let

$$V_n = D(\infty, n) \bigcup \bigcup_{a \notin \Omega} D\left(a, \frac{1}{n}\right)$$

and $K_n = \overline{\mathbb{C}} \setminus V_n$. Then K_n is the quantity sought. □

Corollary 6.1.5. *If Ω is an open set in $\overline{\mathbb{C}}$ and $f \in H(\Omega)$, there are a sequence $\{R_n\}$ of rational functions with poles of $\{R_n\}$ outside Ω such that $\{R_n\}$ uniformly converges to f on each compact subset of Ω.*

Corollary 6.1.6. *Suppose Ω is an open set in $\overline{\mathbb{C}}$ if $\overline{\mathbb{C}} \backslash \Omega$ is connected and $f \in H(\Omega)$, then there are a sequence $\{R_n\}$ of polynomials functions $\{P_n\}$ such that $\{P_n\}$ uniformly converges to f on each compact subset of Ω.*

Uniform Approximation by Polynomials

Theorem 6.1.7 (Mittag–Leffler's Theorem). *Suppose Ω is an open set in $\mathbb{C}$, $A = \{a_j\} \subset \Omega$, A has no limit point in Ω, and to each $a \in A$, there are associated a positive integer $m(a)$ and a rational function*

$$p_a(z) = \sum_{j=1}^{m(a)} c_j (z-a)^{-j}.$$

Then there exists a meromorphic function f in Ω, whose principal part at each $a \in A$ is p_a and which has no other poles in Ω.

Proof. We choose a sequence $\{K_n\}$ of compacts in Ω, as shown in Lemma 6.1.4. Put $A_1 = A \cap K_1$ and $A_n = A \cap (K_n \setminus K_{n-1})$ for $n = 2, 3, \ldots$. Since $A_n \subset K_n$ and A has no limit point in Ω (hence none in K_n), each A_n is a finite set. Put

$$q_n(z) = \sum_{a_j \in K_n} p_{a_j}(z).$$

Each K_n contains finite points such that $a_j \in K_n$, each q_n is a rational function, and $q_{n+1} - q_n \in H(K_n)$. It now follows from Runge's theorem that there exist rational functions R_n, all of whose poles are in $\overline{\mathbb{C}} \setminus \Omega$, such that

$$\sup\{|q_{n+1}(z) - q_n(z) - R_n(z)| : z \in K_n\} < \frac{1}{2^n}.$$

We claim that

$$f(z) = q_1(z) + \sum_{n=1}^{\infty}(q_{n+1}(z) - q_n - R_n(z)) \quad (z \in \Omega)$$

has the desired properties.

Fix N. On K_N, we have

$$f(z) - q_N(z) + \sum_{n=1}^{N} R_n(z) = \sum_{n=N+1}^{\infty}(q_{n+1}(z) - q_n(z) - R_n(z)).$$

Each term in the last sum is less than 2^{-n} on K_N; hence, this last series converges uniformly on K_N to a function which is holomorphic in the interior of K_N. Since the poles of each R_n are outside Ω,

f is meromorphic in the interior of K_N and $f(z) - q_N(z)$ is holomorphic in the interior of K_N. Thus, f has precisely the prescribed principal parts in the interior of K_N, and hence in Ω, since N was arbitrary. $\square$

By Corollary 1.2.5, if $\phi \in C_c^\infty(\mathbb{C})$, then
$$u(z) = -\frac{1}{\pi}\int_\mathbb{C} \frac{\phi(\zeta)d\lambda(\zeta)}{\zeta - z}$$
is a C^∞ solution to $\partial u/\partial \bar{\zeta} = \phi$. By a proof analogous to the preceding proof of Mittag-Leffler's theorem we get an existence theorem for solutions to the inhomogenous Cauchy-Riemann equation.

Theorem 6.1.8. *Suppose that Ω is an open set in $\mathbb{C}$ and $f \in C^\infty(\Omega)$, then there is a $u \in C^\infty(\Omega)$ such that*
$$\frac{\partial u}{\partial \bar{z}} = f \qquad (6.1.6)$$
in Ω.

Proof. Take K_n as before and choose $\phi_n \in C_c^\infty(\text{int}(K_{n+1}))$, such that $\phi_n \equiv 1$ in a neighborhood of K_n. Now, there are $u_n \in C^\infty(\Omega)$ such that $\partial u_n/\partial \bar{z} = \phi_n f$. Observe that $(\partial/\partial \bar{z})(u_{n+1} - u_n) = 0$ in a neighborhood of K_n. Thus, there are rational functions R_n and R_n have their poles outside of Ω, such that $|u_{n+1} - u_n - R_n| < 2^{-n}$ on K_n. Then
$$u = u_1 + \sum_{n=1}^\infty (u_{n+1} - u_n - R_n)$$
is a $C^\infty(\Omega)$-function, since on each fixed compact, there is only a finite number of terms that are not analytic, i.e., for each compact $K \subset \Omega$, there is N, such that $K \subset \text{int}(K_N)$. The tail series
$$u(z) - u_N(z) + \sum_{n=1}^N R_n(z) = \sum_{n=N+1}^\infty (u_{n+1}(z) - u_n(z) - R_n(z))$$
converges uniformly on K_N and thus analytic in $\text{int}(K_N)$. Therefore, we may differentiate termwise, and hence,
$$\frac{\partial u}{\partial \bar{z}} = \frac{\partial u_N}{\partial \bar{z}} = \phi_N f = f$$
on Ω. $\square$

It is clear that if u_0 is a particular solution to equation (6.1.4), then any other solution has the form $u = u_0 + h$, where $h \in H(\Omega)$. Often, it is important to single out a solution with certain properties.

Another way to get further information about a solution is to use some explicit formula. If $f \in L^1(\Omega)$, then

$$u(z) = -\frac{1}{\pi} \int_\Omega \frac{f(\zeta)d\lambda(\zeta)}{\zeta - z}$$

provides a solution to equation (6.1.6). It is also possible to construct such formulas that will work for a large class of f.

Example 6.1.9. If f satisfies that

$$\int_U (1 - |\zeta|^2)|f(\zeta)|d\lambda(\zeta) < \infty,$$

then

$$(Kf)(z) = -\frac{1}{\pi} \int_U \left(\frac{1 - |\zeta|^2}{1 - \bar{\zeta}z}\right) \frac{f(\zeta)d\lambda(\zeta)}{\zeta - z}$$

provides a solution to equation (6.1.6). In fact, one can just apply the arguments above to the function

$$\zeta \mapsto \left(\frac{1 - |\zeta|^2}{1 - \bar{\zeta}w}\right) f(\zeta)$$

and let $z = w$. In this formula, f is thus allowed to grow somewhat at the boundary of U. However, this solution also has a certain minimality property.

Let $B(U) = H(U) \cap L^2(U)$ and $B(U)$ be a closed subspace of $L^2(U)$. Let

$$(Pu)(z) = \frac{1}{\pi} \int_U \frac{u(\zeta)d\lambda(\zeta)}{(1 - \bar{\zeta}z)^2}.$$

One can verify that P is a bounded operator from $L^2(U)$ to $B(U)$. Furthermore, one can check that

$$K(\partial u/\partial \bar{z}) = u - Pu \tag{6.1.7}$$

for $u \in C^1(\overline{U})$. Since the analytic polynomials are dense in $B(U)$, (6.1.7) shows that $Pu = u$ if $u \in B(U)$ is analytic, i.e., P is a projection. Moreover, it is indeed the orthogonal projection $P : L^2(U) \to B(U)$ since its kernel $K(z,\zeta) = (1-\bar\zeta z)^{-2}/\pi$ is hermitian, $\overline{K(z,\zeta)} = K(\zeta,z)$, which in turn implies that P is self-adjoint. The operator P is called the Bergman projection in U and $B(U)$ the Bergman space. Now, if equation (6.1.6) has a solution u in $L^2(U) \cap C^1(\overline{U})$, it follows from (6.1.7) that Kf is the one with minimal norm in $L^2(U)$ since it is orthogonal to $B(U)$.

Let Ω be a region in the extended complex plane. If for any Jordan closed curve γ with $\gamma^* \subset \Omega$, either the interior of γ is entirely contained in Ω or the exterior of γ is entirely contained in Ω, then Ω is called simply connected. For a simply connected region, the following conditions are equivalent.

Theorem 6.1.10. *Let $\Omega \subset \mathbb{C}$ be a plane region and $\Omega \neq \mathbb{C}$, each of the following nine conditions implies all the others:*

(1) *Ω is simply connected.*
(2) *Ω is homeomorphic to the open unit disc U.*
(3) *Any continuous closed curve γ in Ω is homotopic to zero in Ω.*
(4) *Any closed chain γ in Ω is homologous to zero in Ω.*
(5) *$\overline{\mathbb{C}} \setminus \Omega$ is connected.*
(6) *There exists a holomorphic homeomorphism of Ω onto the open unit disc U.*
(7) *For every $f \in H(\Omega)$, there are a sequence of polynomials $\{P_n\}$, such that $\{P_n\}$ converges to f uniformly on compact subset of Ω.*
(8) *For every closed chain γ in Ω and every $f \in H(\Omega)$,*
$$\int_\gamma f(z)dz = 0.$$
(9) *To every $f \in H(\Omega)$ corresponds an $F \in H(\Omega)$ such that $F' = f$.*
(10) *If $f \in H(\Omega)$ and $\frac{1}{f} \in H(\Omega)$, there exists $g \in H(\Omega)$ such that $e^g = f$.*
(11) *If $f \in H(\Omega)$ and $\frac{1}{f} \in H(\Omega)$, there exists $\varphi \in H(\Omega)$ such that $\varphi^2 = f$.*
(12) *If u is a real harmonic function in Ω, there exists $f \in H(\Omega)$ such that $u = \operatorname{Re} f$.*

Proof. From the Riemann mapping theorem, (1) implies (6). From Proposition 1.3.10, (3) implies (4). From Theorem 1.3.4, (4) implies (8). From Lemma 1.2.25, (8) implies (9). It is known from the proof process of Lemma 1.2.25 that (9) implies (10). Taking $\varphi = \exp(\varphi/2) \in H(\Omega)$, then $\varphi^2 = f$, thus (10) implies (11). It is known from the proof process of the Riemann mapping theorem that (11) implies (1). From Theorem 4.1.2, (1) implies (12). From Corollary 6.1.6, (5) implies (7). The implication of (6) to (2) is obvious.

(7) implies (8). If for every $f \in H(\Omega)$, there are a sequence of polynomials $\{P_n\}$, such that $\{P_n\}$ converges to f uniformly on compact subsets of Ω, thus $f \in H(\Omega)$, and for every closed chain γ in Ω, $\int_\gamma P_n(z)dz = 0$, thus $\int_\gamma f(z)dz = 0$.

(12) implies (10). If $f \in H(\Omega)$ and f has no zero in Ω, then $\log|f(z)|$ is a real harmonic function in Ω, there exists $g_0 \in H(\Omega)$ such that $\log|f| = \text{Re} g_0$. Therefore, $|fe^{-g_0}| \equiv 1$ in Ω. Since Ω is a region, there exists $a \in \mathbb{R}$ such that fe^{-g_0} is the constant e^{ia} in Ω, thus $f = e^{g_0 + ia}$ in Ω.

(2) implies (3). Let ψ be a continuous one-to-one mapping of Ω onto U whose inverse ψ^{-1} is also continuous. If $\gamma : [0,1] \to \Omega$ is a closed continuous curve in Ω with parameter interval $[0,1]$, put $H(s,t) = \psi^{-1}(t\psi(\gamma(s)))$. Then $H(s,t)$ is continuous on $[0,1] \times [0,1]$; $H(s,0) = \psi^{-1}(0)$, $H(s,1) = \gamma(s)$ and $H(0,t) = H(1,t)$ because $\gamma(0) = \gamma(1)$. Thus, γ is homotopic to zero in Ω.

(9) implies (5). Assume (5) is false. Then $\overline{\mathbb{C}} \setminus \Omega$ is a closed subset of $\overline{\mathbb{C}}$ which is not connected. It follows that $\overline{\mathbb{C}} \setminus \Omega$ is the union of two non-empty disjoint closed sets K and H. Let K be compact relative to the plane $\mathbb{C}$ and H be the complement of K in $\overline{\mathbb{C}}$ with $\infty \in H$. Let $W = \mathbb{C} \setminus H$, then $W = \Omega \cup K$ is open set and the compact $K \subset W$. Theorem 1.1.2 (with $f = 1$) shows that there exists $\phi \in C_c^\infty(\Omega)$ such that $\phi(z) = 1$ for z in a neighborhood of K, and $0 \leq \phi \leq 1$ in Ω. Taking $\alpha \in K$, then $(z - \alpha)^{-1} \in H(\Omega)$, (9) implies that there exists $f \in H(\Omega)$ such that $f'(z) = (z - \alpha)^{-1}$. Theorem 1.2.4 implies that

$$\pi = \pi\phi(\alpha) = -\int_W \frac{\partial \phi}{\partial \overline{\zeta}} \frac{d\lambda(\zeta)}{\zeta - z} = -\int_W \frac{\partial \phi}{\partial \overline{\zeta}} f'(\zeta) d\lambda(\zeta)$$

$$= \int_W \frac{\partial^2 \phi}{\partial \overline{\zeta} \partial \zeta} f(\zeta) d\lambda(\zeta) = \int_W \frac{\partial \phi}{\partial \zeta} \frac{\partial f}{\partial \overline{\zeta}} d\lambda(\zeta) = 0.$$

This indicates a contradiction. Thus, it follows that (9) implies (5). This completes the proof of the theorem. □

6.2 Uniform Approximation by Polynomials

Let $\operatorname{int} K$ be the interior of a compact K in the complex plane. Let $P(K)$ denote the set of all functions on K which are uniform limits of polynomials in z. Which functions belong to $P(K)$? Two necessary conditions come to mind immediately: If $f \in P(K)$, then $f \in C(K)$ and $f \in H(\operatorname{int} K)$.

The question arises whether these necessary conditions are also sufficient. The answer is negative whenever the complement of K, $\mathbb{C} \setminus K$ is not connected. We see this in the remark of Example 6.1.3. On the other hand, if K is an interval on real axis, the Weierstrass approximation theorem asserts that $P(K) = C(K)$. So, the answer is positive if K is an interval. Runge's theorem also points in this direction, since it states, for compact sets K, if $\mathbb{C} \setminus K$ is connected, then $H(K) \subset P(K)$.

In this section, we prove the theorem of Mergelyan which states, without any superfluous hypotheses, that if $\mathbb{C} \setminus K$ is connected, then $f \in P(K)$ if and only if $f \in C(K)$ and $f \in H(\operatorname{int} K)$.

The principal ingredients of Mergelyan's theorem are Tietze's extension theorem and Runge's theorem.

Lemma 6.2.1. *Suppose $D(z_0, r)$ is an open disc disc of radius of $r > 0$ with the center at z_0, $E \subset D(z_0, r)$, E is compact and connected, $\Omega = \overline{\mathbb{C}} \setminus E$ is connected and the diameter of E is least r, i.e., $\operatorname{diam} E = \sup\{|z_1 - z_2| : z_1 \in E, z_2 \in E\} \geqslant r$. Then there is a function $g \in H(\Omega)$ and a constant b, with the following property: If*

$$Q(\zeta, z) = g(z) + (\zeta - b)g^2(z), \qquad (6.2.1)$$

the inequalities

$$|Q(\zeta, z)| < \frac{100}{r}, \qquad (6.2.2)$$

$$\left| Q(\zeta, z) - \frac{1}{z - \zeta} \right| < \frac{1\,000 r^2}{|z - \zeta|^3} \qquad (6.2.3)$$

hold for all $z \in \Omega$ and $\zeta \in D(z_0, r)$.

Proof. Theorem 6.1.10 implies that Ω is simply connected. By the Riemann mapping theorem, there is therefore a conformal mapping F of the unit disc U onto Ω such that $F(0) = \infty$. $F(w)$ has an expansion of the form

$$F(w) = \frac{a}{w} + \sum_{n=0}^{\infty} c_n w^n \qquad (w \in U).$$

We define

$$g(z) = \frac{1}{a} F^{-1}(z) \qquad (z \in \Omega),$$

where F^{-1} is the mapping of Ω on to U which inverts F, and write

$$b = \frac{1}{2\pi i} \int_{|z-z_0|=r} (z - z_0) g(z) dz + z_0.$$

Theorem 2.4.4 can be applied to (F/a). It asserts that the diameter of the complement of $(F/a)(U)$ is at most 4. Therefore, $\text{diam} E \leqslant 4|a|$. Since $\text{diam} E \geqslant r$, it follows that

$$|a| \geqslant \frac{r}{4}.$$

Since g is a conformal mapping of Ω_0 onto $D(0, 1/|a|)$, thus

$$|g(z)| < \frac{4}{r} \qquad (z \in \Omega).$$

Therefore,

$$|b - z_0| < 4r.$$

If $\zeta \in D(z_0, r)$, $z \in \Omega$, then $|\zeta - z_0| < r$, so

$$|Q(\zeta, z)| \leqslant \frac{4}{r} + 5r \left(\frac{16}{r^2} \right) < \frac{100}{r}.$$

This proves (6.2.2). Fix $\zeta \in D(z_0, r)$. If $z = F(w)$, then $z g(z) = w F(w)/a$, and since $w F(w) \to a$ as $w \to 0$, we have $z g(z) \to 1$ as

$z \to \infty$. Hence, g has an expansion of the form

$$g(z) = \frac{1}{z-\zeta} + \frac{\lambda_2(\zeta)}{(z-\zeta)^2} + \frac{\lambda_3(\zeta)}{(z-\zeta)^3} + \cdots \qquad (|z-\zeta| > 2r).$$

By Cauchy's theorem,

$$\lambda_2(\zeta) = \frac{1}{2\pi i} \int_{|z-z_0|=r} (z-\zeta) g(z) dz = b - z_0 + (z_0 - \zeta).$$

Let

$$\varphi(z) = \left[Q(\zeta, z) - \frac{1}{z-\zeta} \right] (z-\zeta)^2,$$

then $\varphi \in H(\Omega \setminus \{\zeta\})$. $\varphi(z)$ is bounded as $z \to \infty$. Hence, ∞ is a removable singularity of φ. If $z \in (\Omega \setminus \{\zeta\}) \cap D(z_0, r)$, then $|z - \zeta| < 2r$, so (6.2.2) gives

$$|\varphi(z)| < 8r^3 |Q(\zeta, z)| + 4r^2 < 1\,000 r^2. \qquad (6.2.4)$$

By the maximum modulus theorem, (6.2.4) holds for all $z \in \Omega \setminus \{\zeta\}$. This proves (6.2.3). $\square$

Theorem 6.2.2 (Tietze's Extension Theorem). *Suppose K is a compact subset in $\mathbb{C}$ and $f \in C(K)$. Then there exists an $F \in C_c(\mathbb{C})$, such that $F(z) = f(z)$ for all $z \in K$.*

Proof. Assume that f is a real continuous function on K, $-1 \leqslant f \leqslant 1$. Let $W = \{z : d(z, K) < 1\}$, then W is a bounded open set, and $K \subset W$. Put

$$K^+ = \left\{ z \in K : f(z) \geqslant \frac{1}{3} \right\}, \qquad K^- = \left\{ z \in K : f(z) \leqslant -\frac{1}{3} \right\}.$$

Then K^+ and K^- are disjoint compact subsets of K. As a consequence of Proposition 1.1.2, there is a function $f_1 \in C_c(\mathbb{C})$ such that $f_1(z) = \frac{1}{3}$ on K^+, $f_1(z) = -\frac{1}{3}$ on K^-, $-\frac{1}{3} \leqslant f_1(z) \leqslant \frac{1}{3}$ for all $z \in \mathbb{C}$, and the support of f_1, supp $f_1 \subset W$. Thus,

$$|f(z) - f_1(z)| \leq \frac{2}{3} \quad (z \in K), \qquad |f_1(z)| \leq \frac{1}{3} \quad (z \in \mathbb{C}).$$

Repeat this construction with $f - f_1$ in place of f: There exists an $f_2 \in C_c(\mathbb{C})$ such that supp $f_2 \subset W$, so that

$$|f(z) - f_1(z) - f_2(z)| \leq \left(\frac{2}{3}\right)^2 \quad (z \in K), \qquad |f_2(z)| \leq \frac{1}{3} \cdot \frac{2}{3} \quad (z \in \mathbb{C}).$$

In this way, we obtain functions $f_n \in C_c(\mathbb{C})$ such that

$$\left|f(z) - \sum_{k=1}^n f_k(z)\right| \leq \left(\frac{2}{3}\right)^n \quad (z \in K),$$

$$|f_n(z)| \leq \frac{1}{3} \cdot \left(\frac{2}{3}\right)^{n-1} \quad (z \in \mathbb{C}).$$

Put $F(z) = \sum_{k=1}^\infty f_k(z)$. The series $F(z) = \sum_{k=1}^\infty f_k(z)$ converges to $f(z)$ on K, it converges uniformly on $\mathbb{C}$. Hence, F is continuous. Also, supp $F \subset \overline{W}$. Therefore, $F \in C_c(\mathbb{C})$. $\square$

Theorem 6.2.3 (Mergelyan's Theorem). *Let K be a compact set in $\mathbb{C}$, whose complement $\mathbb{C} \setminus K$ is connected. If $f \in C(K)$ and $f \in H(\mathrm{int} K)$, then $f \in P(K)$, i.e., for any $\varepsilon > 0$, there exists a polynomial P such that*

$$|f(z) - P(z)| < \varepsilon. \tag{6.2.5}$$

for all $z \in K$.

Proof. By Tietze's Theorem, f can be extended to a continuous function in the complex plane $\mathbb{C}$, with compact support. We fix one such extension, and denote it again by f.

For any $\delta > 0$, let $\omega(\delta) = \sup\{|f(z_2) - f(z_1)| : |z_2 - z_1| \leq \delta\}$. Since f is uniformly continuous, we have

$$\lim_{\delta \to 0} \omega(\delta) = 0 \tag{6.2.6}$$

for all $z \in K$. By (6.2.6), this proves the theorem.

From now on, $\delta > 0$ will be fixed. We shall prove that there is a polynomial P such that

$$|f(z) - P(z)| < 10\,000\omega(\delta) \qquad (6.2.7)$$

if $0 \leq r \leq \delta$.

Let the function $a(r)$ on $[0, \infty)$ be defined as follows: Put $a(r) = 0$ if $r > \delta$, put

$$a(r) = \frac{3}{\pi\delta^2}\left(1 - r^2/\delta^2\right)^2$$

and define $b(z) = a(|z|)$ for all complex z. It is clear that $b(z) \in C_c^1(\mathbb{C})$. We obtain that, by computing the integrals in polar coordinates,

$$\int_{\mathbb{C}} b(z)d\lambda(z) = \int_0^\delta a(r)r\,dr = 1,$$

$$\int_{\mathbb{C}} \partial b(z)/\partial \bar{z}\,d\lambda(z) = 0$$

and

$$\int_{\mathbb{C}} |\partial b(z)/\partial \bar{z}|\,d\lambda(z) = -2\pi \int_0^\delta a'(r)r\,dr = \frac{24}{15\delta} < \frac{2}{\delta}.$$

Now, define

$$\Phi(z) = \int_{\mathbb{C}} f(z - \zeta)b(\zeta)d\lambda(\zeta) = \int_{\mathbb{C}} f(\zeta)b(z - \zeta)d\lambda(\zeta).$$

Then $\operatorname{supp}\Phi \subset \{z : d(z, \operatorname{supp} f) \leq \delta\}$. Since

$$\Phi(z) - f(z) = \int_{\mathbb{C}} [f(z - \zeta) - f(z)]b(\zeta)d\lambda(\zeta)$$

for all z and $b(\zeta) = 0$ if $|\zeta| > \delta$,

$$|f(z) - \Phi(z)| \leq \omega(\delta). \qquad (6.2.8)$$

Since $b \in C_c^1(\mathbb{C})$, $2|\partial b(z)/\partial \bar{z}| = -a'(|z|)$, we have

$$\partial \Phi(z)/\partial \bar{z} = \int_{\mathbb{C}} f(\zeta)\,(\partial b(z-\zeta)/\partial \bar{z})\,d\lambda(\zeta)$$

$$= \int_{\mathbb{C}} f(z-\zeta)\,(\partial b(\zeta)/\partial \bar{\zeta})\,d\lambda(\zeta)$$

$$= \int_{\mathbb{C}} [f(z-\zeta) - f(z)]\,(\partial b(\zeta)/\partial \bar{\zeta})\,d\lambda(\zeta).$$

Thus,

$$|\partial \Phi(z)/\partial \bar{z}| \leqslant \omega(\delta) \int_{\mathbb{C}} |\partial b(\zeta)/\partial \bar{\zeta}|\,d\lambda(\zeta) < \frac{2\omega(\delta)}{\delta}. \tag{6.2.9}$$

By Theorem 1.2.4,

$$\Phi(z) = -\frac{1}{\pi} \int_{\text{supp }\Phi} \frac{\partial \Phi(\zeta)/\partial \bar{\zeta}}{\zeta - z}\,d\lambda(\zeta). \tag{6.2.10}$$

Put $K_\delta = \{z \in K : d(z, K^c) > \delta\}$, then K_δ is an open set, and

$$\Phi(z) = \int_0^\delta a(r)r\,dr \int_0^{2\pi} f(z - Re^{i\theta})\,d\theta$$

$$= 2\pi f(z) \int_0^\delta a(r)r\,dr = f(z)$$

for all $z \in K_\delta$. By (6.2.10),

$$\Phi(z) = -\frac{1}{\pi} \int_{X(\delta)} \frac{\partial \Phi(\zeta)/\partial \bar{\zeta}}{\zeta - z}\,d\lambda(\zeta), \tag{6.2.11}$$

where $X(\delta) = \{z \in \text{supp}\Phi : d(z, K^c) \leqslant \delta\}$ is a compact set. $X(\delta)$ can be covered by definition of many open discs $D(z_1, 2\delta), \ldots, D(z_n, 2\delta)$ of radius 2δ, whose centers are not in K. Since $\mathbb{C} \setminus K$ is connected, the center z_j of each $D(z_j, 2\delta)$ can be joined to ∞ by a polygonal path in $\mathbb{C} \setminus K$. It follows that each $D(z_j, 2\delta)$ contains a compact connected set E_j, of diameter of at least 2δ, so that $\mathbb{C} \setminus E_j$ is connected and $K \cap E_j = \varnothing$.

We now apply Lemma 6.2.1, with $r = 2\delta$. There exist functions $g_j \in H(\overline{\mathbb{C}} \setminus E_j)$ and constants b_j so that the inequalities

$$|Q_j(\zeta, z)| < \frac{50}{\delta} \tag{6.2.12}$$

and

$$\left| Q_j(\zeta, z) - \frac{1}{z - \delta} \right| < \frac{4000\delta^2}{|z - \zeta|^3} \tag{6.2.13}$$

hold for $z \notin E_j$ and $\zeta \in D(z_j, 2\delta)$, where

$$Q_j(\zeta, z) = g_j(z) + (\zeta - b_j)g_j^2(z).$$

Let $\Omega = \mathbb{C} \setminus (E_1 \cup \cdots \cup E_n)$. Then Ω is an open set which contains K. Put $X_1 = X(\delta) \cap D(z_1, 2\delta)$ and $X_j = (X(\delta) \cap D(z_j, 2\delta)) \setminus (X_1 \cup \cdots \cup X_{j-1})$, $2 \leq j \leq n$. Define

$$R(\zeta, z) = Q_j(\zeta, z) \qquad (\zeta \in X_j, z \in \Omega)$$

and

$$F(z) = \frac{1}{\pi} \int_{X(\delta)} (\partial \Phi(\zeta)/\partial \overline{\zeta}) \, R(\zeta, z) d\lambda(\zeta) \qquad (z \in \Omega).$$

Since

$$F(z) = \sum_{j=1}^{n} \frac{1}{\pi} \int_{X_j} (\partial \Phi(\zeta)/\partial \overline{\zeta}) \, Q_j(\zeta, z) d\lambda(\zeta), \tag{6.2.14}$$

(6.2.14) shows that F is a finite linear combination of the functions g_j and g_j^2. Hence, $F \in H(\Omega)$. By (6.2.9) and (6.2.11), we have

$$|F(z) - \Phi(z)| < \frac{2\omega(\delta)}{\pi\delta} \int_{X(\delta)} \left| R(\zeta, z) - \frac{1}{z - \zeta} \right| d\lambda(\zeta) \qquad (z \in \Omega). \tag{6.2.15}$$

Observe that the inequalities (6.2.12) and (6.2.13) are valid with R in place of Q_j if $\zeta \in X(\delta)$ and $z \in \Omega$. For if $\zeta \in X(\delta)$, then $\zeta \in X_j$ for some j, and then $R(\zeta, z) = Q_j(\zeta, z)$ for all $z \in \Omega$.

Now, fix $z \in \Omega$, when $|\zeta - z| < 4\delta$, and estimate the integrand (6.2.15) by (6.2.12) if $|\zeta - z| < 4\delta$ and by (6.2.13) if $|\zeta - z| \geqslant 4\delta$. Then

$$|F(z) - \Phi(z)| \leqslant \frac{2\omega(\delta)}{\pi\delta} \left\{ \int_{D(z,4\delta)} \left(\frac{50}{\delta} + \frac{1}{|\zeta-z|}\right) d\lambda(\zeta) \right.$$

$$\left. + \int_{\mathbb{C}\backslash \overline{D(z,4\delta)}} \frac{400\delta^2 d\lambda(\zeta)}{|\zeta-z|^3} \right\}$$

$$= \frac{4\omega(\delta)}{\delta} \left\{ \int_0^{4\delta} \left(\frac{50}{\delta} + \frac{1}{\rho}\right) \rho d\rho + \int_{4\delta}^{\infty} \left(\frac{400\delta^2}{\rho^2}\right) d\rho \right\}$$

$$= \frac{4\omega(\delta)}{\delta} \{404\delta + 1\,000\delta\} < 6\,000\omega(\delta).$$

Since $F \in H(\Omega)$, $K \subset \Omega$, and $\mathbb{C} \setminus K$ is connected, Runge's Theorem shows that F can be uniformly approximated on K by polynomials. Hence, (6.2.6) and (6.2.8) show that there exists a polynomial P such that (6.2.7) holds for all z. This completes the proof. □

Exercises VI

1. Suppose that the meromorphic function $f(z)$ in the extended complex plane $\overline{\mathbb{C}}$ has only two poles, with $z = -1$ being a first-order pole and its principal part being $\frac{1}{z+1}$ and $z = 2$ being a second-order pole with principal part $\frac{2}{z-2} + \frac{3}{(z-2)^2}$. Additionally, $f(0) = \frac{7}{4}$. Find the Laurent series of $f(z)$ in the annulus $1 < |z| < 2$.
2. Suppose that the meromorphic function $f(z)$ in complex plane $\mathbb{C}$ has only second-order poles at $z = 1, 2, 3, \ldots$, and the principal part of $z = n$ is $\frac{n}{(z-n)^2}$. Find the general form of $f(z)$.
3. Show that

$$\frac{\pi^2}{\sin^2 \pi z} = \sum_{n=-\infty}^{\infty} \frac{1}{(z-n)^2}.$$

4. Let $\Omega = \{z : |z| < 1, |2z - 1| > 1\}$, $f \in H(\Omega)$. Show there exists a sequence of polynomials $\{p_n(z)\}$ such that $\{p_n\}$ converges uniformly on each compact subset of Ω to f.

5. Show there exists a sequence of polynomials $\{p_n(z)\}$ with $p_n(0)=1$ such that $\{p_n\}$ converges uniformly on each compact subset of $\mathbb{C}\setminus\{0\}$ to 0.
6. Show there exists a sequence of polynomials $\{p_n(z)\}$ such that
$$\lim_{n\to\infty} p_n(z) = \begin{cases} 1, & \operatorname{Re} z > 0, \\ 0, & \operatorname{Re} z = 0, \\ -1, & \operatorname{Re} z < 0. \end{cases}$$

References

[1] Cheng Minde, Deng Donggao, Long Ruilin. *Real Analysis*. Beijing: Higher Education Press, 1993 (in Chinese).
[2] Fang Qiqin. *Complex Function*. Beijing: Peking University Press, 1996 (in Chinese).
[3] Zhang Nanyue, Chen Huaihui (eds.). *Selected Topics in Complex Function Theory*. Beijing: Peking University Press, 1995 (in Chinese).
[4] A.H. Markushevich. *Analytic Functions*. Translated by Huang Zhengzhong. Beijing: Higher Education Press, 1957 (in Chinese).
[5] Li Ruifu, Dai Chongji, Song Guodong (eds.). *Further Topics in Complex Function Theory*. Beijing: Higher Education Press, 1989 (in Chinese).
[6] Gong Sheng (ed.). *Concise Complex Analysis*. Beijing: Peking University Press, 1999 (in Chinese).
[7] Dieter Gaier. *Approximation Theory of Complex Variables*. Translated by Shen Xiechang. Changsha: Hunan Education Press, 1992 (in Chinese).
[8] Liu Shuqin (ed.). *Single-Valued Functions*. Xi'an: Northwest University Press, 1988 (in Chinese).
[9] Lu Shanzhen, Wang Kunyang. *Real Analysis* (2nd Edition). Beijing: Beijing Normal University Press, 2006 (in Chinese).
[10] Kawaada Tatsuo. *Fourier Analysis*. Translated by Zhou Minqiang. Beijing: Higher Education Press, 2004 (in Chinese).
[11] Zhou Minqiang (ed.). *Lecture Notes on Harmonic Analysis*. Beijing: Peking University Press, 1999 (in Chinese).
[12] Yu Jiarong, Lu Jianke (eds.). *Selected Topics in Complex Function Theory*. Beijing: Higher Education Press, 1993 (in Chinese).
[13] Yu Jiarong (ed.). *Complex Function*. Beijing: People's Education Press, 1994 (in Chinese).

[14] M. Andersson. *Topics in Complex Analysis*. Beijing: Tsinghua University Press, 2005.

[15] S. Axler, P. Bourdon, and W. Ramey. *Harmonic Function Theory* (2nd Edition, Graduate Texts in Math., 137, Springer-Verlag, Berlin, 2001.

[16] V. Ahlfors. *Complex Analysis* (3rd Edition). Beijing: China Machine Press, 2004.

[17] R. P. Boas, Jr. *Entire Functions*. Academic Press, New York, 1954.

[18] J. B. Conway. *Functions of One Complex Variable*. Springer-Verlag, New York Heidelberg Berlin, 1978, 2nd Edition. Translated by Lv Yinian and Zhang Nanyue. Shanghai: Shanghai Scientific and Technological Publishing House, 1985.

[19] B.Ya. Levin. *Lectures on Entire Functions*. V.150 Translations of Math, Monographs, 1996.

[20] P.L. Duren. *Theory of H^p-Spaces*. Academic Press, New York, 1997.

[21] Y. Katznelson. *An Introduction to Harmonic Analysis*. (3rd Edition). Beijing: China Machine Press, 2005.

[22] J. Garnett. *Bounded Analytic Functions*. Academic Press, New York, 1981.

[23] M. Rosenblum and J. Rovnyak. *Topics in Hardy Classes and Uninvalent Functions*. Berlin: Birkhäuser Verlag, 1994.

[24] C. Pommerenke. *Univalent Functions*. Vandenhoeck & Ruprecht, Göttingen Press, Princeton. 1975.

[25] I. Privalov. *Boundary Properties of Analytic Functions*. GITTL, Moscow, 1950 (Russian).

[26] W. Rudin. *Real and Complex Analysis*. (3rd Edition). McGraw-Hill Publishing, New Delhi, 1987.

[27] E. M. Stein and R. Shakarchi. *Complex Analysis, Princeton Lectures in Analysis II*. Princeton Univ. Press, Princeton. 2003.

[28] E. M. Stein and G. Weiss. *Real and Complex Analysis*. China Machine Press, Beijing. 2004.

[29] E. C. TiTchmarsh. *The Theory of The Riemann Zeta-Function*. (2nd Edition) revised by D.R. Heath-Brown. Oxford University Press, Oxford. 1986.

[30] K. Yosida. *Functional Analysis*. Springer-Verlag, Berlin, 1978.

[31] A. Zygmund. *Trigonometric Series, Vol. I, II*. (3rd Edition). Cambridge University Press, Cambridge, UK. 1977.

Index

$C_0(\mathbb{R})$, 217
$C_P(\partial U)$, 206
$H^p(\mathbb{C}_+)$ decomposition theorem, 278
$H^p(U)$ factorization theorems, 252
Γ−function, 126
$\bar{\partial}$ problem, 9
$\mathscr{F}$, 296
$\mathscr{F}^{-1}$, 299
$\mathscr{M}([-\pi,\pi])$, 205
$\mathscr{M}(\mathbb{R})$, 217
$\mathscr{M}_P(\partial U)$, 206

A

analytic automorphism group of the unit disk, 54
area theorem, 66, 69
argument principle, 22

B

Banach-Alaoglu theorem, 214, 229
Bernoulli constant, 137
beta function, 134
Bieberbach's theorem, 68
big Picard's theorem, 83, 93
Blaschke product, 250, 252, 276
Bloch's theorem, 84, 87

C

canonical product, 117
Carathéodory's convergence theorem, 76
Carathéodory inequality, 121, 142
Carleman formula, 202, 204
Cauchy inequality, 16
Cauchy formula, 9
Cauchy-Riemann equation, 7, 148
Cauchy theorem, 8, 326
Cauchy type integral, 15, 218
closed chain and index, 26
conformal mapping, 51
continuity to boundary of conformal mappings, 65
convergence of regions, 75
convolution, 218
convolution of measure, 218
curve, 4

D

degree p-Lebesgue point, 3
differentiability and analytic functions, 6
differentiating under the integral sign, 2
differentiating under the summation, 3

Dirichlet problem, 194
distribution function, 205

E

elementary factors, 102
equicontinuous on compact subsets of, 41
Euler summation formula, 130
Euler's constant, 129
Euler's theorem, 140
exponent of convergence, 114
exponential type entire function, 307
extreme value principle, 151

F

F. Riesz–M. Riesz theorem, 257
Fatou theorem, 215, 230
Fourier transform, 291
Fourier-Stieltjes transform, 304
Fubini and Cauchy theorem, 300
Fubini's theorem, 2
fundamental theorem of algebra, 16

G

Gauss formula, 129
generalized Carleman formula, 271
generalized Nevanlinna formula, 270
global Cauchy theorem, 26
Goursat theorem, 11
green formula, 7
Green's function, 194
Green-Pompeiu formula, 8
Gronwall, 66

H

Hadamard factorization theorem, 120
Hadamard's formula, 14
harmonic function, 147
Harnack's theorem, 159
holomorphic fourier transforms, 247
homology, 27
homology version of Cauchy's theorem, 27
homotopy, 32

homotopy version of Cauchy's theorem, 32
Hurwitz, 48

I

infinite products, 99
inverse function theorem, 25
inversion problem of the fourier transform, 293
isolation of zeros, 17

J

Jensen's formula, 247
Jordan closed curve theorem, 4
Jordan closed polygonal line theorem, 4
Joukowski's function, 67

K

Koebe function, 69
Koebe $\frac{1}{4}$ theorem, 78
Koebe's boundary correspondence theorem, 61
Köebe's one-quarter theorem, 69

L

Laplace operator, 147
Laplacian, 147
Laurent's theorem, 18
Lebesgue point set, 3
Lebesgue differential theorem, 3
Lebesgue differentiation theorem, 3
Lebesgue point, 3
Legendre's duplication formula, 140
Liouville's theorem, 16
Little Picard's theorem, 83, 88
Littlewood's theorem, 75

M

Marty's theorem, 50
maximum modulus principle, 61
maximum modulus theorem, 14
mean value formula, 150
Mergelyan's theorem, 327

Index

Minkowski's inequality, 2
Mittag–Leffler's theorem, 319
modulus of integral continuity, 295
Montel, 43
Morera theorem, 10, 300
multiplication formula, 292, 297, 304

N

Nevanlinna representation, 239
non-tangential limit, 215, 230, 275–277
normal family, 45

P

Paley-Wiener theorem, 300
Parseval's equality for fourier transform, 296–297
Parseval's identity for Fourier-Stieltjes transform, 305
path, 26
Phragmén–Lindelöf theorem, 180, 182–183, 186
Picard's theorem, 83
Plancherel theorem, 299, 301
Poisson integral, 229
Poisson kernel, 152, 241
Poisson Representation, 238
Poisson-Stieltjes integral, 218
Poisson-Stieltjes integral of measure μ, 229

R

Rado's theorem, 79
real differentiability, 1
residue, 19
residue theorem, 22, 29
Riemann mapping theorem, 325
Riemann zeta function, 135
Riemann-Lebesgue lemma, 291
Riesz representation theorem, 217
Rouché's theorem, 22, 31
Runge theorem, 315, 324, 331

S

Schottky's inequality, 84
Schwarz formula, 150
Schwarz integral formula, 35
Schwarz lemma, 53, 121
Schwarz's inequality, 300
Schwarz–Pick lemma, 55
simply connected region, 4
Stieltjes inversion formula, 240
Stirling formula, 134

T

Taylor's theorem, 16
the Abel mean of the Fourier-Stieltjes integral of a measure, 306
the distribution function, 217
the Hadamard three-circle theorem, 243
the Harnack inequality, 244
the inversion formula, 299
the minimum principle of harmonic functions, 243
the order and type of an entire function, 106
the Plancherel theorem, 299
the Poisson integral, 206
the Poisson kernel on the unit disc, 152
the polar coordinate form of the Laplace, 157
the Riemann mapping theorem, 56
the Schottky inequality, 90
the Schwar reflection principle, 160
the Weierstrass factorization theorem, 102
theorem of Mergelyan, 324
Tietze's extension theorem, 324, 326
total variation norm, 217

U

uniformly bounded on compact subsets of, 41–43

uniformly convergent on compact
  subsets, 12
uniformly converges to ∞ on compact
  subsets of, 45
uniqueness theorem, 17
univalent, 24

V

Vitali's theorem, 44

W

Weierstrass factorization theorem,
  104, 120
Weierstrass formula, 130
Weierstrass theorem, 12, 43, 76, 165

Z

zeros of holomorphic functions, 99

www.ingramcontent.com/pod-product-compliance
Lightning Source LLC
Chambersburg PA
CBHW070443100125
19686CB00004B/83